Lecture Notes in Electrical Engineering

Volume 105

For further volumes:
http://www.springer.com/series/7818

Nikolaos Voros · Amar Mukherjee
Nicolas Sklavos · Konstantinos Masselos
Michael Huebner
Editors

VLSI 2010 Annual Symposium

Selected papers

Springer

Nikolaos Voros
Department of Telecommunication Systems and Networks
Technological Educational Institute of Messolonghi
National Road
303 00 Nafpaktos
Nafpaktos-Antirrion (Varia)
Greece
e-mail: voros@teimes.gr

Amar Mukherjee
School of Electrical Engineering and Computer Science
University of Central Florida
Central Florida Blvd. 4000
Orlando
FL 32816-2362
USA
e-mail: amar@eecs.ucf.edu

Nicolas Sklavos
Informatics and MM
Technological Educational Institute of Patras
Rhga Feraiou
271 00 Pyrgos
Ileias Region
Greece
e-mail: nsklavos@ieee.org

Konstantinos Masselos
Department of Computer Science and Technology
University of Peloponnese
221 00 Tripolis
Terma Karaiskaki
Greece
e-mail: kmas@uop.gr

Michael Huebner
Institut für Technik der Informationsver
Karlsruhe Institute of Technology
Vinzenz-Priessnitzstr. 1
76131 Karlsruhe
Germany
e-mail: michael.huebner@kit.edu

ISSN 1876-1100
ISBN 978-94-007-1487-8
DOI 10.1007/978-94-007-1488-5
Springer Dordrecht Heidelberg London New York

e-ISSN 1876-1119
e-ISBN 978-94-007-1488-5

Cover design: eStudio Calamar, Berlin/Figueres

Printed on acid-free paper

Springer is part of Springer Science+Business Media (www.springer.com)

Preface

During the last years, significant changes have taken place in markets that traditionally had nothing in common. For example, markets like telecommunication, automotive, consumer electronics and medical equipment converge as far as the underlying systems of their products are concerned. The common needs come from the fact that modern devices contain, in most cases, complex parts relying on advanced hardware equipment. The increasing competition and the market pressure have created the need for hardware products of high reliability with short time-to-market.

The traditional development techniques, where the system development was relying on the experience of highly qualified engineers, are no longer adequate. The complexity of modern hardware systems calls for methodologies and tools supporting them to deal with the increasing market requirements.

The existence of large number of computational intensive structures with diverse features leads also to hardware solutions that are no longer monolithic but able to adapt according to design needs. Moreover, energy aware systems and increased performance requirements are more and more important in an era where the size of most devices decreases while their complexity increases exponentially. Data intensive processing is also receiving renewed attention, due to rapid advancements in areas like multimedia computing and high-speed telecommunications. Many of these applications demand very high performance circuits for computationally intensive operations, often under real-time requirements. Furthermore, their computation power appetite tends to soar faster than Moore's law.

Moreover, the next generation of systems in most markets relies on the "computing everywhere" paradigm, which implies computing chips dedicated by market while at the same time development costs are exploding. This results in an increasing need of flexibility not only at the program level (by software) but also at the chip level (by hardware). So, combining flexibility and performance is now a key enabler for future hardware platforms. In that respect, current solutions are reaching their limits:

- Current computing solutions are out of breath: challenge of computing density and low power.
- Current development and programming tools do not provide the required productivity.

On one hand, the performance of most hardware architectures, in spite of the continuous increase in processors' speed, are, not surprisingly, lagging behind. Processors efficiency is more and more impaired by the memory bandwidth problem of traditional von Neumann architectures.

On the other hand, the conventional way to boost performance through Application Specific Integrated Circuits (ASIC) suffers from sky-rocketing manufacturing costs (requiring high volumes to be amortized) and long design development cycles. In the nanometre era, increasing non recurrent engineering costs could relegate system-on-chip to very few high volume products unless some standardization process is undertaken.

Modern Field Programmable Gate Arrays can implement an entire system-on-chip, but at the cost of large silicon area and high power consumption. Moreover, a huge design productivity issue is raised by the difficulty of embedding algorithms on complex massively parallel architectures, while defining the processing architecture, under time to market pressure. Defining a programming paradigm for new hardware architectures is a difficult problem, where Computer Aided Design technologies call for new design paradigms. Current CAD tools have synthesis capabilities that don't reach the abstraction level required to handle complex hardware implementation.

Although the book *Designing Very Large Scale Integration Systems: Emerging Trends and Challenges* does not intend to provide answers to all the aforementioned open issues, it intends to identify and present in a comprehensive way the trends and research challenges of designing the next generation VLSI systems and systems-on-chip. Throughout the chapters of the book, the reader will have the chance to get an insight to state-of-the-art technology and research results on areas like:

- Emerging devices and nanocomputing,
- Architecture level design of highly complex hardware systems and systems-on-chip,
- Reconfigurable hardware technology, and
- Embedded systems.

All the book chapters are written by experts in the relevant domains and is envisaged to become the starting point for young scientists and practitioners to move science and technology one step further, in an attempt to deal with the ever increasing challenges of modern VLSI systems and systems-on-chip.

N. Voros
A. Mukherjee
N. Sklavos
K. Masselos
M. Huebner

Contents

Part I
Architecture: Level Design Solutions

Chapter 1
Intelligent NOC Hotspot Prediction

Elena Kakoulli, Vassos Soteriou and Theocharis Theocharides

Abstract Hotspots are Network on-Chip (NoC) routers or modules which occasionally receive packetized traffic at a higher rate that they can process. This phenomenon reduces the performance of an NoC, especially in the case wormhole flow-control. Such situations may also lead to deadlocks, raising the need of a hotspot prevention mechanism. Such mechanism can potentially enable the system to adjust its behavior and prevent hotspot formation, subsequently sustaining performance and efficiency. This Chapter presents an Artificial Neural Network-based (ANN) hotspot prediction mechanism, potentially triggering a hotspot avoidance mechanism before the hotspot is formed. The ANN monitors buffer utilization and reactively predicts the location of an about to-be-formed hotspot, allowing enough time for the system to react to these potential hotspots. The neural network is trained using synthetic traffic models, and evaluated using both synthetic and real application traces. Results indicate that a relatively small neural network can predict hotspot formation with accuracy ranges between 76 and 92%.

Keywords Network on-Chip Hotspots · Artificial Neural Networks · VLSI Systems

---

E. Kakoulli · V. Soteriou
Department of Electrical Engineering and Information Technology,
Cyprus University of Technology, Lemesos, Cyprus

T. Theocharides (✉)
Department of Electrical and Computer Engineering KIOS Research Center for Intelligent Systems and Networks, University of Cyprus, Nicosia, Cyprus
e-mail: ttheocharides@ucy.ac.cy

N. Voros et al. (eds.), *VLSI 2010 Annual Symposium*,
Lecture Notes in Electrical Engineering 105, DOI: 10.1007/978-94-007-1488-5_1,

1.1 Introduction

The Network on-Chip paradigm, (NoC) [7] is a distributed, router-based interconnect architecture that is quickly becoming the preferred communication fabric in general purpose Chip Multi-Processors (CMPs) as well as in application-specific Systems-on-Chips (SoCs). NoCs facilitate high communication throughput demands among the various computational, memory and IP cores found in these parallel systems. With technology enabling billions of transistors on a single chip, time-and medium-shared interconnects, such as on-chip buses, are no longer adequate; NoCs on the other hand effectively scale and allow efficient reusability of computational resources. NoCs have already been demonstrated in the Tilera TILE64 CMP [2] and the Intel Teraflops chip [26], among other CMP implementations.

In wormhole flow-control (WFC) [8], commonly employed in NoCs ([3],[11]), communication is organized in the form of packetized messages of arbitrary length, which are broken down into further logical link-width chunks called flow-control units (flits). WFC allows parts of a packet to be forwarded to the next router towards a final destination or consumer node, without having to wait for the remaining flits of the same packet, unlike in store-and-forward flow control, allowing the efficient use of smaller intermediate buffering, that is suitable for power-, thermal-and area-constrained NoCs. The spreading of a packet in a pipelined fashion across several routers at a time, however, makes WFC susceptible to packet blocking and possible deadlocks, as an increased delay at an intermediate router quickly forms backpressure which can be propagated in the reverse direction of the packet's path traversal. Such adverse effect, as outlined later, can further lead to the creation of NoC hotspots. Loss in NoC performance is traditionally alleviated by utilizing several resource utilization-enhancing mechanisms, such as virtual channels [6], control packets that reserve resources in advance [20], specialty architectures or schemes that guarantee communication rates [5], or by purposely inefficiently operating high-performance NoCs at relatively low utilizations to avoid blocking [21].

Hotspots are NoC routers or modules which occasionally receive packetized traffic at a rate higher than their traffic consumption rate. Since interconnecting links and ports are bandwidth-limited, and as the load traffic distribution of real applications is inherently uneven, hotspots are common in popular NoC topologies such as at the diagonal of meshes [19]. Even a single heavily loaded router can cause a hotspot, and thus a hotspot can be created even with the use of infinite bandwidth links. Some factors which can cause hotspots are non-optimal application mapping, lack of traffic balancing with oblivious routing algorithms, application migration and specific application NoC resource demands [4]. Hotspots have a spatial component when a subset of the routers receives the majority of the traffic, and a temporal component as these router nodes receive traffic often. The adverse phenomenon of hotspot formation reduces the performance and the effective throughput of the NoC, and increases average packet latency. This effect is particularly severe as backpressure can cause the buffers of neighboring routers to quickly fill-up in a domino-style effect, leading to a spatial spread in congestion that can cause the network to form a "saturation tree", and even worse, to deadlock. Hotspot formation is unpredictable in general-purpose parallel best-effort on-chip systems such as CMPs, as the application behavior is not exactly known a-priori so as to react accordingly by setting exact spatial–temporal hotspot reduction mechanisms.

This Chapter describes a first attempt towards utilizing artificial intelligence in predicting the formation of traffic hotspots. The Chapter presents a hardware-based artificial neural network (ANN), which is trained and used to detect the formation of hotspots, in an effort to notify the system. The Chapter focuses only on hotspot detection; plans to integrate avoidance and adjustment mechanisms are imminent. The proposed ANN is designed with manageable hardware overheads, implemented as a processing element in the NoC. The ANN monitors the NoC in an effort to detect runtime occurring hotspots, and reactively predicts the location of a potential hotspot that is about to be formed, allowing enough time for the system to adjust and potentially avoid the hotspot.

Section 1.2 presents some approaches in hotspot prevention and Sect. 1.3 presents the Artificial Neural Network-based (ANN) mechanism. Some details on the hotspot model considered, the data processed by the ANN, and the proposed ANN architecture are detailed in Sect. 1.3 as well, while Sect. 1.4 presents the experimental setup, ANN synthesis results and quantitative accuracy results in predicting NoC hotspots in CMPs using real application traces. Section 1.5 concludes this chapter.

1.2 Background and Related work

Hotspot prevention has been explored significantly in the domain of large-scale interconnection networks found in off-chip parallel computer systems, but only recently, and to a much lesser extent, in the area of NoCs. In both domains, most of the work focuses on reducing the number of routed packets destined towards a hotspot, using various methodologies. All of these works assume that the hotspot spatial locations are (1) either known a-priory and pro-actively reduce the possibility of hotspot formation, or (2) spatially and temporarily react actively to reduce the possibility of a hotspot.

There are various hotspot prevention works in the domain of large-scale interconnected parallel computers. References [1, 10, 14, 15, 22] present methods that tackle hotspots in traditional, off-chip networks and multiprocessor systems. Most of the hotspot related work in the domain of off-chip networks however, is unsuitable for NoCs, as the resources are restricted (buffering, virtual channels and routing computation) and packet dropping and re-transmission is unacceptable due to the complexity of the mechanism and the need to attain ultra-high performance. Hotspot prevention is therefore mostly based on spatial techniques via the use of

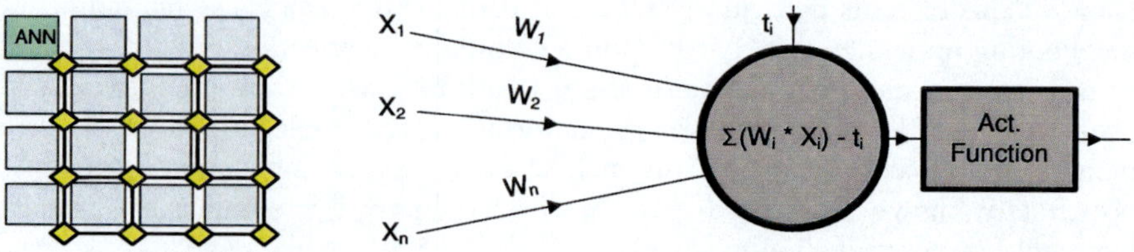

Fig. 1.1 An example placement of an ann prediction engine, as a dedicated pe (*right*) and the mathematical representation of a neuron (*left*)

adaptive routing which aims to route packets around hotspots, such as in [9]. Reference[27] uses a low-cost end-to-end credit-based allocation technique in NoCs to throttle and regulate hotspot-destined traffic, to allow fair sharing of the hotspot resource network and to mitigate the effects of non-hotspot traffic. All existing works however, are based on the premise that hotspots can be predicted due to application-specific behavior. This Chapter presents how an ANN can be used to predict hotspot occurrences in NoCs. ANNs have been used as prediction mechanisms successfully, in several application areas. In particular, they have been used as branch predictors in computer architecture [18, 23], yielding high accuracy, and as resource allocation mechanisms forecasting the computational demands. They have also been used in various other forecasting scenarios, such as weather and stock behavior [17].

An ANN is an information processing paradigm, consisting of computation nodes called neurons and the interconnections between them, called synapses. A neuron takes a set of inputs and multiplies each input by a weight value, accumulating the result until all inputs are received. Weights are determined during the training process. A threshold value (also determined in training) is then subtracted from the accumulated result, and this result is then used to compute the output of the neuron, via the so-called "activation function" (Fig. 1.1 (right)). Typical activation functions include complex mathematical functions such as the hyperbolic tangent function. The neuron output is then propagated to a number of destination neurons in another layer, which perform the same general operation with the newly received set of inputs and their own weight/threshold values, and this is repeated for other layers, depending on the complexity and accuracy required for the application. The activation function used, as well as the connectivity of the neurons, depends on the application as well. Neurons are typically used in multi-layer configurations, where they are trained to perform a desired operation in a similar manner to the way the human brain operates [16]. The ANN operates in two stages; the training stage and the computation stage. Each neuron receives training data that allows it to make necessary decisions based on the input data during execution of a desired operation. A small neural network can be easily implemented in hardware, as neurons can be designed using a Multiplier-Accumulator (MAC) unit, memory to hold the training weights, and a Look-Up Table (LUT) for the activation function. Optimizations, though, are necessary in interconnecting the neurons, however time-division multiplexing between neuron operations allows for resource sharing and minimal hardware needs.

1.3 Artificial Neural Network Hotspot Predictor

As already explained, hotspots are not single routing nodes, but rather a combination of increased utilization in a neighborhood of routers, usually involving two or more routers. Thus, the ANN is designed in such a way as to integrate information of neighboring routers, i.e. "reading" situations happening one and two routers away from the potential hotspot. Furthermore, the hardware needs must be taken into consideration when designing the ANN; this can be done by partitioning the on-chip network into smaller sub-regions, and using a base ANN for each region, in an effort to reduce the hardware resources necessary. Potentially the base ANN can be designed as a Processing Element (PE), as in Fig. 1.1 (left). This Section presents the base ANN mechanism.

1.3.1 Hotspot Modeling

For training and evaluating the ANN-based hotspot prediction mechanism, and to test its effectiveness in forecasting hotspots, a hotspot model with well-defined spatially located hotspot formations centered on randomly chosen routers in an 8×8 mesh network, but of relatively short temporal duration so as to enable testing of the effectiveness of the prediction mechanism, has to be devised. This model enables synthetic generation of network traces that exhibit a range of throughput demands upon the NoC. The synthetic traces allow both training as well as evaluation. The idea is to compare un-even traffic flows (i.e. formed because of a hotspot) versus a base Uniform Random NoC Traffic (URT) pattern. Under URT, all router nodes in an NoC have an equal probability of receiving a packet per unit time, therefore the traffic is almost perfectly balanced and the attainable network throughput is the maximum that can be achieved both practically and theoretically [8]. The developed hotspot model uses the URT model as a base, for most of the time, except when for short pre-specified and periodically occurring time intervals just two arbitrarily selected nodes receive with equal probability all the network traffic from all the remaining sender nodes, with the remaining nodes receiving no traffic. Time Frame Windows (TFW) of 10,000 cycles are then defined, during any time which the two hotspots can occur simultaneously for a short duration of 200 cycles. No other hotspots can occur within a TFW, and during the rest of the duration of the TFW the traffic behaves purely as URT. This model test-stresses the prediction mechanism as hotspots occur both infrequently and for short durations.

Table 1.1 Synthetic hotspot traffic results: prediction accuracy as a function of simulation time at various levels of normalized network throughput

Normalized throughput	20%	40%	60%	80%	100%
0.22*	0.96	0.97	0.97	0.98	0.98
0.33	0.84	0.87	0.88	0.88	0.88
0.44	0.81	0.84	0.86	0.86	0.87
0.56	0.85	0.88	0.89	0.89	0.91
0.67*	0.098	0.98	0.98	0.98	0.99
0.78	0.86	0.88	0.88	0.89	0.91
0.89	0.85	0.86	0.88	0.90	0.92
0.98*	0.95	0.96	0.96	0.96	0.96

Asterisks (*) indicate training data

Table 1.2 Synthetic hotspot traffic results: percentage of false positive hotspot identifications

Normalized throughput	False positive hotspots (%)
0.22*	4
0.33	8
0.44	5
0.56	6
0.67*	5
0.78	6
0.89	7
0.98*	4

Asterisks (*) indicate training data

1.3.2 ANN Training

The ANN was trained using part of the synthetic traffic traces suite, consisting of buffer utilization data collected over half a million cycles. Three different traffic scenarios were targeted during this phase; moderate, average, and high hotspot temporal intensity, expressed in terms of the NoC's normalized saturation throughput (for hotspot traffic) at 0.22, 0.67 and 0.98 (asterisks in Tables 1.1 and 1.2), through the duration of the simulation. Moreover, training data includes multi-router hotspot scenarios. Using the synthetic traces, utilization rates of all buffers in all routers were collected by measuring the average utilization rate of each router during 50-cycle intervals. This data was fed as training input to the ANN.

The Matlab ANN toolbox and supervised learning algorithms [12, 13] were used for the training phase. The acceptable error rate was set to less than 10%, given the large data size and considerations in training memory for the ANN weights, as well as the large range variations and sparsity of the training set data. Error rates less than 5% cannot be obtained in realistic training time, thus are considered impractical for the purposes of the presented network; however, potential optimization of the training set can probably improve the accuracy of the network and enable training for less than 5% acceptable error rate [25]. As the output of the hotspot prediction mechanism represents a probability that a hotspot will be formed in location (x, y) inside a 2D NoC, the acceptable error rate represents the probability that a hotspot is not created. If, for example, the ANN is trained with a resulting accuracy of 85%, this essentially represents that there is an 85% probability of a hotspot happening in location (x, y).

The training data obtained consists of 8-bit weight values, biased to yield positive integers for ease of hardware implementation. During the training stage, the bit-width precision of the weights has to be determined. By varying this in intervals of 4 bits, it was found that 4-bit weight values yield high error rates (>20%), whereas for 8-bit and 12-bit precision, the error rate is ~9.2 and ~8.4% respectively. This indicates that 8 bits are adequate for the weight data. The hyperbolic tangent can be used as the activation function, as its advantage lies in its function properties; it is an odd function with mirror symmetry:

$$\tanh(-x) = -\tanh(x) \textit{ and } \tanh(\overline{z}) = \overline{\tanh(z)}.$$

The function is also asymptotic, with the output of the function considered to be 1 (or-1) for input values above or below a certain threshold. A Look-Up Table (LUT) implementation requires only positive values ranging from zero to the threshold value that yields 1 (or-1). The activation function can be implemented as a multi-ported LUT, to parallelize accesses to it and enable parallel neuron computation.

The base ANN hotspot prediction mechanism consists of two perceptron layers which monitor 2×2 and 3×3 blocks of routers in the NoC, and fully connected hidden and output layers that combine the two perceptron layers and return the location of a potential hotspot router. The ANN operates by receiving buffer utilization statistical data from each router that it monitors, in discrete time intervals, and, using training weights, attempts to detect a pattern that can potentially lead to a hotspot located at one of the routers that the network monitors. The prediction time needs to be early enough for any adjustment mechanisms to interfere and potentially prevent the hotspot from forming. Therefore, the ANN must be accurate, and must also be fast as it needs to output a prediction early ahead of the next timing interval that it is about to receive new monitoring data.

The ANN can be scaled to monitor larger NoCs, by hierarchically connecting the base ANN together to an additional layer of neurons and training appropriately. Furthermore, more layers can be added as the network size grows. The hardware overheads necessary for larger ANNs will increase significantly, and the latency will also increase drastically, as the top ANN needs to receive data from each of the smaller ANNs and compute the final output; this adds significant delays in the overall operation of the ANN. However, the ANN idea can potentially be used in various network sizes and topologies, taking the hardware budgets into consideration.

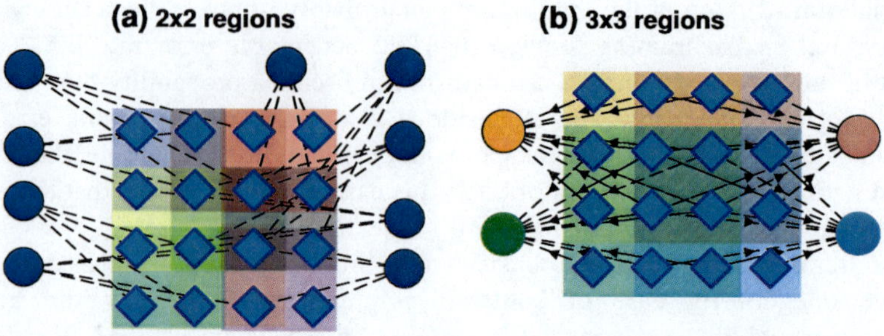

Fig. 1.2 Partitioning a 4 × 4 mesh noc—the 2 × 2 and 3 × 3 partitions and the respective input layer neurons

1.3.3 Neural Network Architecture

The base ANN is directed towards a 2-D Mesh architecture, and was trained and designed to support up to 4 × 4 mesh NoC infrastructures. It consists of an input layer which partitions the routers being monitored into 9 segments of 2 × 2 routers, and 4 segments of 3 × 3 routers. The segmentations are shown in Fig. 1.2. The segmentation is done in such a way so as to detect hotspots not as single routers, but as a combination of events affecting routers one and two hops next to the probable hotspot location. The network receives the average buffer utilization from each router in the region that is responsible for monitoring, processes the information and returns as output a binary vector containing the location of a potential hotspot router, including hotspots where two or more routers become hotspots.

The first ANN (input) layer consists of 9 neurons responsible for monitoring regions of 2 × 2 routers and 4 neurons responsible for monitoring regions of 3 × 3 routers (Fig. 1.2). Each of the input layer neurons simply acts as a MAC, that multiplies the incoming utilization rate from each of the routers the neuron is responsible for. When the results are accumulated, they are passed through the activation function (hyperbolic tangent), and propagated to a hidden layer of 16 neurons, all fully connected. The hidden layer neurons similarly process the 13 inputs from each of the input layer neurons, and propagate their result to an output neuron which returns a 16-bit vector indicating whether or not a hotspot is predicted. If the output neuron predicts a hotspot, the location of a hotspot is encoded in the 16-bit output vector. The overall base ANN system is shown in Fig. 1.3.

The base ANN is designed with emphasis on hardware reuse and with the fact that results should be computed prior to the next timing interval that the NoC will re-transmit new utilization data. An overview of the base ANN architecture is shown in Fig. 1.4. The primary hardware component of the ANN is an 8-bit × 8-bit MAC unit; given however that the second layer of neurons could be computed in parallel, the network is designed with 16 MAC units, to boost the computation, reusing them on demand during the first layer of computation nodes.

Fig. 1.3 The 3-layer ANN structure. neuron connections from input layer to hidden layer are shown for only two neurons, for readability purposes

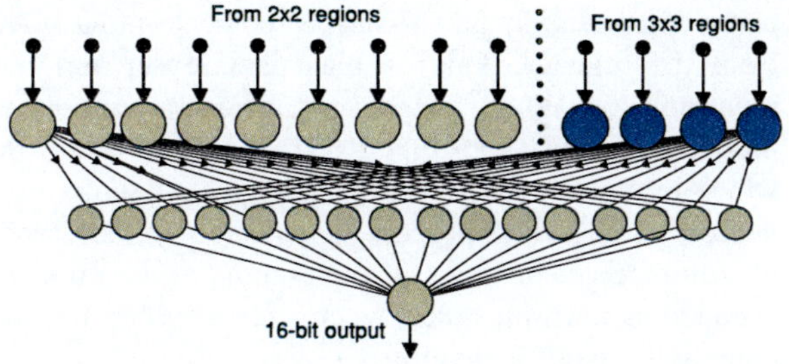

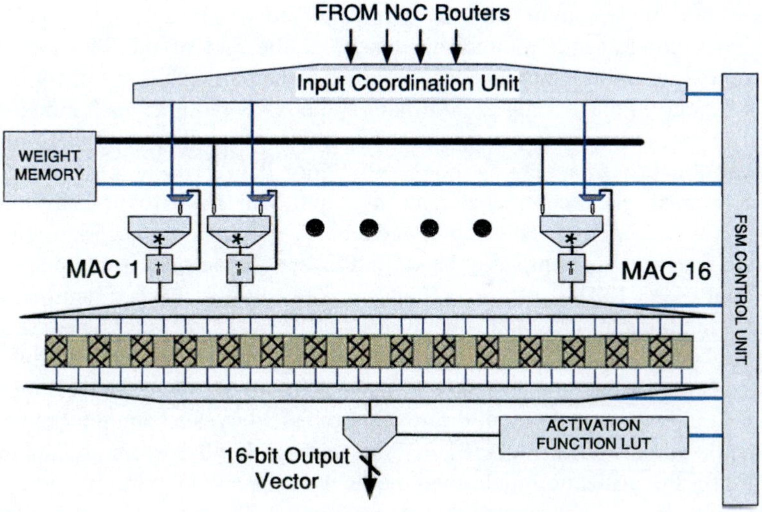

Fig. 1.4 The ANN prediction engine in hardware. It consists of 16 MAC units (only 2 are shown for readability purposes), a memory which holds the ANN weights, registers used to hold partial neuron sums, and the activation function LUT. The entire process is controlled through an FSM control unit and an input coordination unit

The neuron weights are stored in a RAM, with a 128-bit bus, transmitting at most sixteen 8-bit weight values in parallel during the multiplication stage. A FSM control unit synchronizes the entire ANN computation, where input values from each router arrive and are directed towards the appropriate MAC unit, depending on the neuron that they belong to. It is assumed that buffer utilization data for each router arrive as individual packets; using one packet per cycle as incoming data rate, the ANN receives 4 buffer utilization rates for a single router per cycle. Given the overlap of the 2×2 and 3×3 regions, the values are used at least twice, therefore, they are directed to the corresponding MAC unit. The MAC units are also interconnected to a set of accumulation registers, where each register holds the value of each neuron necessary for the computation. There are 30 registers; 13 for the input layer neurons, 16 for the output layer neurons and 1 for the output neuron. When the next input arrives at the ANN and is directed to a neuron already being computed, its partial sum is fed back into a MAC unit using the stored value from the register. This enables data reuse, and reduces the overall hardware requirements. When a neuron finishes its computation cycle, its final sum of products is used as an input to the LUT, to return the activation function value, which in turn is reused as input to the MAC units for computing the second layer neuron values. The process is repeated for all second layer neurons. The last neuron receives as input all the outputs of the 16 s layer neurons, and its result is encoded as a 16-bit binary vector representing the location(s) of the routers with potential hotspot formation.

The computation starts when the first buffer utilization values arrive from one of the routers under monitoring. Data is collected within a 50-cycle TFW; if a router does not transmit monitoring data until the end of the TFW, its buffer utilization is set to maximum, as it is assumed to be part of a hotspot. By the time the ANN receives the last router information, the ANN requires 5 additional cycles to finalize computation of the first layer neurons (given that computation of the first layer of neurons happens as soon as utilization data arrive at the ANN). This is in part because each completed neuron output will have to pass through the activation function LUT. This can be reduced by parallelizing access to the LUT. Once the activation is complete, the second layer (16 neurons) is computed; there are a total of 13 MAC operations, hence the results for the last neuron will be available in 26 cycles (13 MAC operations and 13 LUT accesses). The last neuron performs 16 parallel MAC operations and a mapping of the result to the 16-bit output vector, needing an additional 3 cycles (1 for all MACs, 1 for the summation and 1 for the mapping). In total, the ANN requires 34 cycles from the time the last router utilization data is received until its output. If all routers are on time in transmitting the utilization data, then the network needs an extra 16 cycles to fill up the computation pipeline (1 cycle per router), bringing the total minimum number of cycles to 50, which is the targeted time interval between outputs.

1.4 ANN Hotspot Prediction Results

1.4.1 Experimental Setup

The ANN was trained and modeled in two steps; training and evaluation. During the training stage, synthetic traffic utilization figures were used, as explained in Sect. 1.3.2. The evaluation stage consisted of two separate steps: (1) evaluation using synthetic traffic models with varied hotspot rates in an 8×8 2D mesh network using 4 base ANN engines, with each such ANN monitoring a non-overlapping 4×4 mesh sub-network, and (2) evaluation using the Raw CMP benchmarks [24], over the 4×4 Raw network (see Sect. 1.4.3). The output of the ANN was compared to the actual trace, both for synthetic traffic and the Raw benchmarks, and the ANN accuracy is therefore evaluated. This accuracy was measured during constant simulation intervals, to observe the behavior of the network as the amount of data increases.

A detailed cycle-accurate simulator is used to provide the buffer utilization data for training and evaluation of the ANN. The simulator supports k-ary 2-mesh topologies with 3 GHz 5-stage pipelined router cores, each with two virtual channels (VCs). Packets are composed of five 32-bit flits with each flit transported in 1 link cycle over links of 96Gbps bandwidth. Each router consists of eight unidirectional channels (four incoming and four outgoing).

1.4.2 Synthetic Traffic Prediction Results

Various hotspot spatio-temporal intensities are used as input to the ANN. These intensities are modeled as a function of the normalized saturation throughput, defined as the throughput at which the latency value is three times the zero-load latency of the network. The output of the ANN is then compared with the actual behavior of each trace, in discrete intervals of 20% increments during the simulation process. A hotspot prediction is considered to be successful if it can be foreseen at least 50 cycles ahead of its occurrence in the NoC, and that its coordinate locality is correct. Table 1.1 shows the accuracy of the ANN for the synthetic traffic models. Each model exhibits different hotspot behavior; Table 1.1 also illustrates the training traces used for training (marked with asterisks (*)); these traces were also used as evaluation benchmarks to validate the operation of the ANN. The accuracy ranges between 88% and 92%, values which are relatively good given the 10% acceptable error margin used during training. Table 1.2 shows the number of false positive predictions (i.e. predicted hotspots which never occurred in the actual trace). While false positives might seemingly not be a problem, their effect can only be studied when the overall cost of a hotspot correction/avoidance mechanism is implemented that utilizes the prediction information. The accuracy can be further improved, by improving the training set, and by potentially using more bits to represent the weights.

1.4.3 Real-System Traffic Prediction Results

The Raw CMP [24] is a system for general-purpose and embedded computing, consisting of 16 identical tiles interconnected via a 4 $\times$ 4 mesh array. Every tile contains its own pipelined RISC processor, computational resources, memory, and programmable routers. The Raw architecture contains four parallel networks: two dynamic networks to support unpredictable inter-communication requirements among the tiles (e.g. cache misses), and two static networks to allow implementation software-directed routing among tiles with ordered and flow-controlled transfer of operands and data streams between functional units. The switching of the routers in the two static networks is pre-programmed by the compiler,

Table 1.3 Results for the raw benchmarks: prediction accuracy results as a function of simulation time

Benchmark	20%	40%	60%	80%	100%
8b_encode	0.82	0.82	0.80	0.80	0.81
802.11a	0.88	0.89	0.88	0.88	0.89
adpcmRAW	0.84	0.86	0.85	0.86	0.85
Fft	0.76	0.78	0.78	0.80	0.80
Mpeg2	0.91	0.90	0.91	0.91	0.92
Streams	0.95	0.94	0.95	0.95	0.95
vpr	0.92	0.90	0.91	0.91	0.90

Table 1.4 Results for the raw benchmarks: percentage of the correctly identified hotspots at least 50 system clock cycles in advance

Benchmark	Percentage of predicted hotspots at least 50 cycles in advance (%)
8b_encode	64
802.11a	69
adpcmRAW	82
Fft	80
Mpeg2	72
Streams	84
vpr	68

collectively configuring the entire network on a cycle-by-cycle basis to enable predictable flow of data.

Seven traces extracted from the Raw's static network, through binaries compiled by the Raw compiler on the Raw cycle-accurate simulator were also used to evaluate the base ANN. The traces accurately match the hardware timing. The same methodology as with the synthetic benchmarks was followed; the ANN prediction accuracy for each of the seven Raw benchmark traces was obtained as before. Table 1.3 shows the accuracy rates obtained during discrete simulation intervals, until the simulation was completed at half a million cycles, while Table 1.4 shows the percentage of those identified hotspots that were detected at least 50 cycles in advance of their future occurrence. It is evident that the ANN remains accurate throughout the simulation, even if for the fast Fourier transform (fft) benchmark, the prediction accuracy is relatively low. The ANN does exceptionally well with three benchmark traces (mpeg2, streams and vpr). The streaming nature of the mpeg2 and stream applications follows a repetitive pattern, therefore if the ANN is properly trained, streaming applications could largely benefit from this.

1.4.4 Hardware Synthesis Results

The ANN is evaluated also in terms of CMOS area requirements and hardware overheads. Hence, the ANN hardware architecture shown in Fig. 1.4 was implemented and synthesized using Synopsys Design Vision, targeting a 65 nm commercial CMOS technology library. The targeted frequency was 500 MHz, operating at 1 V power supply voltage. Synthesis results indicate that an NoC router with 2 VCs, lookahead routing and speculative switch allocation requires ~73,240 gates, while an estimated amount of 67,600 gates is required for the base ANN implementation that controls a 4 × 4 region, making the ANN predictor smaller in area than the aforementioned pipelined router. Thus, the base ANN hardware overhead is 5.8% with respect to the base 16-router region it controls. These results indicate that the ANN predictor can be easily integrated as a PE in a typical NoC architecture. Synopsys' PrimePower is also used to provide an estimate of the power consumption. Using 50% switching activity probability, the ANN consumes an estimated 0.016 mW when computing one hotspot prediction. The estimated power requirements of the ANN predictor are considered negligible when compared to existing real architectures such as Teraflops [2]. Both these results show that the overall hardware and power overheads of the ANN predictor for hotspot management are very low, and thus allow for feasible hardware implementations in an NoC, given the benefits of performance hotspot prediction in on-chip interconnected systems.

1.5 Conclusions and Future Work

This Chapter presented an intelligent ANN hotspot prediction mechanism. The ANN uses buffer utilization data to dynamically monitor the interconnect fabric, and reactively predicts the location of an about to-be-formed hotspot. The ANN is trained using synthetic traffic models, and evaluated using both real and synthetic application traces. Results show that a relatively small neural network architecture can predict hotspot formation with accuracy ranges between 76 and 92%.

The promising results encourage further research in using intelligent mechanisms for managing NoCs. Further optimized ANNs can be explored in predicting hotspots in NoCs more accurately. Furthermore, the coupling of proactive ANN prediction mechanisms with reactive hotspot reduction mechanisms to avoid the formation of unforeseen hotspots occurring in NoCs can be very beneficial. Additionally, other system information such as link utilization and topology information can be used to enhance the training of the ANN, in order to improve the accuracy and efficiency of the predictor.

References

1. Baydal E et al (2005) A Family of mechanisms for congestion control in wormhole networks. In IEEE TPDS 16(9):772–784 Sept 2005
2. Bell S et al (2008) TILE64 Processor: A 64-Core SoC with mesh interconnect. In: ISSCC, pp 88–598 Feb 2008

3. Bertozzi D, Benini L (2004) Xpipes: A Network-on-Chip architecture for gigascale Systems-on-Chip. In: IEEE Circ Syst 4(2):18–31, Second Quarter
4. Bjerregaard T, Mahadevan S (2006) A survey of research and practices of Network-on-Chip. In ACM CSUR 38(1):1–51 March 2006
5. Bolotin E et al (2004) QNoC: QoS architecture and design process for Network on Chip. In Elsevier JSA 50(2–3):105–128 Feb 2004
6. Dally WJ (1992) Virtual-channel flow control. In IEEE TPDS 3(2):94–205 March 1992
7. Dally WJ, Towles B (2001) Route packets, not wires: on-Chip interconnection networks. In: DAC, pp 684–689 June 2001
8. Dally WJ, Towles B (2004) Principles and practices of interconnection networks. Morgan kaufmann publishers Inc. ISBN 9780122007514
9. Daneshtalab M et al (2006) NoC hot spot minimization using antNet dynamic routing algorithm. In: ASAP, pp 33–38 Dec 2006
10. Duato J et al (2005) A new scalable and cost-effective congestion management strategy for lossless multistage interconnection networks. In: HPCA, pp 108–119 Feb 2005
11. Goossens K et al (2005) AEtherealn Network on chip: concepts, architectures, and implementations. In: IEEE DTC, pp 414–421 Sept-Oct 2005
12. Hashem S et al (1999) A novel approach for training neural networks for long-term prediction. In IJCNN 3:1594–1599 July 1999
13. Hashemi KS et al (1991) On the number of training points needed for adequate training of feedforward neural networks. In: IFNNPS, pp 232–236 July 1991
14. Ho WS, Eager DL (1989) A novel strategy for controlling hot-spot congestion. In: IEEE ICPP, pp 14–18
15. Gaughan PT, Yalamanchili S (1993) Adaptive routing protocols for hypercube interconnection networks. In IEEE Computer 26(5):12–23 May 1993
16. Jain AK et al (1996) Artificial neural networks: a tutorial. In IEEE Computer 29(29):31–44 March 1996
17. Maqsood I et al (2004) An ensemble of neural networks for weather forecasting. In Neural Computing & Applications 13(2):112–122 June 2004
18. McCoy A et al (2007) Multistep-Ahead Neural-Network Predictors for Network Traffic Reduction in Distributed Interactive Applications. In: ACM TOMACS, 17(4):1–30
19. Nilsson E et al (2003) Load Distribution with the Proximity Congestion Awareness in a Network on Chip. In: DATE, pp 11126–11127 March 2003
20. Peh L-S, Dally WJ (2000) Flit-Reservation Flow Control. In: HPCA, pp 73–84 Jan 2000
21. Pande PP et al (2005) Performance evaluation and design trade-offs for Network-on-Chip interconnect architectures. In IEEE TPDS 54(8):1025–1040 Aug 2005
22. Sarbazi-Azad H et al (2001) An analytical model of fully-adaptive wormhole-routed k-ary n-cubes in the presence of hot spot traffic. In IEEE TOC 50(7):623–634 July 2001
23. Steven G et al (2001) Dynamic branch prediction using neural networks. In: DSD, pp 178–185 Sept 2001
24. Taylor MB et al (2004) Evaluation of the raw microprocessor: an exposed-wire-delay architecture for ILP and streams. In: ISCA, pp 2–13
25. Teixeira A et al (2000) A multi-objective optimization approach for training artificial neural networks. In: IEEE SBRN, pp 168–172 Jan 2000
26. Vangal S et al (2007) An 80-tile 1.28TFLOPS Network-on-Chip in 65 nm CMOS. In: ISSCC, pp 98–99 Feb 2007
27. Walter I et al (2007) Access regulation to hot-modules in wormhole NoCs. In: NoCs, pp 137–148 May 2007

Chapter 2
Accurate Asynchronous Network-on-Chip Simulation Based on a Delay-Aware Model

Naoya Onizawa, Tomoyoshi Funazaki, Atsushi Matsumoto and Takahiro Hanyu

Abstract A performance-evaluation simulator, such as a cycle-accurate simulator, is a key tool for exploring appropriate asynchronous Network-on-Chip (NoC) architectures in early stages of VLSI design, but its accuracy is insufficient in practical VLSI implementation. In this paper, a highly accurate performance-evaluation simulator based on a delay-aware model is proposed for implementing an appropriate asynchronous NoC system. While the unit delay between circuit blocks at every pipeline stage is constant in the conventional cycle-accurate simulator, which causes poor accuracy, the unit delay between circuit blocks in the proposed approach is determined independently by its desirable logic function. The use of this "delay-aware" model makes it accurate to simulate asynchronous NoC systems. As a design example, a 16-core asynchronous Spidergon NoC system is simulated by the conventional cycle-accurate and the proposed simulator whose results, such as latency and throughput, are validated with a highly precise transistor-level simulation result. As a result, the proposed simulator achieves almost the same accuracy as one of the transistor-level simulators with the simulation speed comparable to the cycle-accurate simulator.

N. Onizawa (✉) · T. Funazaki · A. Matsumoto · T. Hanyu
Research Institute of Electrical Communication, Tohoku University, Sendai, Japan
e-mail: onizawa@ngc.riec.tohoku.ac.jp

T. Funazaki
e-mail: funazaki@ngc.riec.tohoku.ac.jp

A. Matsumoto
e-mail: matumoto@ngc.riec.tohoku.ac.jp

T. Hanyu
e-mail: hanyu@ngc.riec.tohoku.ac.jp

N. Voros et al. (eds.), *VLSI 2010 Annual Symposium*,
Lecture Notes in Electrical Engineering 105, DOI: 10.1007/978-94-007-1488-5_2,

2.1 Introduction

With circuit size reaching billions of transistors, traditional shared-bus architectures become unusable for a complex System-on-Chip (SoC). Network-on-Chip (NoC) design paradigm [1] has been recently proposed to provide scalable on-chip global communication [2]. Especially, NoC based on Globally Asynchronous Locally Synchronous (GALS) system [3] takes advantage of the benefits of asynchronous circuits such as low power consumption and communication robustness [4].

On-chip communication performance, such as latency and throughput, in NoC is evaluated based on static performance analyses or system-level simulations for design space exploration of NoC architectures at early stages before implementation [5, 6]. To validate the fast evaluation techniques, a cycle-accurate simulator is useful for performance evaluation of synchronous NoC architectures because behavior of circuits within one clock cycle is precisely simulated [7, 8]. However, on-chip communication in the GALS-NoC is controlled by request-acknowledge based handshaking between modules without a global clock, the asynchronous operations are not performed at a constant period. Therefore, the GALS-NoC performance cannot be accurately evaluated by using the cycle-accurate simulator. For early design exploration of the GALS-NoC architectures, an accurate simulator supporting the asynchronous operation is required to validate the fast evaluation techniques.

This paper presents a highly accurate simulator based on a delay-aware model for evaluation of the on-chip communication performance, such as throughput and latency, in the asynchronous NoC architectures. In the proposed delay-aware model, each asynchronous operation between modules is modeled by a delay module. Since the delay information is selected dependent on conditions by a circuit-delay table, asynchronous on-chip communication operating at different speed can be simulated. The circuit delay is determined by transistor-level simulation results of post-layout asynchronous NoC routers and communication links, which improves the accuracy of the simulation. In addition, as the asynchronous operation is started when the acknowledge information of the previous operation is returned to the module, the delay module is scheduled to be active when inputs to the module satisfy the specific condition. The scheduling makes it possible to simulate local synchronization by the handshake operation between modules.

To evaluate the accuracy of simulations, the proposed and the cycle-accurate simulators are validated using a highly precise transistor-level simulation result in a 16-core asynchronous Spidergon [9] NoC architecture. As a result, the proposed simulator obtains the minimum latency with about 5% error with respect to that with about 35% error from the cycle-accurate simulator at comparable speed.

The rest of this paper is organized as follows. Section 2.2 describes related works and our motivation. Section 2.3 describes a delay-aware model for the asynchronous router. Section 2.4 describes the simulation accuracy of our simulator. Section 2.5 concludes this paper.

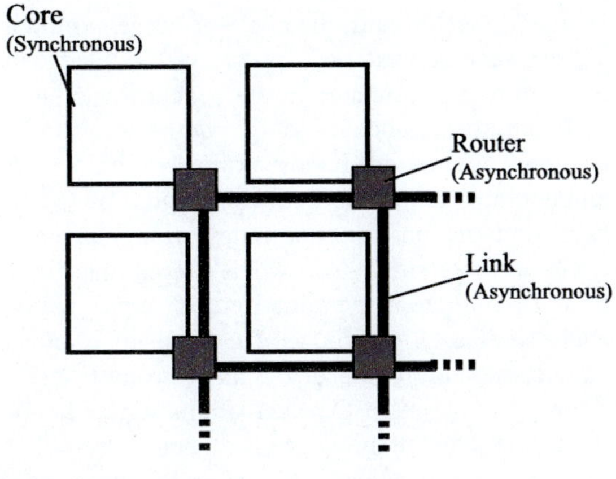

Fig. 2.1 NoC architecture based on a 2D-mesh topology

2.2 Background and Motivation

2.2.1 GALS-NoC Architecture

Figure 2.1 shows a typical NoC architecture based on a 2D-mesh topology, which consists of switching routers, communication links between the routers and processing cores. The processing cores transmit/receive data using the inter-core communication network. The switching routers compute where to transmit an incoming data and arbitration between potential concurrent data and finally transmit the selected data on the selected link.

The GALS-NoC architecture is that NoC is implemented based on a GALS system, where processing cores at different clock frequencies transmit/receive data through the asynchronous switching routers [10, 11] and communication links [12, 13]. In general, the asynchronous inter-core communication network is realized by using quasi delay-insensitive(QDI) circuits which are robust against a delay variation because of its timing assumption [14].

2.2.2 Related Work and Contributions

NoC performance, such as throughput and latency are varied with topology, flow control, routing algorithm and flit size. In order to obtain an efficient NoC architecture for specific applications, performance evaluation techniques are extremely important at early stages before implementation. The performance evaluation techniques includes static performance analyses [5, 15] and simulations, where the simulations are done based on non-cycle accurate high-level modeling [16, 6] and cycle-accurate modeling [7, 8]. In general, the static analyses and the non-cycle accurate simulations are used to explore an optimum NoC architecture at early stages because of fast evaluation. To validate the fast evaluation techniques, the cycle-accurate simulation is used. Since behavior of the circuit, such as routers, is precisely modeled within a clock cycle, on-chip communication performance in the synchronous NoC can be accurately evaluated.

In contrast, since on-chip communication is performed based on request-acknowledge based handshaking without a global clock in the GALS-NoC architecture, the asynchronous circuits are not operated at constant speed. The asynchronous on-chip communication cannot be accurately simulated by the cycle-accurate simulator, which causes inaccurate results of performance evaluation. To validate fast evaluation techniques using the static analyses and the non-cycle simulation techniques for the asynchronous NoC design space exploration, an accurate simulation of the asynchronous NoC architecture is required.

In [17], a System C-based simulation technique for performance evaluation of the GALS-NoC architecture has been proposed. The simulator supports the asynchronous transfer delay which is independent of a clock cycle by using the circuit delay in packet switching. However, this is not normally used in the NoC because of hardware overhead, where a complete packet is stored in the router. The circuit delay is not accurate because the asynchronous routers and communication links are not implemented. In addition, the simulator does not support the acknowledge operation, which cannot accurately simulate the handshake operation.

This paper presents a highly accurate simulator to evaluate the on-chip communication performance, such as throughput and latency, in the GALS-NoC architectures. Our contributions of this work are: a) to model the acknowledge operation for the accurate simulation of the asynchronous circuit; b) to use highly precise delay information extracted from transistor-level simulation results of post-layout asynchronous NoC circuits for improving the accuracy; and c) to support wormhole switching which is normally used in NoC because of small hardware.

2.3 Delay-Aware Model for an Asynchronous NoC Simulation

2.3.1 Asynchronous Router

In the GALS-NoC architecture, every core communicates with each other by packet-based communication through asynchronous inter-core network whose elements are asynchronous switched routers and communication links. In wormhole switching, a packet is divided into several flow control digits (flits) which are transferred through the network. Figure 2.2 shows a format of wormhole-switched packets in a deterministic routing algorithm. The packet consists of three types of flits (*FlitType*): a header flit, a body flit and a tail flit. The header flit contains a static routing path in an address field in order to perform the packet routing from one processing core to a destination processing core. The address field contains some sub-addresses, such as Addr1, Addr2... and AddrN. Each router uses one sub-address (Addr1) in the address field and shifts the address field for the following routers.

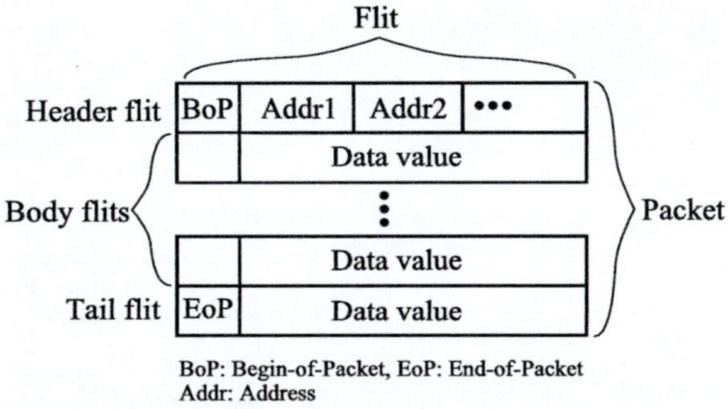

Fig. 2.2 Format of a wormhole-switched packet

Figure 2.3 shows a block diagram of an asynchronous router which is divided into an input unit, a routing unit, an arbitration unit and an output unit by pipelatches (PLs). The input unit is first-in-first-out (FIFO) buffers for storing incoming flits from other switched routers via communication links or a processing core. The routing unit selects the link based on the address field of the header flit in routing controller (RC) and then transfers the flit to the selected arbitration unit using shifter (SH). The arbitration unit determines a flit which can be transferred to the output unit in arbitration controller (AC). When multiple flits simultaneously request the same output unit, a selected flit is transferred to the output unit, while the other flits remain in the routing unit. The output unit selects a flit from the arbitration unit in multiplexer (MX) and then transfers the flit to other switched routers via communication links or a processing core.

In the asynchronous router, a flit in each unit is asynchronously transferred by request-acknowledge based handshaking between PLs which are connected with the input and the output of its unit. In general, the asynchronous router is designed by using a QDI logic circuit based on a 4-phase protocol. A flit with the request information is encoded as data represented by a 1-of-N signal, while spacer is inserted into two time-consecutive data to distinguish them [14], where data and spacer are defined by *PhaseType* shown in Table 2.1. PL is implemented by C-elements [18] and an OR gate. The C-element is a standard asynchronous storage element, whose output is low (high) when all inputs are low (high), and which otherwise holds its value. When the 1-of-N signal (data) is stored in the C-elements, one of inputs to the OR gate becomes high, which asserts an output of the OR gate. In contrast, the output of the OR gate is deasserted when spacer is stored in the C-elements. The output signal of the OR gate means acknowledge information of the data or spacer. *PhaseType* of the acknowledge information is DataPhase when the output signal of the OR gate is high, and SpacerPhase when low shown in Table 2.1.

Figure 2.4 shows an example of an asynchronous router operation. Firstly, a header flit with the request information (data) is stored in PL of the input unit. Secondly, PL of the input unit asserts ***ACK_OUT*** as acknowledge information of data whose *PhaseType* is DataPhase. This leads to changing ***DATA_IN*** from data to spacer whose *PhaseType* is Spacer Phase. Thirdly, spacer is stored in PL of the

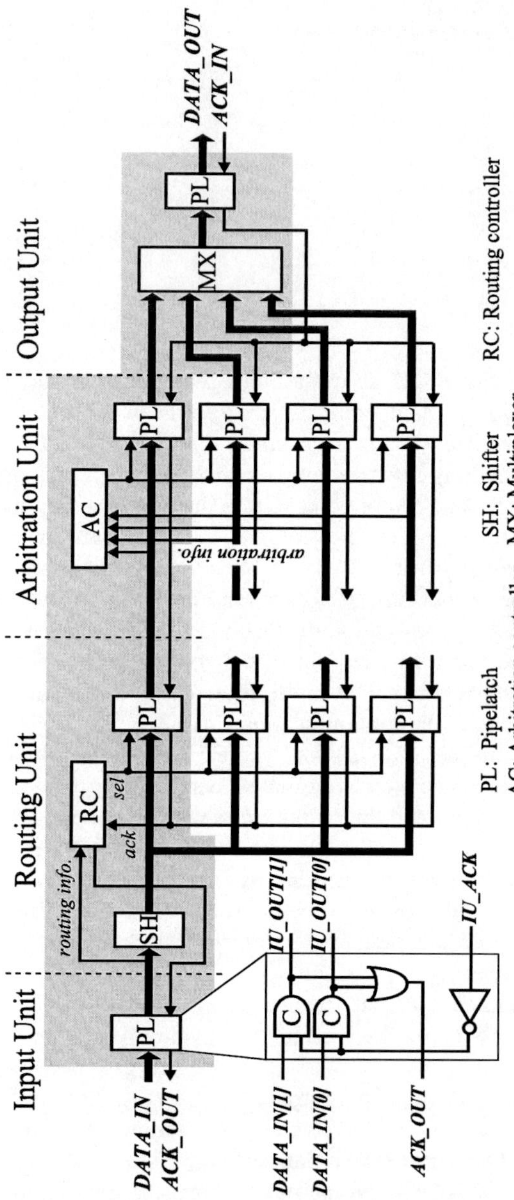

Fig. 2.3 Block diagram of an asynchronous router

input unit when the acknowledge information of data is arrived to PL of the input unit from the next unit (router unit). This means that a flit (data or spacer) is stored in PL when the flit (***DATA_IN***) and the acknowledge information (***IU_ ACK***) have different *PhaseType*. Finally, PL of the input unit deasserts ***ACK_OUT*** as acknowledge information of spacer.

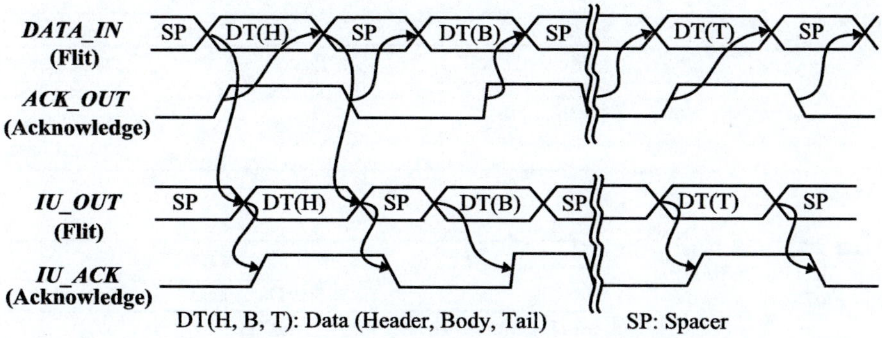

Fig. 2.4 Example of an asynchronous router operation

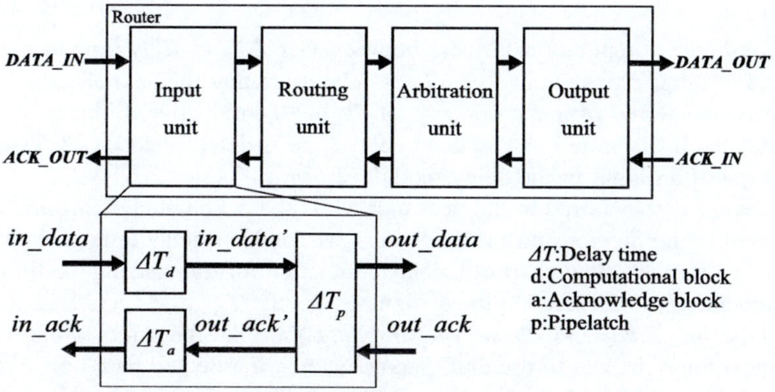

Fig. 2.5 Asynchronous router based on the delay-aware model

This represents one cycle of the asynchronous operation based on the 4-phase protocol. Namely, the asynchronous unit is operated based on a one-cycle delay time which is varied with *FlitType* and *PhaseType*, since the operations are different on the conditions.

2.3.2 Delay-Aware Model

To simulate accurately latency and throughput of a specific asynchronous NoC architecture, the proposed delay-aware model captures the functionality and timing information of each unit operating at different speed. Figure 2.5 shows the asynchronous router based on the delay-aware model. Each unit between PLs is modeled by a delay module based on the delay information of its desirable logic function. The delay module includes computation delay time (ΔT_d), acknowledge delay time (ΔT_a) and pipelatch delay time (ΔT_p). ΔT_d is delay time that the flit is

Table 2.1 *PhaseType* of the asynchronous signal

		PhaseType
Flit	Data	DataPhase
	Spacer	SpacerPhase
Acknowledge	High	DataPhase
	Low	SpacerPhase

Table 2.2 Delay time dependent on *PhaseType* and *FlitType*

		PhaseType	
		DataPhase	SpacerPhase
FlitType	Header flit	ΔT_{hd}	ΔT_{hs}
	Body flit	ΔT_{bd}	ΔT_{bs}
	Tail flit	ΔT_{td}	ΔT_{ts}

transferred over computational blocks between PLs. ΔT_p is delay time that the flit is stored in the C-elements of PL. ΔT_a is delay time that the acknowledge information is transferred over the OR gate of PL and combinational blocks between PLs after the flit is stored in PL. Each delay time is determined by *FlitType* and *PhaseType* of an input in the delay module shown in Table 2.2

As the flit is transferred to the next unit while the acknowledge information is transferred to the previous unit in each unit, two kinds of delay time is simulated. Figure 2.6a shows a flow chart of a delay simulation for transferring the flit in the delay module. T_x is the arrival time of signal x. Firstly, T_{in_data} is determined when *PhaseType* (*in_ data*) and *PhaseType* (*in_ data'*) are not the same, which means that a new flit is arrived to the unit. Secondly, ΔT_d is selected based on *FlitType* and *PhaseType* of *in_ data* shown in Table 2.2. Then, $T_{in\_data'}$ is calculated by adding the ΔT_d to T_{in_data}. Thirdly, $T_{in\_data'}$ and T_{out_ack} are compared and then the larger one is selected when *PhaseType* (*in_ data*) and *PhaseType* (*out_ ack*) are not the same. Finally, T_{out_data} is calculated by adding ΔT_p.

Figure 2.6b shows a flow chart of a delay simulation for transferring the acknowledge information in the delay module. Two flow charts shown in Fig. 2.6 are almost the same. $T_{out\_ack'}$ can be simulated as T_{out_data} shown in Fig. 2.6a. Then, ΔT_a is determined by *FlitType* and *PhaseType* of *out_ ack* and is then added to $T_{out\_ack'}$.

2.4 Evaluation

2.4.1 Simulation Environment

Figure 2.7 shows simulation environment for performance evaluation of asynchronous NoC architectures. It is assumed that a target asynchronous NoC architecture is simulated by a testbench which calculates latency and throughput from the

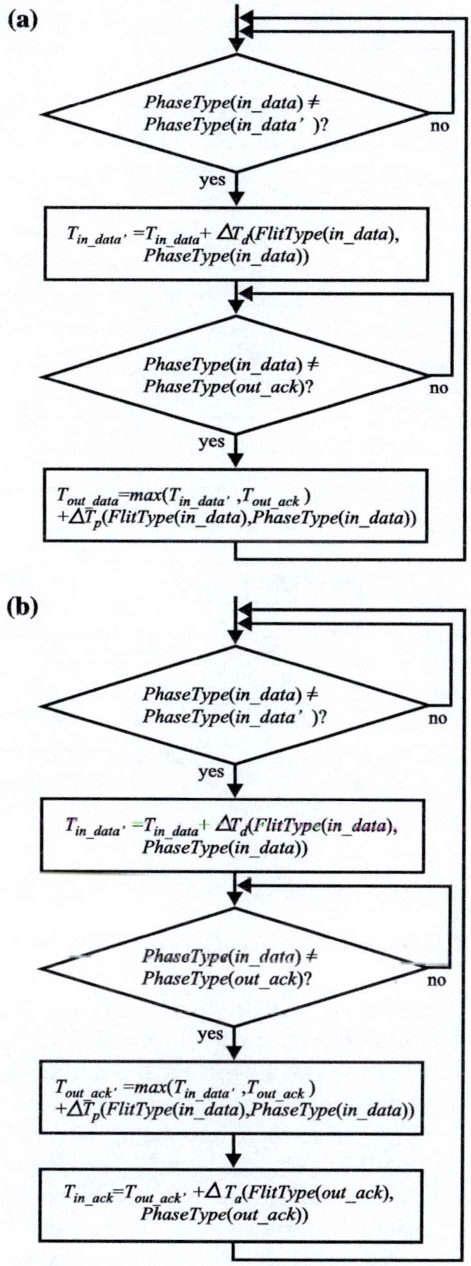

Fig. 2.6 Flow charts of a delay simulation for transferring: **a** the flit and **b** acknowledge information

simulation result. The target NoC architecture includes asynchronous routers (Rs) modeled as Sect. 2.3.2 and interconnect wires between routers. The interconnect wire is modeled as a delay element based on its wire length. Packet Transceivers (PTs) and

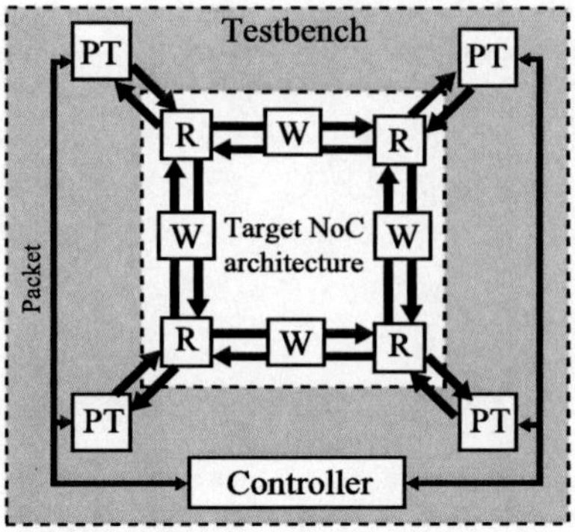

Fig. 2.7 Simulation environment

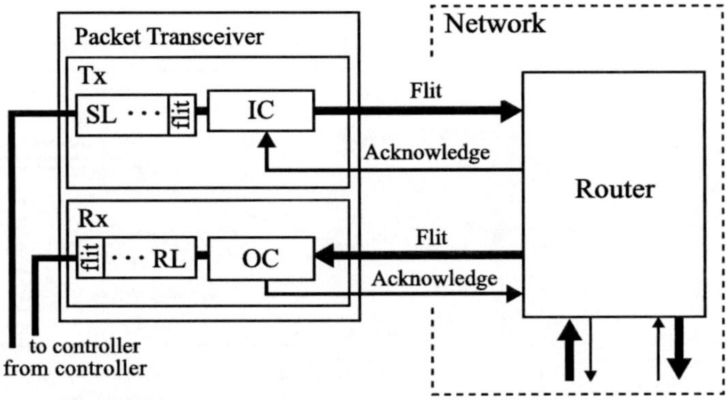

Fig. 2.8 Block diagram of a Packet Tranceiver (PT)

Controller are included in the testbench. The target asynchronous NoC architecture and the testbench are described using Verilog HDL.

In CT, packets are generated based on parameters, such as packet size, flit size and offered traffic. The packets are partitioned into flits, which are in turn encoded by 1-of-*N* signals, where information of a static routing path between source and destination PTs is embedded in the header flit. The encoded flits are transmitted to the source and the destination PTs based on the information of the static routing path.

Figure 2.8 shows a block diagram of PT which is composed of transmitter (Tx) and receiver (Rx). Tx includes sending list (SL) and input controller (IC). Rx includes receiving list (RL) and output controller (OC). The encoded flits transmitted from CT are stored in SL when the header flit indicates that PT is the source of the routing path, or are stored in RL when the header flit indicates that PT is the destination of the routing path.

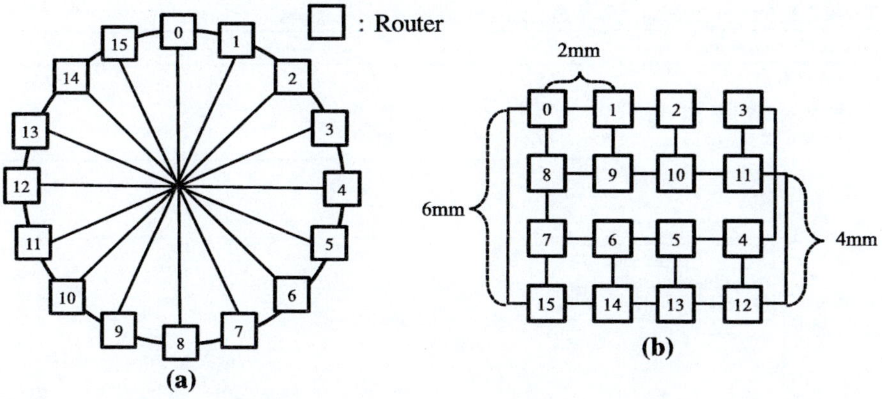

Fig. 2.9 16-core Spidergon NoC: **a** topology, **b** physical layout

In Tx, since the asynchronous operation based on the handshake protocol is not operated at constant speed, the flits should be transferred to a router in the target asynchronous NoC architecture when IC detects that the acknowledge signal from the router is changed. Time to transfer the flit is recorded in SL for calculation of latency and throughput. In Rx, the flits from other PTs are received. When the received flit is the same as the previously stored flit in RL, time to receive the flit is recorded in RL for calculation of latency and throughput. Then, OC changes the acknowledge signal of the flit from the router.

2.4.2 Simulation Accuracy

To evaluate the accuracy of simulators, a 16-core asynchronous Spidergon [9] NoC architecture shown in Fig. 2.9 is simulated by the conventional cycle-accurate simulator and the proposed simulator whose results, such as latency, are validated with a highly precise transistor-level simulation result. Each node (router) has links to its clockwise and counter-clockwise neighboring nodes, and a direct link to its diagonally opposite neighbor shown in Fig. 2.9a. In the physical layout shown in Fig. 2.9b, the physical connections between routers only need to cross at one point. It is assumed that each processing core with its router is connected by three types of wires whose lengths are 2, 4 and 6 mm.

In the 16-core asynchronous Spidergon NoC architecture, asynchronous routers and the communication links are designed by the QDI logic circuit based on a 4-phase dual-rail encoding. The packet is partitioned into 12-bit flits. BoP and EoP are represented respectively by 1 bit, while data value or address is represented by 10 bits depicted in Fig. 2.2. The transistor-level simulation results are obtained by NanoSim under a ROHM 0.18 μm CMOS technology. The conventional cycle-accurate and the proposed simulators are described by Verilog HDL.

Table 2.3 Router latency versus the number of packet collisions

# of packet collisions	0	1	2	3
Proposed [ns]	55.2	83.5	111.7	139.9
Cycle-accurate [ns]	62.8	94.3	125.8	157.2
Transistor-level [ns]	57.6	86.9	115.6	145

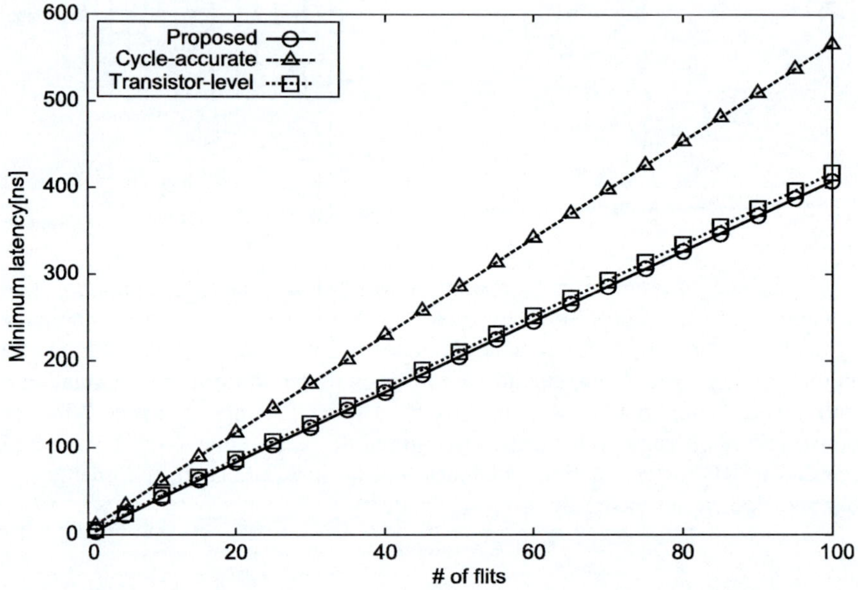

Fig. 2.10 Minimum latency in the 16-core asynchronous Spidergon NoC architecture

Table 2.3 shows latencies of the asynchronous router versus the number of packet collisions where one packet consists of 15 flits. The latencies are D_t, D_c and D_p from the transistor-level, the cycle-accurate and the proposed simulators, respectively, where an error is defined as a ratio of an absolute difference $|D_t - D_c|$ or $|D_t - D_p|$ to D_t. The latencies in the cycle-accurate simulator differ from those in the transistor-level simulator by about a 10% error because all units are simulated at a constant period limited by the worst-case delay. In contrast, as the proposed simulator simulate each unit operating at different speed by using the delay-aware model, the latencies can be obtained within a 5% error in comparison with one of the transistor-level simulators.

Figure 2.10 shows the minimum latencies through the 16-core asynchronous Spidergon NoC architecture which is dependent on the number of flits in one packet. Actually, the flit transmission in the asynchronous NoC is performed at different speed determined by the routing path. However, in the cycle-accurate simulator, the flit transmission in every unit is simulated at low speed dependent on the longest (6 mm) wire, which causes about a 35% error. In contrast, the flit transmission in each unit is simulated at different speed in the proposed simulator. Therefore, the minimum latencies can be obtained within 5% error in comparison with one of the transistor-level simulators.

The proposed and the cycle-accurate simulator run at almost the same time which is about 4 orders of magnitude faster than the transistor-level simulator.

2.5 Conclusion

In this paper, a highly accurate asynchronous NoC simulator has been proposed for validation of fast evaluation techniques, such as static performance analyses and system-level simulators which are used for early design space exploration of NoC architectures. Since the asynchronous operation is modeled by a delay module, where the delay time is changed based on a flit type, the non-periodic operation based on the handshake protocol can be accurately simulated. As a result, the proposed simulator achieves almost the same accuracy as the transistor-level simulator with the comparable speed to the cycle-accurate simulator.

The future direction includes building a fast system-level asynchronous NoC simulator, whose accuracy is validated with the proposed simulator upon different topologies.

Acknowledgements This research was supported by JST, CREST. This simulation is supported by VLSI Design and Education Center (VDEC), the University of Tokyo in collaboration with Synopsys, Inc. and Cadence Design Systems, Inc.

References

1. Benini L, De Micheli G (2002) Networks on chips: a new SoC paradigm. IEEE Comput 35(1):70–78
2. Vangal S, Howard J, Ruhl G, Dighe S, Wilson H, Tschanz J, Finan D, Iyer P, Singh A, Jacob T, Jain S, Venkataraman S, Hoskote Y, Borkar N (2007) An 80-tile 1.28TFLOPS network-on-chip in 65 nm CMOS. In: IEEE ISSCC Digest of technical papers, pp 98–99 Feb 2007
3. Chapiro DM (1984) Globally-asynchronous locally-synchronous sysmtes. Ph.D. Thesis, Stanford University, Stanford, CA, Oct 1984
4. Lattard D, Beigne E, Clermidy F, Durand Y, Lemaire R, Vivet P, Berens F (2008) A recongurable baseband platform based on an asynchronous network-on-chip. IEEE J Solid-State Circuits 43(1):223–235
5. Bakhouya M, Suboh S, Gaber J, El-Ghazawi T (2009) Analytical modeling and evaluation of on-chip interconnects using network calculus. In: The 2009 3rd ACM/IEEE international symposium on networks-on-chip, NOCS '09, pp 74–79
6. Banerjee A, Wolkotte PT, Mullins RD, Moore SW, Smit GJM (2009) An energy and performance exploration of network-on-chip architectures. IEEE Trans Very Large Scale Integr VLSI Syst 17:319–329
7. Loghi M, Angiolini F, Bertozzi D, Benini L, Zafalon R (2004) Analyzing on-chip communication in a MPSoC environment. In: Design, automation and test in Europe conference and exhibition, DATE ' 04, vol. 2. pp 752–757

8. Wolkotte PT, Holzenspies PK, Smit GJM, (2007) Fast, accurate and detailed NoC simulations. In: The first international symposium on networks-on-chip, NOCS '07, pp 323–332
9. Moadeli M, Shahrabi A, Vanderbauwhede W, Ould-Khaoua M (2007) An analytical performance model for the spidergon NoC. In: The 21st international conference on advanced information networking and applications, AINA '07, pp 1014–1021
10. Beigne E, Clermidy F, Vivet P, Clouard A, Renaudin M (2005) An asynchronous NOC architecture providing low latency service and its multi-level design framework. In: The 11th IEEE international symposium on asynchronous circuits and systems, ASYNC '05, pp 54–63
11. Bainbridge J, Furber S (2002) Chain: a delay-insensitive chip area interconnect. IEEE Micro 22(5):16–23
12. Mizusawa K, Onizawa N, Hanyu T (2008) Power-aware asynchronous peer-to-peer duplex communication system based on multiple-valued one-phase signaling. IEICE Trans Electron E91-C(04)):581–588
13. Otake Y, Onizawa N, Hanyu T (2009) High-performance asynchronous intra-chip communication link based on a multiple-valued current-mode single-track scheme. In: IEEE international symposium on circuits and systems, ISCAS '09, pp 1000–1003
14. Sparsø J, Furber S (2001) Principles of asynchronous circuit design. Kluwer Academic Publisher, Dordrecht
15. Ogras UY, Marculescu R (2007) Analytical router modeling for networks-on-chip performance analysis. In: Design, automation and test in Europe conference and exhibition, DATE '07, pp 1–6
16. Kahng AB, Li B, Peh L-S, Samadi K (2009) ORION 2.0: a fast and accurate NoC power and area model for early-stage design space exploration. In: Design, automation and test in Europe conference and exhibition, DATE '09, pp 423–428
17. Ling X, Chiu-Sing C (2007) A network-on-chip system-level simulation environment supporting asynchronous router. In: The 7th international conference on ASIC, ASICON '07, pp 1241–1244
18. Shams M, Ebergen JC, Elmasry MI (1998) Modeling and comparing CMOS implementations of the C-element. IEEE Trans Very Large Scale Integr VLSI Syst 6(4):563–567

Chapter 3
Trust Management Through Hardware Means: Design Concerns and Optimizations

Apostolos P. Fournaris and Daniel M. Hein

Abstract Trust in security demanding software platforms is a very important feature. For this reason, Trusted computing group has specified a TPM hardware module that can enforce and guaranty a high trust level to all the platform's involved entities. However, the TPM's features can not be fully exploited in systems under extreme physical conditions. To solve this problem, the use of a special purpose hardware module, physically connected to a host security system's device acting as a local trusted third party, has been proposed in literature. In this chapter, we describe the hardware structure of such a hardware module, called Autonomous Attestation Token (AAT) and discuss hardware resource constrains, security bottlenecks that can stem from improper design of its various components integrated in the AAT's structure. We conclude that the efficiency of the AAT system is closely related to the efficiency of its public key encryption–decryption unit (RSA encryption–decryption module). In this book chapter, we address these issues by describing a design methodology toward a low hardware resources (small chip covered area) and side channel attack resistant RSA hardware architecture. The described hardware architectures' implementations provide very optimistic results of very low chip covered area and high computation speed thus verifying the efficiency of the proposed algorithms and architecture design approach.

A. P. Fournaris (✉)
Department of Electrical and Computer Engineering,
University of Patras, Rio Campus, Patras, Greece
e-mail: apofour@gmail.com

D. M. Hein
Institute for Applied Information Processing and Communications,
Graz University of Technology, Graz, Austria
e-mail: daniel.hein@iaik.tugraz.at

N. Voros et al. (eds.), *VLSI 2010 Annual Symposium*,
Lecture Notes in Electrical Engineering 105, DOI: 10.1007/978-94-007-1488-5_3,

3.1 Introduction

Trust in Information Systems constitutes a fundamental security issue in most sensitive data handling applications. However, achieving a high level of trust in the entities of such systems is not an easy task. There are some computer communication systems where the nature of the handled information is so sensitive that untrusted behaviors can not be tolerated. In such systems, trust is ensured by hardware means [1].

Trusted Computing is an initiative concerned with establishing trust in commodity-off-the-shelf computing platforms that is propagated by the Trusted Computing Group (TCG). Trusted Computing provides a facility to establish the load-time software integrity of a computing platform and attest this platform configuration to interested third parties. The Trusted Computing concept relies heavily on a specialized security chip, the *Trusted Platform Module* (*TPM*) [2]. A TPM is physically bound to the computing platform it protects. Nowadays, many Personal Computing platforms are already shipped with a TPM by default. The TPM provides security functions such as 2048 bit RSA cryptographic unit and key generator, a SHA-1 hash engine, secure non-volatile storage, and a true random number generator. The TPM supports the integrity protection of the host platform's software with two facilities: a shielded storage for platform software measurements, so-called Platform Configuration Registers (PCRs) and a way to cryptographically corroborate that these measurements are indeed authentic and protected by a genuine TPM.

Many researchers [3, 4] have remarked that Trusted Computing can be effectively applied to distributed software applications like mobile agent systems. Trusted Computing can help protecting the integrity of the distributed components in a mobile agent system. Using a transitive chain of trust model Trusted Computing and the TPM can ensure that a malicious user cannot hide tampering with, or misusing of a host computer in the mobile agent system. For this Trusted Computing relies on a process called "Remote Attestation", where a remote trusted third party decides if the applying computer has a trusted software configuration. The third party can then grant access to protected services based on this decision, for example by issuing a required cryptographic key. The remote trusted third party approach is not always viable, especially if the availability of such a service cannot be guaranteed. This is especially true if there is no reliable communication channel to this remote entity. A solution to this problem is "local attestation" [5, 6]. The functionality of this operation is based on an Autonomous Attestation Token (AAT) on which a number of cryptographic credentials are stored in a secure way. The retrieval of these credentials can only be done after a successful local attestation session.

In this book chapter, we explore the hardware structure of such a module, we analyze its functionality, discuss the security algorithms that are required to ensure trust and propose solutions on how the hardware units of the module can cooperate efficiently without compromises in the provided trust level. We introduce a local trusted third party hardware module architecture and describe how this architecture must function in order to match the trust specification of a security demanding system. After analyzing the AAT hardware structure, hardware efficiency and security bottlenecks are examined. One important bottleneck of the AAT hardware structure is the efficient design of its public key encryption–decryption unit (RSA unit). As a result, an RSA encryption–decryption unit design methodology is presented that takes into account the security demanding nature of the AAT by providing resistance against Fault and Simple power side channel attacks. The Chinese Remainder Theorem (CRT) RSA algorithm is adopted, employing a Fault and simple power attack resistant modular exponentiation (FSME) algorithm that uses values in carry-save logic. The proposed architecture is structured around an FSME unit that is reused for finding the two CRT RSA exponents in order to minimize the required chip covered area. To prove this area efficiency, the proposed FSME unit is implemented in FPGA technology and the implementation results are presented thus proving that SCA resistance in the proposed structure can be achieved with few compromises in performance.

The book chapter is organized as follows. In Sect. 3.2, the protocols behind the AAT are described and in Sect. 3.3 the AAT hardware structure is proposed and analyzed. In Sect. 3.4 an analysis is made on the problems-constrains of the AAT and possible solutions are discussed. In Sect. 3.5, algorithms for the AAT RSA encryption–decryption module are described. In Sect. 3.6 hardware architectures are proposed for CRT RSA. The described units performance characteristics are presented in Sect. 3.7.

3.2 Attestation Through Security Hardware Module

As a distributed information processing system a mobile agent system relies on the fact that each participating party executes the agents as intended by the operator of the agent platform. This requires a secure environment that can be trusted to execute the distributed code unhindered. One way to achieve his goal is to use special tamper proof hardware with support for trusted third parties [7]. The ubiquity of powerful commodity-of-the-shelf computer hardware, makes it desirable to use these relatively cheap platforms for distributed computing applications. This requires a way to determine if such a platform can be trusted to execute distributed code. Special purpose hardware chips have been proposed to determine the trusted state of such a platform [8].

One promising approach in this direction uses the *Autonomous Attestation Token* (*AAT*) [5]. The AAT restricts access to cryptographic credentials depending on the trusted software configuration of the host platform. Thus, the AAT does not need to provide a full tamper-proof execution environment, but leverages the host platform by ensuring its software integrity. The AAT is a hardware plug-in device for Local Attestation and the AAT concept allows for a variety of possible interfaces and form factors for its connection to a host platform. Possible candidates are SD cards, or USB tokens. The AAT contains a secure non-volatile memory to protect cryptographic keys and credentials. This information is only released to the host platform if the host platform has a trusted software configuration. The AAT works in conjunction with a

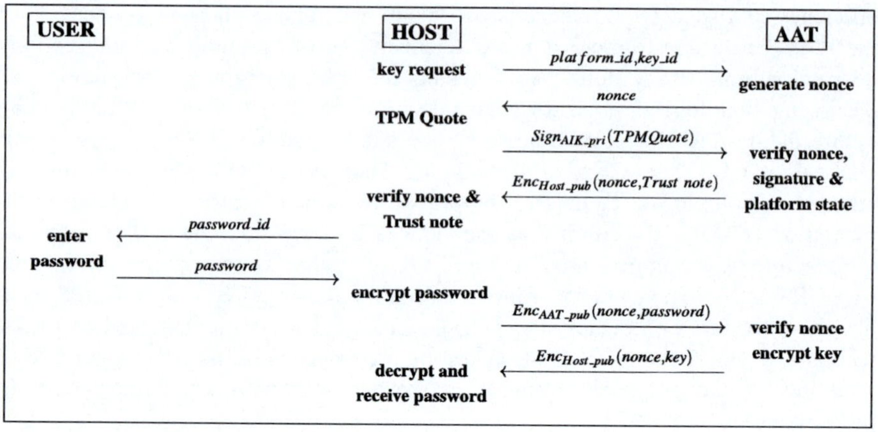

Fig. 3.1 The Host-AAT local attestation protocol

TPM to establish the host platform's trusted software configuration. The TPM is permanently bound to the host platform and is used to store integrity measurements of the host platform. These measurements take the form of SHA-1 hashes stored in the TPM's PCRs. A special TPM operation, the so-called *TPM Quote*, is used to attest the measured platform state. This is done by creating a SHA-1 hash of selected PCRs and signing it with a special TPM protected key, which can only be used for this purpose. Based on the information contained by a TPM Quote the AAT can decide to release protected information to the host.

In this section we introduce a protocol (cf. Fig. 3.1) for local attestation between the AAT and the host platform. The protocol is based on a local attestation protocol variant with increased Man-in-the-Middle resistance [6]. The goal of the protocol is to protect against an attacker who has full control on the communication channel between the host and the AAT (Dolev Yao threat model). The three entities in the protocol are the user who wants to employ a mobile agent system, the host platform used to connect to this distributed service and the AAT which guards the required service credentials in its protected storage. For the required credentials to be released the host platform must be in a trusted software configuration and the user must supply a password. To protect the communication messages between the AAT and the host platform previously exchanged RSA key pairs are used for encryption. In this protocol knowledge of a specific public key is used to corroborate the authenticity of both the AAT and the host. Therefore, both the private and the public communication keys must be kept secret. In the host this is ensured by using a TPM protected key pair for the communication ($Host_{pair}$) and storing the public AAT key (AAT_{pub}) in a file encrypted with the private part of the host communication key pair ($Host_{pri}$). On the AAT all keys for communication ($AAT_{pair}, Host_{pub}$) are kept in a secure storage location. For signing the TPM Quote a second TPM RSA key pair (Attestation Identity Key AIK_{pair}) is required. The public part of this key (AIK_{pub}) must be previously stored in the secure storage of the AAT as well.

The key release protocol is initiated when the host requests a key from the AAT by providing the *platform_id* and the *key_id*. The *platform_id* is necessary so that the AAT can select the correct encryption key $Host_{pub}$. As a response, the AAT sends a nonce. The host platform creates a TPM Quote containing the nonce, and a SHA-1 hash representing the platform software configuration. The TPM Quote is signed with AIK_{pri} by the TPM. The hosts then sends the signed TPM Quote to the AAT. The AAT verifies the signature using AIK_{pub}, checks that the nonce is correct and validates the given platform software configuration against the acceptable configurations stored in the AAT's shielded memory. The AAT then generates a Trust note that contains the result of the TPM Quote verification and the *password_id*. In case the verification fails the *password_id* is replaced with a random number. The Trust note along with the nonce is encrypted using the host's public key $Host_{pub}$ and send to the host. When the message is received, the host decrypts it and reads the Trust note. If the Trust note acknowledges a trusted platform configuration the host reads the *password_id* and prompts the user to insert the password. The password along with the nonce is then encrypted using the AAT's public key AAT_{pub} and send to the AAT. The AAT decrypts the message and verifies the nonce and the password. If both are correct the AAT encrypts the nonce and the requested key using $Host_{pub}$ and sends the result to the host. The host retrieves the key by decrypting the AAT message with $Host_{pri}$ and by verifying the nonce.

Each message exchange in the protocol is protected from replay attacks through the use of a random nonce generated by the AAT. This nonce is also concatenated to every protocol message to ensure that this message is related to the current run of the protocol. Also, after decryption, the message is expected to include the nonce value thus the message receiver can verify that the message is a correctly encrypted value and not some garbage data inserted into the channel. The channel is secured by encrypting all messages including sensitive information along with the nonce generated by the AAT. Knowledge of the host's public key by the AAT attests that the hardware device connected to this host is a legitimate AAT. Likewise the hosts identity is corroborated by knowledge of AAT_{pub} and the signature on the TPM Quote, because AIK_{pri} is bound to the host platform.

Apart from key release, the communication between a host and the AAT may involve a series of other actions. Those actions are related to the addition or deletion of keys and valid host software configurations. For this application establishing a symmetrically encrypted session is highly desirable. We suggest the use of AES with the Needham-Schroeder-Lowe [9] key agreement protocol.

3.3 AAT Hardware Structure

We envision the AAT as a synchronous System on Chip (SoC) device. The hardware structure of the AAT can be determined by the functions that it must fulfill. The AAT due to its potential connection with a TPM has a TPM-like structure and includes an RSA signature unit, a control processor unit, a non

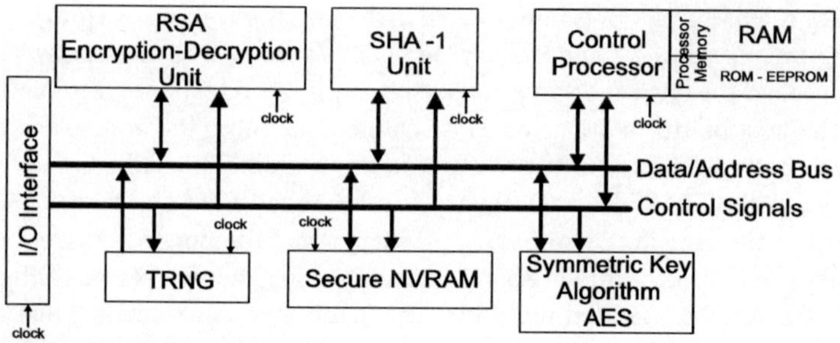

Fig. 3.2 The proposed AAT hardware structure

volatile memory unit for key storage, a true random number generator unit (TRNG), a SHA-1 hash function unit and a symmetric key encryption/decryption and key generation unit (AES algorithm).

The suggested hardware structure of an AAT chip is presented in Fig. 3.2. The system is structured around a data bus where all the data values are transferred for reading by or writing to a requesting unit of the AAT. There is also an address bus connected to the memory unit for a successful memory data reading and writing. An additional bus is also connected to all the units of the chip which is responsible for passing all the control signals to those units. Signals of this bus are in general managed by the processor.

The processor unit is responsible for controlling the whole AAT system and realizing the AAT functionality by enabling-controlling (chip select, CS, signal) AAT component units and performing operations that don't require the involvement of other units (i.e. comparisons or memory search). For this task, the processor has stored in its ROM memory a firmware program implementing the various AAT commands and a series of data required by those commands.

There are three memory modules included in the AAT hardware structure. The first module is a RAM unit that is employed for temporary value storage during a single AAT-Host session. The second module is a ROM–EEPROM unit that is mainly used for storing and updating the AAT firmware that realizes the AAT functionality. The third module is an NVRAM (flash memory) unit which constitutes the main storage area for all the sensitive information involved in the AAT transactions. The three memory units are protected by special hardware structures that deter attackers from deciphering memory data.

The SHA-1 unit is implementing the SHA-1 hash function and the RSA encryption–decryption unit is responsible for performing the arithmetic operation of modular exponentiation ($m^e modN$) as defined in RSA public key scheme. The SHA-1 unit has a data input/output and a control signal indicating the beginning of a hash function operation. The RSA unit has as inputs the modulus N value (part of the RSA public key), the message m to be encrypted-decrypted and the public or private key e along with a control signal indicating the beginning of a modular exponentiation encryption or decryption.

The AES encryption/decryption unit is responsible for the key generation, encryption and decryption of an established session's data that are transmitted to and from the AAT. It has control signals that indicate an encryption, decryption and key generation operation. The AES generated session key is only saved within the AES unit and is changed when is replaced by a newly generated session key.

The TRNG unit is connected to the data path through its data output and has a 9 bit control signal that determines the bit length of the generated random number. The NVRAM has a read/write control signal while is connected to the data and address bus in order to read the address and use it to write or read the data values in or out of it. The system has a clock generator unit for managing the AAT different clocks and a a reset signal along with chip select (CS) signals to enable or disable each AAT unit. The system works in a synchronous way.

Note, that in order to ensure a high security level, the RSA keys used in the AAT should be of length 2048 bits. As a result, the data related to the RSA encryption and decryption will have similar bit length. However, there is no feasible processing system able to operate with buses of such bits. Therefore, the 2048 bit values are broken into several blocks (to match the bus bit length) and reconstructed inside the AAT units (in example the RSA encryption–decryption unit).

3.4 Hardware and Security AAT Issues and Possible Solutions

Designing a state of the art security chip that matches recent and future market needs is a challenging act. The cryptographic level that should be provided by the chip must conform to the strictest security rules and the adopted cryptographic algorithms should be widely accepted but also have a place in future applications. In this chapter we address some of the possible drawbacks that the AAT can have, under the above described needs. RSA algorithm is considerably adopted in recent application but serious speculations arise as to how the algorithm can compete against more modern public key schemes like Elliptic Curves. It seems that as long as the key length is kept high ($\geq$2048 bits), RSA can be considered very secure from cryptanalytic point of view. However, the algorithm's life expectancy and future is determined by its computational efficiency and hardware resources use as well as its protection against hardware specific security attacks (Side Channel Attacks).

3.4.1 RSA Efficiency and Security Issues

The AAT architecture presented in the previous section posses several design challenges, however, the biggest one is related to the RSA encryption–decryption unit. Public key cryptographic algorithms due to their high computational complexity are considered difficult to implement, have significant computational time and consume considerable number of hardware resources (chip covered area, power dissipation). So, in order to design an efficient AAT architecture that will be able to process information quickly and fit in a relatively constrained environment, chip area optimization of the RSA encryption–decryption unit is necessary. Furthermore, since the AAT, being a removable smart card like device, operates in a hostile environment where it can be stolen or manipulated by untrusted users, security attacks on its hardware structure can not be excluded. Such attacks called side channel attacks exploit an architecture's hardware characteristics leakage (power dissipation, computation time, electromagnetic emission etc.,) to extract information about the processed data use them to deduce cryptographic keys and messages [10]. While RSA is considered very secure against traditional cryptanalysis, side channel attacks (SCA) have been successful in determining RSA keys using information leaking from a straightforward implementation of the algorithm. For the above reasons, it can be concluded that the RSA encryption–decryption unit can be a performance and security bottleneck and special attention need to be given to its design in order to avoid the above indicated problems.

Among the most effective SCA launched on RSA are Fault attacks (FA) and simple power attacks (SPA). The fault attack (FA) goal is to disturb a hardware device during cryptographic operation execution, analyze the faulty behavior of the disturbed device and as a result deduce sensitive information. Combining such attack with a simple power attack (SPA), where a hardware device's power trace is measured and exploited for secret information leakage [10], a crypto-system attacker can relatively easy deduce a cryptographic key of a secure hardware device. Many researchers have proposed solutions on protecting RSA from FA and SPA with relative success [11–14]. However, those solutions are focused on an FA-SPA resistance on algorithmic level without taking into account the implementation cost for one FA SPA secure RSA encryption–decryption operation. This cost is associated with the arithmetic operations required for RSA and can be very high-restrictive when applying the existing FA SPA resistant RSA solutions in real security devices. SCA countermeasures can be generic, on circuit level, (more effective against power attacks) [15, 16] or specialized, focused on specific cryptographic algorithms, on an algorithmic level. The second approach can be more effective in general, since it utilizes techniques that better negate a cryptoalgorithm's specialized SCA weaknesses.

In RSA, the target of the FA and SPA is the modular exponentiation unit, that constitute the core component in message encryption–decryption. SPA resistance is achieved by making the arithmetic operations during the exponentiation algorithm execution undiscriminated from an external observer [13]. This countermeasure can be further enhanced by blinding the modulus N using a random number. Fault attack countermeasures are based on techniques of detecting single fault injection and blocking further processing thus prohibiting the release of secret information. Giraud in [11] proposed a FA-SPA resistant modular exponentiation algorithm and later Kim and Quisquater [14] proposed an attack on the algorithm along with a way to thwart it. Recently, Fournaris [17] proposed a modification of Giraud's and Kim's RSA algorithm [14] that works using Montgomery modular multiplication algorithm and results in very optimistic performance characteristics but was not designed as a CRT RSA module. The RSA modified algorithms of [11, 17] and [14] guaranty FA resistance by introducing two values, s_0 and s_1 and checking if a known equation between them is always true. If this connection between s_0 and s_1 is disturbed then a fault attack is detected and the cryptographic processes is canceled.

3.5 Proposed FA-SPA CRT RSA Algorithm

In our approach about a FA-SPA resistant RSA module we adopt the methodology described in [17] for Montgomery modular multiplication and adapt it to the system at hand. The algorithm in [17] seems very promising in terms of security and hardware performance. It is based on the Montgomery multiplication algorithm that is adapted for Giraud's modular exponentiation methodology resulting in a CRT FA-SPA resistant RSA algorithm. However, the work of [17] only focus on how FA-SPA modular exponentiation can be done and not how a fully functional CRT based FA-SPA unit can operate. In our approach, the ideas of [17] are applied in the AAT architecture in order to propose an efficient and secure FA-SPA RSA unit.

Introducing a random number a in the computational sequence, as suggested in [14], an FA-SPA resistant modular exponentiation algorithm (FSME) can be presented. This algorithm employs Montgomery modular multiplication as a structural element, chosen among other approaches due to its efficient realization in hardware. Following the above directive, we adopt an optimized version of the Montgomery modular multiplication algorithm (CSMMM), as described in [17, 18], that employs carry-save logic (C-S logic) in all its inputs, outputs and intermediate values along with precomputation. The complete FSME algorithm is described below.

FA-SPA Montgomery Modular Exponentiation (FSME) algorithm
Input: $m, a, e = (1, e_{t-2}, \ldots e_0), N$
Output: $(a + m^{e-1}) modN, (a + m^e) modN$
Initialization: $T = R^2 modN, a_R = a \cdot RmodN$ where $R = 2^{n+2}$

1. $s_0 = CSMMM(T, m, N)$
2. $s_1 = CSMMM(s_0, s_0, N)$
3. **For** $i = t - 2$ **to** 1
   a. **If** $e_i = 1$ **then**
      i. $s_0 = CSMMM(s_0, s_1, N)$
      ii. $s_1 = CSMMM(s_1, s_1, N)$

b. **else**
   i. $s_1 = CSMMM(s_0, s_1, N)$
   ii. $s_0 = CSMMM(s_0, s_0, N)$
4. $s_1 = CSMMM(s_0, s_1, N)$
5. $s_0 = CSMMM(s_0, s_0, N)$
6. $s_1 = CSMMM(s_1 + a_R, 1, N)$
7. $s_0 = CSMMM(s_0 + a_R, 1, N)$
8. **If** (Loop Counter i and exponent e are not modified) **then** return (s_0, s_1) **else** return error

The above algorithm can be considered only part of a CRT RSA module. Let $N = p \cdot q$ be the RSA modulus described as the product of two large secret prime numbers p and q. Also, let r be the public RSA exponent and d be the private RSA exponent satisfying the equation $r \cdot d \equiv (1)mod(p-1) \cdot (q-1)$. We denote as d_p the value $d_p = dmod(p-1)$ and as d_q the value $d_q = mod(q-1)$. Using the above, the CRT FA-SPA RSA algorithm will have the following form:

FA-SPA CRT RSA algorithm

Input: $m, a, p, q, d_p, d_q, i_q = q^{-1}modp, N$

Output: $m^d modN$

1. $(s_0^p, s_1^p) = FSME(m, a, d_p, p)$
2. $(s_0^q, s_1^q) = FSME(m, a, d_q, q)$
3. $\acute{S} = s_0^q + q \cdot ((s_0^p - s_0^q) \cdot i_q modp)$
4. $S = s_1^q + q \cdot ((s_1^p - s_1^q) \cdot i_q modp)$
5. $\acute{S} = (m \cdot \acute{S} + a)modN$
6. $S = (S + a \cdot m)modN$
7. **If** $(S = \acute{S})$ and p, q not modified **then** return $(S - a - a \cdot m)modN$ **else** return error

Observing the algorithms reveals that an RSA encryption–decryption process consists of two execution of the FSME algorithm. That algorithm is run using d_p and d_q exponents and p, q modulus respectively. We can assume without lost of generality that p, q are of similar bit length and therefore about half the bit length of N. Thus, using CRT the tedious n bit modular exponentiation operation is broken into two parallel $n/2$ bit modular exponentiations that can give results faster. The price for this action, is a final CRT operation (steps 3 and 4) for both fault secure streams of data. Note, that the use of the random number a prohibits the discovery of the RSA decryption result before the fault attack check is performed (step 7).

3.6 Proposed Hardware Architectures

Designing an FA-SPA resistant RSA encryption/decryption poses several challenges in a system with constrained resources like the AAT. Our primary goal is to minimize the chip covered area without compromises in security. For this reason, we employ one FSME unit responsible for modular exponentiation using either p or q. The FSME unit is structured around two Montgomery Modular Multipliers (CSMMU) using Carry-Save logic input and output signals. The CSMMU is a 5 parallel $(n/2+1)$ bit and 2 serial (1) bit input architecture that generates a Montgomery modular multiplication product in Carry-Save format every $n/2+2$ clock cycles. The functionality and full architecture of the CSMMU is described in [17, 18]. Observing FSME algorithm, reveals that there are several algorithmic steps that can be performed in parallel. Parallelism can be applied on steps 3ai and 3aii, steps 3bi and 3bii, steps 4 and 5 as well as steps 6 and 7. Furthermore, one part of the step pair always performs Montgomery modular multiplication between s_0 and s_1 in parallel with the other part of the step pair performing Montgomery modular squaring either of s_0 or s_1. We assign each parallel operation in one CSMMU. Therefore, only two such units are required for all necessary calculations of the FSME algorithm. A problem can occur using this approach and it has to do with the correct inputs in the Montgomery squaring unit. Those inputs are taken either from the output of the same CSMMU or the output of the multiplication CSMMU. The correct input choice of the squaring CSMMU is not related only to the i-th bit of e in round i but it also has to do with the operations performed in the previous algorithmic round (round i-1) and therefore it is also related to the (i-1)th e bit. Thus, in the proposed architecture, the data path control in squaring CSMMU is related to the value difference between the ith and (i-1)th bit of e and is estimated through a XOR gate.

The FSME unit needs to be used twice for one CRT RSA decryption session since it implements steps 1 and 2 of FSME in a serial fashion. In the first use of the FSME unit the outcome of $m^{d_p} mod p$ is calculated in carry-save format and stored in two registers while on the second use of the FSME unit the outcome $m^{d_q} mod q$ is calculated and stored in a similar way. Note, that following the FSME algorithm there are two calculation streams for an exponentiation outcome, s_0 and s_1, and therefore each utilization of the FSME unit requires 4 storage registers (2 for s_0 and 2 for s_1). Thus, a total of 8 registers is needed. The stored values are chosen through a multiplexer as inputs for the CRT Transformation unit that is responsible for implementing the operations of steps 3 and 4 of the CRT FA SPA RSA algorithm. The CRT transformation unit will be used twice to provide outcome for both those steps. In the first use of CRT transformation unit, the inputs provided by the multiplexer are s_0^q and s_0^p in carry-save format while on the second use of the CRT transformation unit the inputs would be s_1^q and s_1^p. The output of the CRT transformation unit is an n-bit value in carry-save format stored in two n-bit registers. After the two utilization of the CRT transformation unit the 2 outcomes are inserted into the Fault detection unit that realizes the functionality of step 7 in the CRT FASPA RSA algorithm. The outcome of this unit is either an error code provided by an all zero value or the correct encrypted-decrypted message of the RSA algorithm. The proposed CRT FASPA RSA architecture is presented in Fig. 3.3.

As explained in Sect. 3.4 the data bus is constrained in specific bit lengths that at the moment cannot exceed 32 or 64 bits. For this reason, we develop a

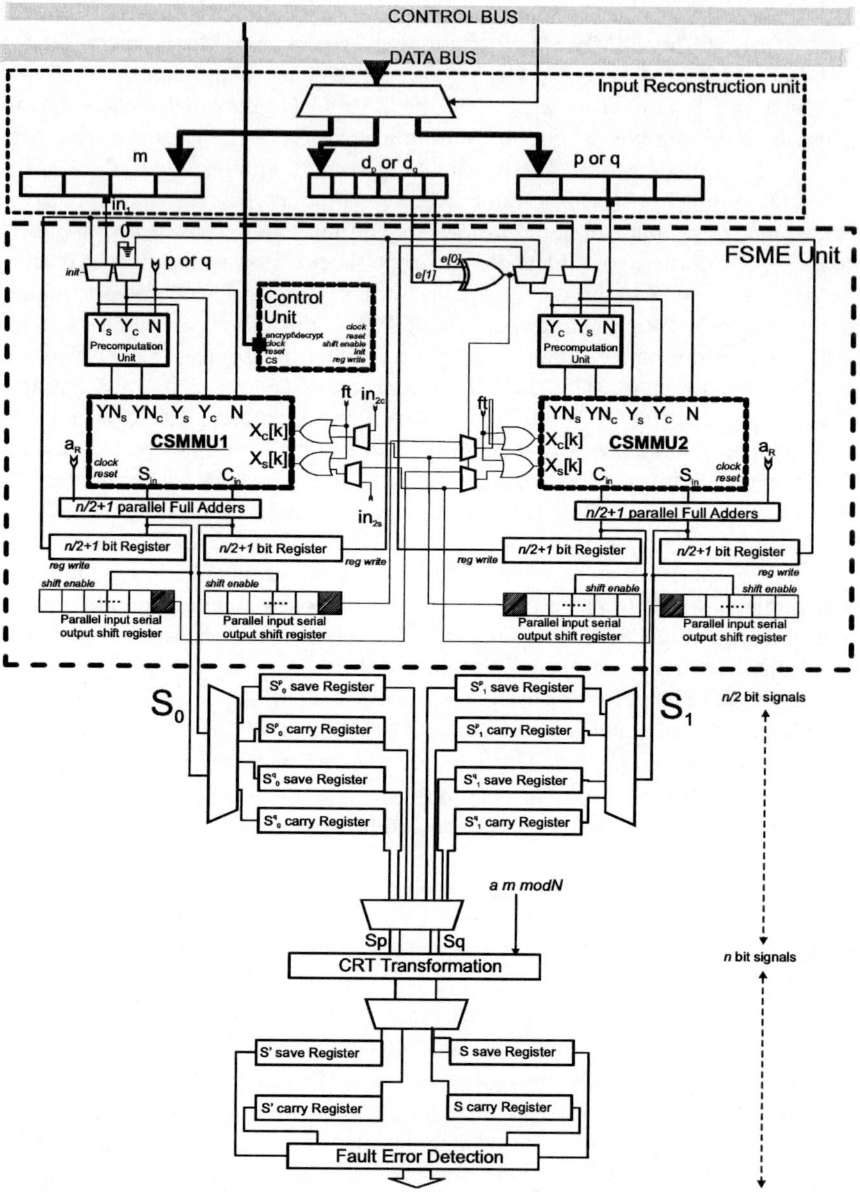

Fig. 3.3 The CRT FA-SPA resistant RSA architecture

mechanism for that bus data insertion into the RSA architecture so that the arbitrary input RSA algorithm where the bit length of the processed values is of higher bit length than the one provided by the data bus, can fully work. The data bus inputs are multiplexed using the input reconstruction unit. There exist 2 digit input parallel output shift registers (for the message m and the p or q modulus) and one digit input serial output shift register (for the d_p or d_q exponent). As the data bus is accessed in digits analogous to the AAT control processor word length (w bits), they are stored to the appropriate register through a multiplexer structure.

Initially, the modulus register is filled out, after $n/2w$ clock cycles. Then, the message register is filled out, after $n/2w$ consecutive clock cycles. The data bus words are inserted in those registers in each clock cycle and then right shifted w bits until the whole register length is full. The register's value is outputted in parallel fashion using a *parallel out* control signal. When the message and modulus registers are filled with the message and p values, the storage of the exponent may begin. Input is provided to the exponent register in w bit words and output is performed in a serial fashion (the two least significant bits of the stored value are outputted). Note, that the exponent register has two modes of operation, data insertion and data output processing. In the first mode of operation the register data are right shifted w bits in order to make space for the next data bus word to be stored while in the second mode of operation, register data are right shifted 1 bit. The data insertion can be completed in $n/2w$ clock cycles however, RSA operations can begin after the first clock cycle of that time since only the 2 least significant bits of the exponent are needed in every modular exponentiation round.

3.7 Performance

Assuming that n bit RSA encryption–decryption needs to be performed and that the data bus clock speed is considerably higher than the proposed CRT RSA unit clock speed, data insertion to the RSA unit's input registers (modulus, exponent and message registers) can be done with small speed overhead to the whole system. Thus, ignoring the cost of data insertion the proposed CRT FA SPA RSA architecture requires $C_{CRT\,RSA} = C_{FSME} + C_{CRT} + C_{FA}$ clock cycles for one encryption–decryption operation where C_{FSME} is the clock cycle number for Modular exponentiation using the FSME algorithm, C_{CRT} is the clock cycle number for CRT transformation and C_{FA} is the clock cycle number for the Fault detection. The $C_{CRT\,RSA}$ is dominated by the FSME unit since C_{FSME} has $O(nt)$ complexity where t is the Hamming weight of the FSME exponent. The C_{CRT} and C_{FA} numbers are of $O(n)$ complexity. Thus, we can focus our analysis on the FSME unit. The CSMMU needs $\frac{n}{2} + 2$ clock cycles to come up with a result. The FSME algorithm is concluded after $t - 2$ algorithmic rounds where the CSMMM algorithm is executed in parallel two times in each round. Also, in FSME, 2 execution of CSMMM are needed for initialization (FSME steps 1 and 2) and 2 parallel CSMMM executions for post processing calculations (FSME parallel steps 4–5 and 6–7). As a result, the total number of clock cycles for one modular exponentiation using FSME is $(t + 2) \cdot (\frac{n}{2} + 2)$. Since through the CRT FA SPA RSA algorithm (steps 1 and 2) there are two FSME executions, all modular exponentiations are concluded after $(t + 2) \cdot (n + 4)$ clock cycles.

Table 3.1 Modular Exponentiation Comparisons for $n = 1024$

Arch.	Technology	Area (slices)	Freq. (MHz)	FA-SPA
Prop.	XC5VLX50T	4340	324.4	Yes
[19]	XC2V3000	12537	152.5	No
[20]	XC2V6000	23208	96	No
[18]	XC2V3000	7873	129	No

To further evaluate our proposed system we realized the FSME architecture in FPGA technology (xilinx virtex 5) using VHDL language for $n = 1024$ RSA operations. Assuming that p and q are approximately $n/2$ bit length, a 512 bit FSME architecture was implemented and measurements in chip covered Area (FPGA slices) and clock frequency (MHz) were taken. Those measurements are presented in Table 3.1 and comparison with other existing RSA works are made.

Acknowledgements The work reported in this paper is supported by the European Commission through the SECRICOM FP7 European project under contract FP7 SEC 218123

References

1. Sklavos N, Zhang X (2007) Wireless security and cryptography: specifications and implementations. CRC Press Inc, Boca Raton
2. Group TC(2007) TCG TPM specification version 1.2. URL https://www.trustedcomputinggroup.org/specs/TPM/
3. Xiaoping Wu ZS, Zhang H (2008) Secure key management of mobile agent system using tpm-based technology on trusted computing platform. Computer science and software engineering, International conference on 3, pp 1020–1023. doi:http://doi.ieeecomputersociety.org/10.1109/CSSE.2008.256
4. Tan HK, Moreau L (2001) Trust relationships in a mobile agent system. In: Mobile agents, number 2240 in LNCS, Springer, Heidelberg, pp 15–30
5. Hein D, Toegl R (2009) An autonomous attestation token to secure mobile agents in disaster response. In: The first international ICST conference on security and privacy in mobile information and communication systems (MobiSec 2009), Torino
6. Fournaris AP (2010) Trust ensuring crisis management hardware module. Inf Secur J: A Global Perspect 19(2):74–83
7. Uwe G. Wilhelm SS, Buttya'n L (1999) Introducing trusted third parties to the mobile agent paradigm. In: Secure internet programming: security issues for mobile and distributed objects. Springer, Heidelberg, pp 471–491
8. Jonathan M. McCune Adrian Perrig AS, van Doorn L (2007) Turtles all the way down: research challenges in user-based attestation. In: Proceedings of the workshop on hot topics in security (HotSec). URL http://www.truststc.org/pubs/286.html
9. Lowe G (1995) An attack on the needham-schroeder public-key authentication protocol. Inf Process Lett 56(3):131–133
10. Kocher P, Jaffe J, Jun B (1999) Differential power analysis. In: Advances in cryptology proceedings of crypto 99, Springer, Heidelberg, pp 388–397
11. Giraud C (2006) An rsa implementation resistant to fault attacks and to simple power analysis. IEEE Trans Comput 55(9):1116–1120
12. Vigilant D (2008) Rsa with crt: a new cost-effective solution to thwart fault attacks. In: Oswald E, Rohatgi P (eds.) CHES, Lecture notes in computer science, vol 5154. Springer, Heidelberg, pp 130–145

13. Joye M, Yen SM (2003) The montgomery powering ladder. In: CHES '02: Revised papers from the 4th international workshop on cryptographic hardware and embedded systems, Springer, London, pp 291– 302
14. Kim CH, Quisquater JJ (2007) Fault attacks for crt based rsa: new attacks, new results, and new countermeasures. In: Sauveron D, Markantonakis C, Bilas A, Quisquater JJ (eds.) WISTP, Lecture notes in computer science, vol 4462. Springer, Heidelberg
15. Bhattacharya K, Ranganathan N (2008) A linear programming formulation for security aware gate sizing. In: GLSVLSI '08: Proceedings of the 18th ACM Great Lakes symposium on VLSI. ACM, New York, pp 273–278
16. Tiri K, Verbauwhede I (2006) A digital design flow for secure integrated circuits. IEEE Trans CAD Integr Circuits Syst 25(7):1197–1208
17. Fournaris AP (2010) Fault and simple power attack resistant rsa using montgomery modular multiplication. In: Proceedings of the IEEE international symposium on circuits and systems (ISCAS2010). IEEE (30 May 2002, June 2010)
18. Fournaris AP, Koufopavlou OG (2005) A new rsa encryption architecture and hardware implementation based on optimized montgomery multiplication. In: ISCAS (5), IEEE, pp 4645–4648
19. Shieh MD, Chen JH, Wu HH, Lin WC (2008) A new modular exponentiation architecture for efficient design of rsa cryptosystem. IEEE Trans Very Large Scale Integr Syst 16(9):1151–1161
20. McIvor C, McLoone M, McCanny J (2004) Modified montgomery modular multiplication and rsa exponentiation techniques. IEE Proc-Comput Digital Tech 151(6):402–408

Chapter 4
MULTICUBE: Multi-Objective Design Space Exploration of Multi-Core Architectures

Cristina Silvano, William Fornaciari, Gianluca Palermo, Vittorio Zaccaria, Fabrizio Castro, Marcos Martinez, Sara Bocchio, Roberto Zafalon, Prabhat Avasare, Geert Vanmeerbeeck, Chantal Ykman-Couvreur, Maryse Wouters, Carlos Kavka, Luka Onesti, Alessandro Turco, Umberto Bondi, Giovanni Mariani, Hector Posadas, Eugenio Villar, Chris Wu, Fan Dongrui, Zhang Hao and Tang Shibin

Abstract Given the increasing complexity of Chip Multi-Processors (CMPs), a wide range of architecture parameters must be explored at design time to find the best trade-off in terms of multiple competing objectives (such as energy, delay, bandwidth, area, etc.). The design space of the target architectures is huge because it should consider all possible combinations of each hardware parameter (e.g., number of processors, processor issue width, L1 and L2 cache sizes, etc.).

This project is supported by the EC under grant MULTICUBE FP7-216693.

C. Silvano (✉) · W. Fornaciari · G. Palermo · V. Zaccaria · F. Castro
Politecnico di Milano, Milano, Italy
e-mail: silvano@elet.polimi.it

M. Martinez
Design of Systems on Silicon (DS2), Valencia, Spain

S. Bocchio · R. Zafalon
STMicroelectronics, Catania, Italy

P. Avasare · G. Vanmeerbeeck · C. Ykman-Couvreur · M. Wouters
IMEC vzw, Leuven, Belgium

C. Kavka · L. Onesti · A. Turco
ESTECO, Trieste, Italy

U. Bondi · G. Mariani
ALaRI - Universita' della Svizzera Italiana, Lugano, Switzerland

H. Posadas · E. Villar
University of Cantabria, Cantabria, Spain

C. Wu
STMicroelectronics Beijing, Beijing, China

F. Dongrui · Z. Hao · T. Shibin
Institute of Computing Technology, Chinese Academy of Sciences, Beijing, China

N. Voros et al. (eds.), *VLSI 2010 Annual Symposium*,
Lecture Notes in Electrical Engineering 105, DOI: 10.1007/978-94-007-1488-5_4,

In this complex scenario, intuition and past experience of design architects is no more a sufficient condition to converge to an optimal design of the system. Indeed, Automatic Design Space Exploration (DSE) is needed to systematically support the analysis and quantitative comparison of a large amount of design alternatives in terms of multiple competing objectives (by means of Pareto analysis). The main goal of the MULTICUBE project consists of the definition of an automatic Design Space Exploration framework to support the design of next generation many-core architectures.

4.1 Introduction

The main goal of MULTICUBE project is the definition of an automatic framework to support multi-objective Design Space Exploration (DSE) of multi and many core SoC architectures. The framework enables the tuning of several architectural parameters to minimize multiple metrics (such as energy and latency) while meeting system-level constraints (such as throughput, bandwidth and QoS). The project focuses on the definition of an automatic modeling and optimization methodologies for improving the conventional SoC design flow. In such a design flow, the optimization process is still done manually, based on the past experience of the designer to leverage existing simulation models to explore the design space and find out power-performance trade-offs (Pareto analysis).

The project targets an **efficient** and **automatic** exploration of parallel embedded architectures in terms of several design parameters such as available parallelism (e.g., number of cores, processor issue width), cache-subsystem (e.g. cache size and associativity) and Network-on-Chip (NoC) related parameters (e.g., channel buffer size). When dealing with complex many-core architectures, the design space exploration can easily become very huge, making a full-search simulation-based exploration unfeasible. Automatic exploration techniques based on optimisation heuristics should be used to figure out an approximate Pareto set in a reasonable time. The exploration time can even be reduced by decreasing the number of design points to be simulated. This can be done by using analytical models to predict the system behavior corresponding to design points not yet simulated. These analytical models are defined from a training set of simulations.

4.2 The MULTICUBE Design Methodology

The MULTICUBE design flow (shown in Fig. 4.1) consists of two frameworks:

Power/Performance Estimation Framework includes the simulation tools for power and performance estimation of the target architectures to be explored. Each configurable simulation model accepts an architectural configuration as input and

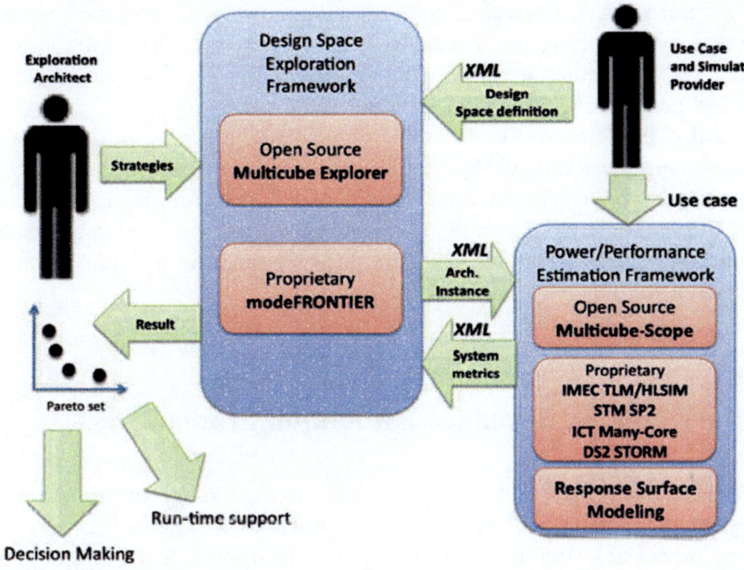

Fig. 4.1 The MULTICUBE design flow

generates, at the end of simulation, the corresponding system-level metrics running the reference application. The combination of architecture and reference application is defined, from here on, as *use case*. The simulators used in the context of the project are either open source (MULTICUBE SCOPE) or proprietary (such as the IMEC TLM platform, the STMicroelectronics SP2 simulator or the Institute of Computing Technology's (ICT) many core simulator). The tool interface has been standardized with the definition of the MULTICUBE XML interface so as other simulators can be plugged in with minimal effort. Besides event-based simulation, a set of **Response Surface Models** has been developed to further speedup the evaluation of the system-level metrics through analytical expressions derived from a training set of simulations.

Design Space Exploration Framework contains the tools for automatic DSE and interacts with the power/performance estimation framework by iteratively generating different instances of the design space and reading back power and performance evaluation metrics (*system metrics*). This framework includes several optimization heuristics for identifying power/performance trade-offs and generating the approximated Pareto frontier which may be more or less close to the actual Pareto front. The DSE tools developed in the project are: the open source (MULTICUBE Explorer tool) and the proprietary (ESTECO's modeFRONTIER) optimization tool (widely used tool in multi-disciplinary optimization).

The Use Case and Simulator Provider. (S)he is the system architect providing the simulator with a set of configurable parameters and performance and power estimation model for the target use case (*model setup*).

The Exploration Architect. (S)he is in charge of optimizing the configurable architecture by exploiting the automatic DSE framework. The exploration architect can directly interact with the DSE engine to set up exploration strategies, metrics and constraints.

The final exploration results (Pareto frontier) can be further pruned by a set of decision-making mechanisms. The solution set can be ordered according to some ranking criteria to derive the final candidate solution for the implementation. Besides, the Pareto frontier can then be used to support run-time management of the resources. This last feature has been already tested on an industrial use case and described in [1].

4.3 Power and Performance Estimation Framework

The MULTICUBE project can easily manage a wide range of simulation technologies ranging from full-system cycle-accurate to application-level back-annotated functional simulation by using a standardized XML-based interface. The capability to span across several abstraction levels is fundamental to model cross-validation between different abstraction levels (e.g., high-level vs. TLM) or to provide mixed-level optimization strategies [2]. In the optimization literature, the simulator is also referred to as the *solver*.

The power and performance estimation tools can be grouped as *open-source* and *proprietary*:

4.3.1 Open Source Estimation Framework

The open source prototype tool is MULTICUBE SCoPE and it is based on existing technology (SCoPE) [3] developed by University of Cantabria for performing HW/SW co-simulation. MULTICUBE SCoPE enables the definition of SystemC platform template models to evaluate performance and power consumption. MULTICUBE SCoPE efficiency comes from the fact that performance and power estimations of the software side are performed at the application source code level by using back-annotation. The back-annotated software components are then linked to the hardware components by using standard SystemC interfaces. This modeling style is called *Timing Approximate*. Software back-annotation avoids instruction set simulation therefore decreasing several orders of magnitude simulation time by and at the same time maintaining a fairly good accuracy with respect to cycle-accurate simulations. MULTICUBE SCoPE also provides some hooks for enabling C/C++ software code to invoke operating system primitives compliant with POSIX and MicroC/OS.

4.3.2 Proprietary Estimation Framework

These modeling and simulation tools span several abstraction levels. To simulate IMEC multimedia architecture based on the ADRES processor [4], two simulators are provided: a High-Level Simulator (HLSim) and a SystemC-based platform simulator based on CoWare[1] virtual prototyping environment. The target platform is composed of a variable number of processor nodes and memory nodes. All processor nodes contain an ADRES processor, IMEC proprietary VLIW processor and its scratch-pad local data (L1) memory. The processing nodes are connected to the memory nodes by a configurable communication infrastructure. It can be either a multi-layer AHB bus, which provides a full point-to-point connectivity, or a NoC model built with the CoWare AVF cross-connect IP.

From one side, IMEC HLSim provides fast simulation at the source code level of parallel applications by providing support for timing annotation, thread spawning/joining, DMA transfers, FIFO communication and loop synchronization modeling. HLSim takes as an input the platform architecture description which contains information on the processors, the memory hierarchy structure, the timing figures and the power consumption of the other IP cores. For example, the timing information of the computation-intensive loops in the threads is measured on the Instruction Set Simulator and is used as an input in HLSim. This means that the timing figures are instruction-accurate because they take all the compiler optimizations into account. On the other side, IMEC's CoWare simulator is based on cycle-accurate *Transaction-Level Modeling* (TLM) and provides a multi-processor platform template leveraging a NoC as interconnection model. The template links compute ADRES nodes (processing elements) and memory nodes together using a central communication backbone. This model has been used for creating back-annotation timing data used in the HLSim simulator and performing cross-validation. Overall, we found [5] an acceptable relative accuracy with a significant speed-up in simulation time (see Fig. 4.2, where the relative accuracy and speed of MULTICUBE SCoPE technology has been added for comparison).

An instruction set simulator has been used for SP2 superscalar processor provided from STMicroelectronics and for the many-core architecture from ICT. Both simulators expose program execution time and power consumption as system level metrics. More in detail, the ICT many-core architecture is a tiled MIMD machine composed of a bi-dimensional grid of homogeneous, general-purpose compute elements, called *tiles*. A 2D-mesh network architecture is used for connecting the cores to a non-shared memory sub-system.

Finally, as an example of control-oriented architecture, the DS2's STORM platform has been integrated in the framework. The platform is used to model a PLC (Programmable Logic Controller) technology with several implementation choices. For this platform, both Ethernet QoS and internal communication are considered as metrics.

---

[1] CoWare is now a part of Synopsys Inc.

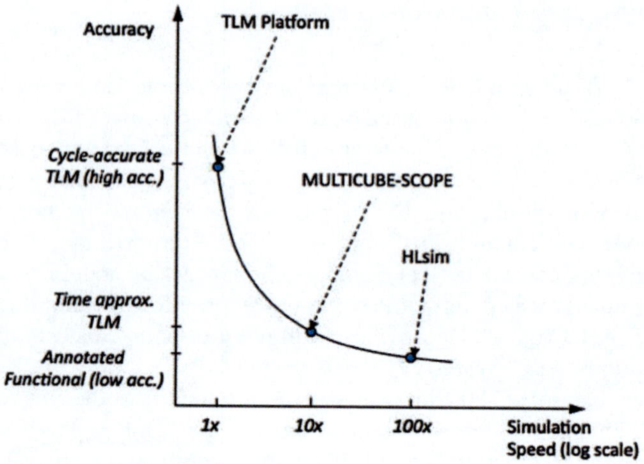

Fig. 4.2 Accuracy versus speed trade-off for several abstraction level simulators

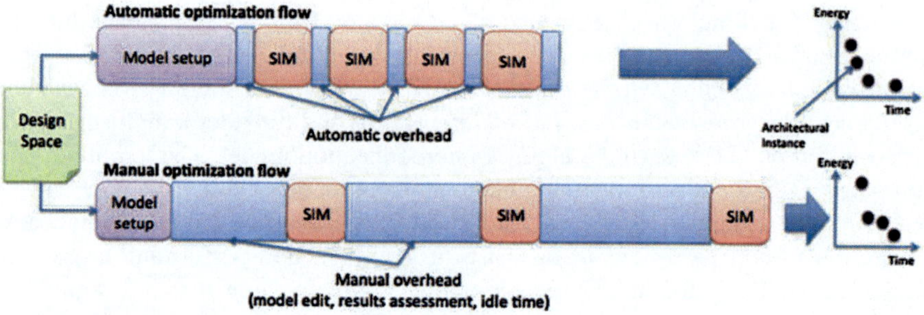

Fig. 4.3 Comparison between a manual (or conventional) design space exploration approach and an automatic approach

4.4 Advantages of Automatic DSE

Multi-objective optimization heuristics and high-level simulation techniques can successfully contribute to dramatically decrease the exploration time, while guaranteeing a certain level of 'closeness' to the actual Pareto frontier. Reducing the exploration time reduces the time-to-market of the product and therefore the risk to miss market deadlines (with the associated economic loss).

Both conventional and automatic DSE starts from a definition of *design-space* (see Fig. 4.3) representing the set of feasible architectural configurations. In a conventional DSE approach, the starting point is the definition of an initial model (*model setup phase*) of the target architecture. The enabling simulation technologies used for creating such a model can range from cycle-accurate SystemC up to annotated functional simulation, with variable ratio of accuracy with respect to speed. The exploration is done iteratively by subjective assumptions of the designer, who will edit manually the architectural simulator and modify a limited number of parameters per evaluation. The model simulation corresponds to a limited portion of the time of the overall exploration procedure. A larger amount of time is spent by the designer editing the configuration parameters and/or the model structure and analyzing the results. There is also an idle time (from the point of view of the use of computational resources) that lasts from the end of the simulation till the moment in which human operator is informed about it and handles the simulation tools to get the results.

In this case, the overall quality of the DSE is based on the designer's ability and past experience to assess the results and to move towards the next instance of the model to be simulated based on aggregate information resulting from the simulation campaign. The outcome (solution set) is thus highly dependent on the skills and past experience of the designer.

An automatic DSE flow uses numerical objective criteria combined with efficient heuristics to drive the exploration. The basic assumption is that the model should be a *configurable template* model which can be automatically manipulated to generate any instance of the target design space. The model setup stage, thus, may be longer than in the conventional case. Given the configurable model template, the DSE tool will change systematically all the parameters at each step and will evaluate the best result based on robust optimisation heuristics. The automatic selection of the next configuration to be simulated (model selection) is typically faster than the conventional one because it does not involve any overhead due to human intervention. The final output of the automatic DSE flow is a set of *dominant* configurations in the design space which are, with high probability, close to the actual Pareto set. Besides, all data concerning previous evaluations are stored in a structured database which can be automatically analyzed by using mathematical/statistical tools to derive aggregate information about the *population* of analyzed design points.

4.5 Design Tool Integration Based on the MULTICUBE XML

In the MULTICUBE project we addressed the formalization of the interaction between the simulator and the DSE tools, that is essentially an automatic program-to-program interaction (see Fig. 4.1):

1. The DSE tool generates one feasible system configuration whose system metrics should be estimated by the simulator.
2. The simulator generates a set of system metrics to be passed back to the DSE tool.

To automatically link the use case simulator to the DSE tool, a design space definition file should be released by the use case and simulator provider together with the executable model of the use case (simulator). This file describes the set of configurable parameters of the simulator, their value range and the set of evaluation metrics that can be estimated by the simulator. This file describes also how to invoke the simulator as well as an optional set of rules with respect to the generated parameter values should be compliant with. The rules are only used by the exploration tool to avoid the generation of invalid or unfeasible solutions during the automated exploration process. The above interaction has been addressed by creating a specification based on an XML based grammar for writing both the design space definition file and the simulator interface files. The grammar is defined by using the XSD schema language.

4.5.1 Design Space Definition

The definition of the design space is done by using an XML file that is composed of a preamble, which defines the name-space and supported version. The current release of the MULTICUBE XML specification is R1.4.

```
<?xml version="1.0" encoding="UTF-8"?>
<design_space xmlns="http://www.multicube.eu/" version="1.4">
   <simulator> ... </simulator>
   <parameters> ... </parameters>
   <system_metrics> ... </system_metrics>
   <rules> ... </rules>
</design_space>
```

The remaining part of the file describes the simulator invocation method (**<simulator>** tag), the set of parameters of the simulator which can be configured (**<parameters>** tag), the system metrics which can be estimated by the simulator (**<system_metrics>** tag) and the rules which have to be taken into account by exploration engine in order to generate feasible configurations.

4.5.1.1 Simulator Invocation

The **<simulator_executable>** marker is used for specifying the complete path name of the executable:

```
<simulator>
   <simulator_executable path="/path/my_simulator_executable" />
</simulator>
```

4.5.1.2 Parameters Definition

The **<parameters>** tag is used by the use case and simulator provider to specify the names, the types and the ranges of the parameters that can be explored by the DSE tool. The section contains a list of **<parameter>** markers:

```
<parameter>
   <parameter name="il1_cache_block_size_bytes"
      description="..." type="exp2" min="8" max="64"/>
   <parameter name="bpred" description="b.p.␣type" type="string">
      <item value="nottaken"/>
      <item value="taken"/>
      <item value="perfect"/>
      <item value="bimod"/>
      <item value="2lev"/>
      <item value="comb"/>
   </parameter>
   ...
</parameters>
```

For each parameter an unique name must be provided. The parameter types can be divided into two categories: scalar types, variable vector types. Scalar types can be **integer**, **boolean** (a subset of integers), **exp2** (progression of power of 2) and **string** (a type for defining categorical variables). Vector types can be used to describe combination of boolean values (*on-off-masks* or *permutations*). In particular, on-off-masks can be useful for describing the space of active processors while permutations can be used to describe the mapping of tasks on the available processors.

4.5.1.3 System Metrics Definition

The <**system_metrics**> section is used by the use case and simulator provider to specify the names, the types and the units of the system metrics that can be estimated by the simulator:

```
<system_metrics>
   <system_metric name="cycles" type="integer" unit="cycles" />
   <system_metric name="instructions" type="integer" unit="insts" />
   <system_metric name="powerconsumption" type="float" unit="W" />
   <system_metric name="area" type="float" unit="mm2" />
</system_metrics>
```

A complex expression of the system metrics can be defined as one of the objectives of the exploration algorithm.

4.5.2 Simulator Input/Output XML Interface

The simulator input file contains a preamble and a sequence of <**parameter**> sections where, for each parameter, the name and the value is specified. The number of <**parameter**> sections and the name of the parameters should be the same as defined in the XML Design Space description file. Similarly the simulator output file contains a preamble and a sequence of <**system_metric**> sections where, for each metric, the name and the value is specified. Besides, an appropriate error reporting syntax has been described in the specification.

4.6 Design Space Exploration Framework

The structure of the exploration framework is composed of an open-source optimization tool (MULTICUBE Explorer) and a proprietary multi-objective optimization tool (ESTECO's modeFRONTIER). The open-source tool has been designed from scratch to address embedded system exploration, while the proprietary tool is an existing tool for multi-disciplinary optimization which has been re-targeted to address discrete embedded system exploration. In general, DSE is an optimization process that takes into account a typical set of IP parameters that are associated with the memory system structure (e.g., cache size), the inherent parallelism of the processor (e.g., number of simultaneous tasks and the instruction issue width) and the on-chip interconnect configuration. The optimization problem involves either the maximization or minimization of *multiple objectives* (such as execution time, power consumption, area, etc.) making the definition of optimality not unique [6].

In our context, the multi-objective optimization targets a set of n system configuration parameters grouped by a configuration vector:

$$a = \begin{bmatrix} a_1 \\ \dots \\ a_n \end{bmatrix} \in A, \tag{4.1}$$

where A is usually a finite, discrete domain (subset of N_0^n). The multi-objective optimization problem is defined as a set of m objective functions to be minimized (or maximized):

$$\min_{a \in A} \phi(a) = \begin{bmatrix} \phi_1(a) \\ \dots \\ \phi_m(a) \end{bmatrix} \in \boldsymbol{R}^m, \tag{4.2}$$

subject to a set of k constraints which, without loss of generality, can be expressed as:

$$\boldsymbol{\chi}(a) = \begin{bmatrix} \chi_1(a) \\ \dots \\ \chi_k(a) \end{bmatrix} \leq \begin{bmatrix} 0 \\ \dots \\ 0 \end{bmatrix}. \tag{4.3}$$

The set of *feasible* solutions of an optimization problem is defined as the *feasible region*:

$$\Phi = \{a \in A | \boldsymbol{\chi}_i(a) \leq 0, 1 \leq i \leq k\}$$

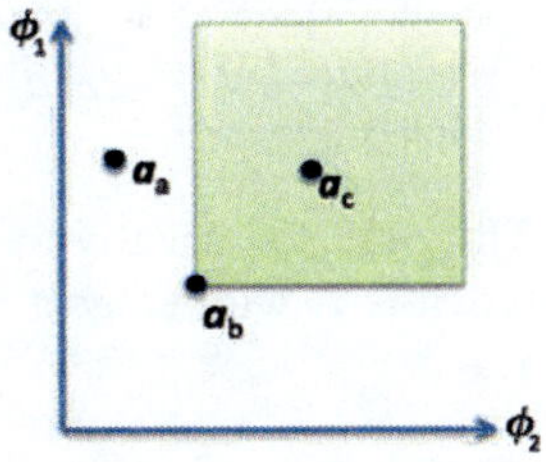

Fig. 4.4 Pareto Dominance, $\Omega = \{a_a, a_b, a_c\}$. Point a_c is dominated by point a_b

In single-objective optimization problems, the feasible set is *totally ordered* w.r.t. the function $\phi(a)$, thus a single exact solution a exists. When several objectives are defined, the feasible set is partially ordered through *dominance relation* and, thus, multiple exact solutions exist. The solution set is defined as the *exact Pareto set* of the problem. We say that ϕ *dominates* γ when:

$$\phi \prec \gamma = \begin{cases} \phi_i \leq \gamma_i, & \forall i = 1, \ldots, m \text{ and} \\ \phi_i < \gamma_i & \text{for at least one } i \end{cases} \tag{4.4}$$

Given a subset of feasible configurations $\Omega \subseteq A$, we define the Pareto set Ψ associated to Ω as:

$$\Psi(\Omega) = \left\{ a \middle| \begin{pmatrix} a \in (\Phi \cap \Omega) \wedge \\ \neg\exists b \in \Psi(\Omega) \text{s.t.} \phi(b) \prec \phi(a) \end{pmatrix} \right\}. \tag{4.5}$$

The Pareto set $\Psi(\Phi)$ is the *exact Pareto set* of the problem. Based on set theory, the projection of $\Psi(\Phi)$ in the objective space is called the *exact Pareto front*.

Whenever the considered solution set Ω is a subset of the feasible solution space Φ, the Pareto set $\Psi(\Omega)$ is called an *approximate Pareto set* of the problem.

Figure 4.4 shows a feasible set of solutions $\Omega = \Phi = \{a_a, a_b, a_c\}$ for an unconstrained minimization problem for $[\phi_1, \phi_2]$. Point a_c is dominated by point a_b since both $\phi_1(a_c)$ and $\phi_2(a_c)$ are greater than $\phi_1(a_b)$ and $\phi_2(a_b)$. In this case we thus have that $\Psi(\Omega) = \{a_a, a_b\}$.

The overall goal of multi-objective optimization heuristics is to identify approximate Pareto sets which are as *near* as possible to the exact Pareto set. For evaluating the quality of the approximate Pareto sets, a measure of the distance between the exact and the approximate Pareto sets can be introduced. In literature, many quality functions have been proposed to tackle this problem [7]. In the MULTICUBE project we leveraged the *Average Distance from Reference Set* (ADRS). The ADRS is used to measure the distance between the exact Pareto set $\Pi = \Psi(\Phi)$ and the approximate Pareto set $\Lambda = \Psi(\Omega)$ [8]:

$$\text{ADRS}(\Pi, \Lambda) = \frac{1}{|\Pi|} \sum_{a_R \in \Pi} \left(\min_{a_A \in \Lambda} \{\delta(a_R, a_A)\} \right), \tag{4.6}$$

where δ is a measure of the normalized distance in the objective function space of two configurations:

$$\delta(a_R, a_A) = \max_{j=1,\ldots,m} \left\{ 0, \frac{\phi_j(a_A) - \phi_j(a_R)}{\phi_j(a_R)} \right\} \tag{4.7}$$

The ADRS is usually measured in terms of percentage; the higher the ADRS, the worst is Λ with respect to Π.

4.6.1 Multi-Objective Optimization Heuristics

In this section we present the multi-objective optimization heuristics implemented in MULTICUBE project and an analysis of the features carried out on an industrial architecture. So far, the following optimisation algorithms have been implemented and analyzed:

- **Standard algorithms**: In this group we can find well-known multi-objective optimisation algorithms that have been implemented by following the original specification. The group is composed of the NSGA-II [9] and MOGA-II [10] algorithms.
- **Enhanced algorithms**: In this group we can find enhanced versions of standard algorithms in order to deal with the specific discrete parameters addressed in MULTICUBE. The group is composed of the MOSA [11], ES and MOPSO [12] algorithms.
- **New algorithms**: In this group we can find all the algorithms that have been specifically defined in the MULTICUBE project for multi-objective optimization in the context of SoC design optimization. In this group we can find the MFGA (Evolutionary) [13] and APRS (Randomized) algorithms.

4.6.1.1 Evaluation of the Algorithms

The algorithm validation shown here is based on the Low-Power Processor Use Case (Superscalar Processor SP2) delivered by STMicroelectronics-China. The executable model for the design space exploration consists of sp2sim simulator, which is an instruction set simulator for the SP2 microprocessor design. The benchmark application selected is the 164.gzip application, based on the popular gzip compression algorithm. The design space is defined as a combination of 11 configuration parameters, among which out-of-order execution parameters (issue width, reorder buffer length, rename register number), cache sub-system (instruction and data cache configurations and load/store queue sizes), branch prediction (type and table size). For improving the validation speed, the number of input parameters was reduced to 7 by fixing the remaining ones to a constant value. The final design space of the problem is composed of 9,216 designs while the three objectives to be minimized are the total number of cycles, the power dissipation and the area.

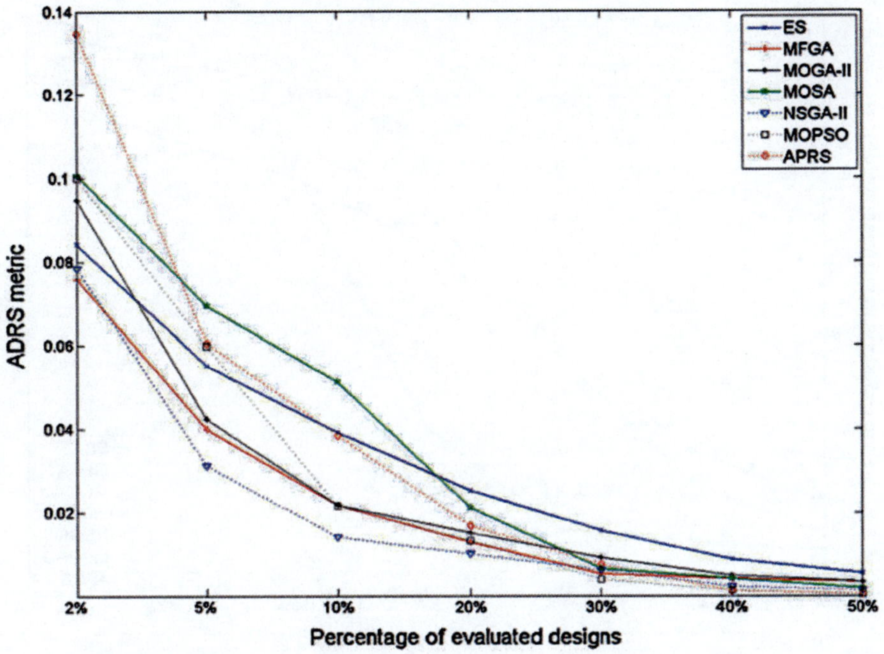

Fig. 4.5 ADRS found plotted against the percentages of evaluated designs

A fair evaluation of non-deterministic algorithms requires several repeated runs without changing any parameter besides the random generator seed. Notwithstanding the relative small search space consisting of only 9,216 designs, very large variations can be observed in the algorithms behavior and a rigorous study is needed to analyze also this aspect. Besides, preliminary tests were performed in order to identify non-significant parameters that have been then fixed to a constant value. Algorithms configuration parameters are usually problem-dependent. Typically, they depend on the user expectations: the optimal choices for the parameters that control the ratio between exploration and exploitation phase (e.g., the temperature schedule in the MOSA heuristic) are strictly related to the number of evaluations the user can afford. It was decided to tune these parameters considering the largest target (i.e. 50% of the design space, as described below) and accepting possible worse results in the partial samples. The evaluation process then proceeds checking at fixed numbers of explored points the quality of the non-dominated front found so far. The steps selected for the evaluation are: 184 designs (corresponding to about 2% of the design space), 461 (5%), 922 (10%), 1,843 (20%), 2,765 (30%), 3,686 (40%) and 4,608 (50%). Only the requests of evaluation of new designs were counted, since sometimes the algorithm could require the evaluation of one or more already evaluated designs due to the inherent behavior of their random engines.

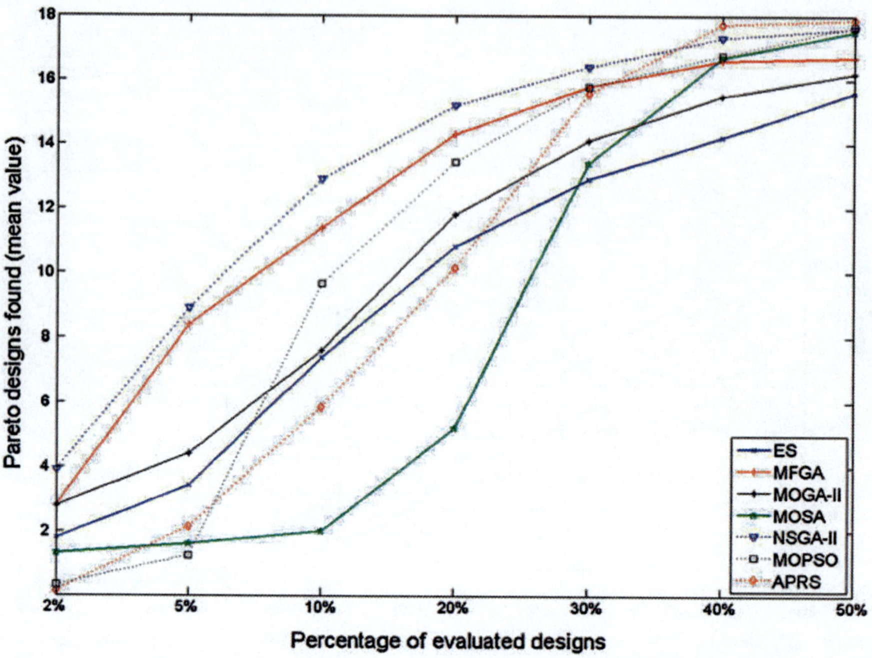

Fig. 4.6 Number of Pareto points found plotted against the percentages of evaluated designs

Concerning the target architecture and the considered algorithms, we found that they easily reach an ADRS value below 2% evaluating 30% of the design space (lower ADRS corresponds to a better solution, see Fig. 4.5). In particular, NSGA-II presented the best ADRS over all the considered range. Besides, we can note variations in the slope of the lines for some algorithms that are due to different 'behavioral' phases of the specific optimization process. We point out also that a semi-automatic support is also provided by modeFRONTIER tool to choose the most suitable algorithm for the target design space.

Figure 4.6 shows the number of true Pareto points found by the algorithms at the specified percentage of design space exploration. Also in this case, NSGA-II has scored the highest numer of Pareto designs found over all the considered range.

4.6.2 Response Surface Modeling

To further increase the efficiency of the exploration process, *Response Surface Modeling* (RSM) techniques have been introduced in the project. The basic idea is to reduce the number of simulations by defining an analytical response model of the system metrics based on a subset of simulations used as *training set*. Then the analytical expressions are used to predict the unknown system response.

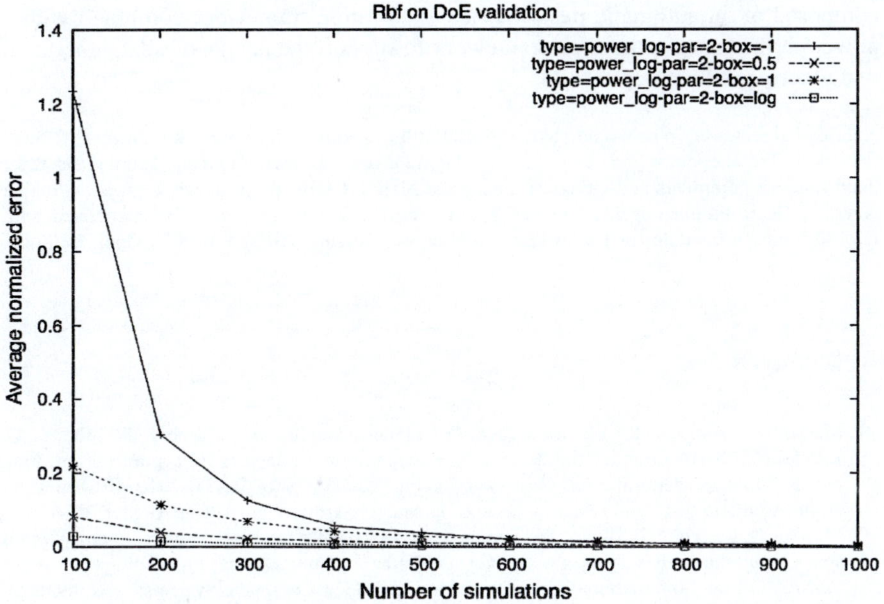

Fig. 4.7 Radial Basis Function prediction accuracy versus the number of training samples (100 samples $\simeq 9\%$ of the design space)

More in detail, RSMs provide an approximation $\boldsymbol{\rho}(a)$ of the given vector system-level metrics $\boldsymbol{\phi}(a)$. Each response $\rho(a)$ is trained with a set of observations derived from an actual simulation of a configuration a. In the project, several RSM techniques have been implemented, among them Radial Basis Functions [14], Linear Regression [15, 16], Artificial Neural Networks [17] and Shepard's Interpolation. Every RSM presented dramatic speed-up in terms of evaluations. Besides, we found that a peculiar mathematical transformation of input training set known as Box-Cox λ transform [15] had a great impact on the prediction accuracy. Figure 4.7 shows the average prediction accuracy of the Radial Basis Function response surface model on the system metrics of the ICT many-core architecture by varying the number of training samples, the model order and the λ transform. We can note that, for a logarithmic or squared Box λ transform (box = log and box = 0.5), we have less than 10% relative error with as few as 9% of the design space used as a training set.

4.7 Conclusions

In this chapter we presented the fundamental problems addressed by the European Project MULTICUBE and the main achievements obtained at the end of the project. In particular, we presented the definition of the MULTICUBE design flow composed of an automatic design space exploration framework combined with a power/performance estimation framework to support the design of next generation many-core architectures.

Acknowledgements We would like to gratefully acknowledge our EC Project Officer, Panagiotis Tsarchopoulos and our reviewers: Alain Perbost, Andrzej Pulka and Kamiar Sehat for their valuable comments and guidance during the MULTICUBE project review process.Prabhat Avasare, Geert Vanmeerbeeck, Chantal Ykman and Maryse Wouters are also associated with Interdisciplinary Institute for BroadBand Technology, Belgium (IBBT), B-9050 Gent, Belgium.

References

1. Mariani G, Avasare P, Vanmeerbeeck G, Ykman-Couvreur C, Palermo G, Silvano C, Zaccaria V (2010) An industrial design space exploration framework for supporting run-time resource management on multi-core systems. In: Proceedings of DATE 2010: IEEE design, automation and test conference in Europe. Dresden, Germany, pp 196–201, Mar 2010
2. Mariani G, Palermo G, Silvano C, Zaccaria V (2009) Multiprocessor system-on-chip design space exploration based on multi-level modeling techniques. In: Proceedings of IEEE IC-SAMOS'09—International Conference on Embedded Computer Systems: Architectures, MOdeling, and Simulation. Samos, Greece, pp 118–124, July 2009
3. Posadas H, Castillo J, Quijano D, Fernandez V, Villar E, Martinez M (2010) SystemC platform modeling for behavioral simulation and performance estimation of embedded systems. Behav Model Embedded Syst Technol: App Des Implementation pp 219–243
4. Mei B, Sutter B, Aa T, Wouters M, Kanstein A, Dupont S (2008) Implementation of a coarse-grained reconfigurable media processor for avc decoder. J Signal Process Syst 51(3):225–243
5. Avasare P, Vanmeerbeeck G, Kavka C, Mariani G (2010) Practical approach to design space explorations using simulators at multiple abstraction levels. In: Design Automation Conference (DAC) User Track Sessions, Anaheim, USA, June 2010
6. Hwang CL, Masud ASM (1979) Multiple objective decision making—methods and applications: a state-of-the-art survey, vol 164. Lecture notes in economics and mathematical systems. Springer, Heidelberg
7. Okabe T, Jin Y, off B (2003) A critical survey of performance indices for multi-objective optimization. In: Proceedings of the IEEE congress on evolutionary computation, pp 878–885
8. Jaszkiewicz A, Czyak P (1998) Pareto simulated annealing—a metaheuristic technique for multiple-objective combinatorial optimisation. J Multi-Criteria Decis Anal (7):34–47
9. Deb K, Agrawal S, Pratab A, Meyarivan T (2000) A fast and elitist multi-objective genetic algorithm: NSGA-II. In: Proceedings of the parallel problem solving from nature VI conference, pp 849–858
10. Poloni C, Pediroda V (1998) GA coupled with computationally expensive simulations: tools to improve efficiency. In: Quagliarella D, Périaux J, Poloni C, Winter G (eds) Genetic algorithms and evolution strategies in engineering and computer science. Recent advances and industrial applications, Chap. 13, Wiley, Chichester, pp 267–288
11. Smith KI, Everson RM, Fieldsend JE, Murphy C, Misra R (2008) Dominance-based multiobjective simulated annealing. Evol Comput, IEEE Trans 12(3):323–342
12. Palermo G, Silvano C, Zaccaria V (2008) Discrete particle swarm optimization for multi-objective design space exploration. In: Proceedings of DSD 2008: IEEE Euromicro conference on digital system design architectures, methods and tools, Parma, Italy, pp 641–644, Sep 2008

13. Turco A, Kavka C (2010) MFGA: a genetic algorithm for complex real-world optimization problems. In: Proceedings of BIOMA 2010, the 4th international conference on bioinspired optimization methods and their applications,.Lubiana, Slovenia, May 2010
14. Joseph PJ, Vaswani K, Thazhuthaveetil MJ (2006) A predictive performance model for superscalar processors. In: MICRO 39: Proceedings of the 39th annual IEEE/ACM international symposium on microarchitecture. IEEE Computer Society. Washington, DC, pp 161–170
15. Joseph PJ, Vaswani K, Thazhuthaveetil MJ (2006) Construction and use of linear regression models for processor performance analysis. The twelfth international symposium on high-performance computer architecture. pp 99–108
16. Lee BC, Brooks DM (2006) Accurate and efficient regression modeling for microarchitectural performance and power prediction. In: Proceedings of the 12th international conference on architectural support for programming languages and operating systems 40(5):185–194
17. Bishop C (2002) Neural networks for pattern recognition. Oxford University Press, Oxford

Chapter 5
2PARMA: Parallel Paradigms and Run-time Management Techniques for Many-Core Architectures

C. Silvano, W. Fornaciari, S. Crespi Reghizzi, G. Agosta, G. Palermo, V. Zaccaria, P. Bellasi, F. Castro, S. Corbetta, A. Di Biagio, E. Speziale, M. Tartara, D. Melpignano, J.-M. Zins, D. Siorpaes, H. Hübert, B. Stabernack, J. Brandenburg, M. Palkovic, P. Raghavan, C. Ykman-Couvreur, A. Bartzas, S. Xydis, D. Soudris, T. Kempf, G. Ascheid, R. Leupers, H. Meyr, J. Ansari, P. Mähönen and B. Vanthournout

Abstract The 2PARMA project focuses on the development of parallel programming models and run-time resource management techniques to exploit the features of many-core processor architectures. The main goals of the 2PARMA project are: definition of a parallel programming model combining component-based and single-instruction multiple-thread approaches, instruction set virtualisation based on

---

The project is supported by the EC under grant 2PARMA-FP7-248716.

C. Silvano (✉) · W. Fornaciari · S. C. Reghizzi · G. Agosta · G. Palermo · V. Zaccaria · P. Bellasi · F. Castro · S. Corbetta · A. Di Biagio · E. Speziale · M. Tartara
Dipartimento di Elettronica e Informazione – Politecnico di Milano, Milano, Italy
e-mail: silvano@elet.polimi.it

D. Melpignano · J.-M. Zins
STMicroelectronics, Grenoble, France

D. Siorpaes
STMicroelectronics, Milano, Italy

H. Hübert · B. Stabernack · J. Brandenburg
Fraunhofer HHI, Fraunhofer, Germany

M. Palkovic · P. Raghavan · C. Ykman-Couvreur
IMEC vzw Belgium and IBBT, Leuven, Belgium

A. Bartzas · S. Xydis · D. Soudris
Institute of Communication and Computer Systems – National Technical University of Athens, Athens, Greece

T. Kempf · G. Ascheid · R. Leupers · H. Meyr · J. Ansari · P. Mähönen
RWTH–Aachen University, Aachen, Germany

B. Vanthournout
Synopsys Belgium, Leuven, Belgium

N. Voros et al. (eds.), *VLSI 2010 Annual Symposium*,
Lecture Notes in Electrical Engineering 105, DOI: 10.1007/978-94-007-1488-5_5,

portable byte-code, run-time resource management policies and mechanisms as well as design space exploration methodologies for many-core computing architectures.

5.1 Introduction

The main trend in computing architectures consists of integrating small processing cores in a single chip where the cores are connected by an on-chip network. Given the technology trend, we would expect in the coming years moving from multi- to many-core architectures. Multi-core architectures are nowadays prevalent in general purpose computing and in high performance computing. In addition to dual- and quad-core general-purpose processors, more scalable multi-core architectures are widely adopted for high-end graphics and media processing, e.g., IBM Cell BE, NVIDIA Fermi, SUN Niagara and Tilera TILE64. To deal with this increasing number of processing cores integrated in a single chip, a global rethinking of software and hardware design approaches is necessary.

The 2PARMA project focuses on the design of a class of parallel and scalable computing processors, which we call Many-core Computing Fabric (MCCF) template. This template is composed of many homogeneous processing cores connected by an on-chip network. The class of MCCF promises to increase performance, scalability and flexibility only if appropriate design and programming techniques will be defined to exploit the high degree of parallelism exposed by the architecture.

Benefits of MCCF architectures include finer grained possibilities for energy efficiency paradigms, local process variations accounting, and improved silicon yield due to voltage/frequency island isolation possibilities. To exploit these potential benefits, effective run-time power, data and resource management techniques are needed. Moreover the MCCF offers customisation capabilities to extend and to configure at run-time the architectural template to address a variable workload.

The 2PARMA project aims at overcoming the lack of parallel programming models and run-time resource management techniques to exploit the features of many-core processor architectures focusing on the definition of a parallel programming model combining component-based and single-instruction multiple-thread approaches, instruction set virtualisation based on portable bytecode, run-time resource management policies and mechanisms as well as design space exploration methodologies for MCCFs.

The research objectives of the project are intended to meet some of the main challenges in computing systems:

- To improve performance by providing software programmability techniques in order to exploit the hardware parallelism;

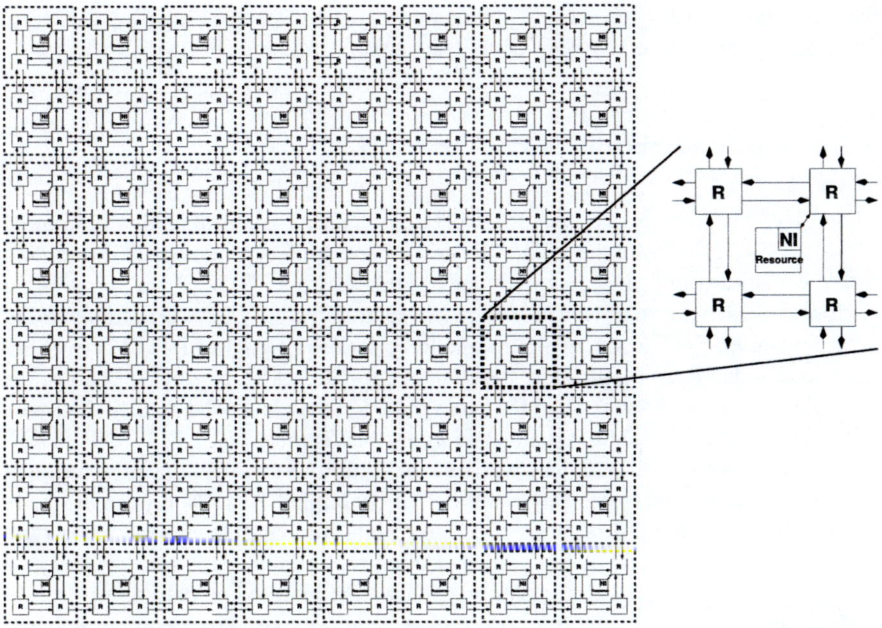

Fig. 5.1 2PARMA Many-core computing fabric template

- To explore power/performance trade-offs and to provide run-time resource management and optimisation;
- To improve system reliability in terms of lifetime and yield of hardware components by providing transparent resource reconfiguration and instruction set virtualisation;
- To increase the productivity of the process of developing parallel software by using semi-automatic parallelism extraction techniques and extending the OpenCL programming paradigm for parallel computing systems.

The rest of this paper is organised as follows. Section 5.2 provides an introduction to the target architectural template. Section 5.3 describes the 2PARMA design flow and the design methodologies employed, while Sect. 5.4 introduces the applications targeted in the project. Finally, Sect. 5.5 draws conclusions and outlines the future work.

5.2 MCCF Architecture Template

The 2PARMA project focuses on the MCCF template, which is composed of many homogeneous processing cores connected by an on-chip network, as shown in Fig. 5.1.

The project will demonstrate methodologies, techniques and tools by using innovative hardware platforms provided and developed by the partners, including the "Platform 2012"—an early implementation of MCCF provided by STMicroelectronics—and the many-core COBRA platform provided by IMEC.

5.2.1 STMicroelectronics Platform 2012

The P2012 program is a cooperation between STMicroelectronics and Commissariat a l'Energie Atomique (CEA) to design and prototype a regular computing fabric capable to improve manufacturing yield. Platform 2012 (P2012) is a high-performance programmable accelerator whose architecture meets requirements for next generation SoC products at 32 nm and beyond. The goal of P2012 is twofold: from one side, it is to provide flexibility through massive programmable and scalable computing power; from the other side, to provide a solid way to deal with increasing manufacturability issues and energy constraints.

Organised around an efficient Network-on-Chip communication infrastructure, P2012 enables connecting a large number of decoupled STxP70 processors SMP clusters, offering flexibility, scalability and high computation density. The Platform 2012 computing fabric is composed of variable number of 'tiles' that can be easily replicated to provide scalability. Each tile includes a computing cluster with its memory hierarchy and a communication engine. The computing fabric operation is coordinated by a fabric controller and is connected to the SoC host subsystem through a dedicated bridge, with DMA capabilities. Clusters of the fabric can be isolated to reduce power consumption (or to switch-off a particular faulty element) and frequency/voltage scaling can be applied in active mode.

The P2012 computing fabric is connected to a host processor such as the ARM Cortex A9, via a system bridge. In this way, the fabric is exposed to legacy operating systems like the GNU/Linux OS.

Many P2012 platform design choices are still open to be explored, and the 2PARMA Consortium, which is one of the very early adopters of this technology, effectively contributes to the platform architecture specification and relevant optimisations.

5.2.2 IMEC ADRES-based COBRA Platform

The IMECs COBRA platform is an advanced platform template targeting 4G Giga-bit per second wireless communication. This platform can be customised to handle very high data rates as well as low throughputs in a scalable way. This platform largely consists of four types of cores. *DIFFS*, an ASIP processor tuned towards sensing and synchronisation, and optimised for very low power. It is tuned towards average duty cycle. *ADRES* [1], a coarse-grained reconfigurable core template [2] consisting of a number of functional units connected in a given interconnect network. The core has been tuned to be capable of doing inner modem processing of various standards efficiently. *FlexFEC* [3], a flexible forward

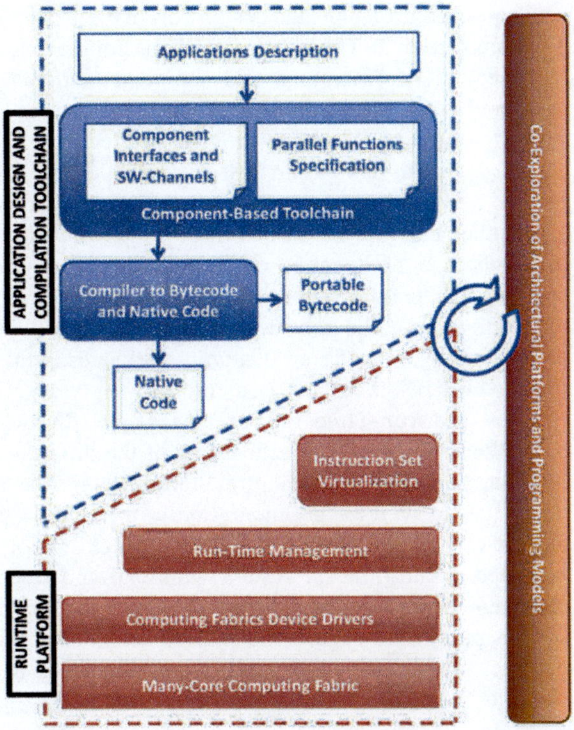

Fig. 5.2 2PARMA Design Flow and Tools

error correction ASIP that is capable of doing different outer modem processing. It is a SIMD engine template where the instruction set, bit width of the data-path and the number of SIMD slots can be chosen based on the set of requirements of the standard to be run. An *ARM* host processor for controlling the tasks on the platform (e.g. the run-time manager task).

The first three cores (DIFFS, ADRES and FlexFEC) can be instantiated for a mix of targeted standards that need to be supported. Also all parts of the plattorm are programmable in C (ADRES, ARM) or assembly (FlexFEC, DIFFS). The communication is ensured by customised InterConnect Controller (ICC) cores that are programmable at assembly level as well. In this platform, besides the type and the size of each core, the number of each type of core can be selected based on the different standards that need to be supported on the platform.

5.3 Design Flow and Tools

In this section, we provide an overview of the 2PARMA design flow, which aims at supporting development, deployment and execution of applications on MCCFs, as well as a more in-depth coverage of design decisions and early results for critical parts of the design flow.

The tool environment and design flow of the 2PARMA project is shown in Fig. 5.2. The compilation tool chain starts with the component-based application source code (C-based) to be assembled and compiled to byte-code and further dynamically translated to machine code. Then the machine code execution and deployment will be supported by an OS layer to provide isolated logical devices efficiently communicating (device-to-device and host-to-device). The GNU/Linux operating system will be used as the software reference common ground for the host processor.

Another main goal consists in developing methodologies and tools to support the application/architecture co-exploration. More in detail, the project focuses on profiling the parallel applications aimed at finding the bottleneck of the target platform and on the robust design space co-exploration of static and dynamic parameters by considering dynamic workloads, while identifying hints/guidelines for dynamic resource management.

Then, the Run-Time Resource Manager (RTRM) provides adaptive task and data allocation as well as scheduling of the different tasks and the accesses to the data for many-core architectures. Furthermore, the adequate power management techniques as well as the integration to the Linux OS will be provided. Based on the set of operating points given by the DSE tool at design time and the info collected at run-time on system workload and resource utilisation, the run-time management techniques will optimise data allocation and data access scheduling, task mapping and scheduling and power consumption. The rest of this section provides more detail on the techniques employed in the design flow.

5.3.1 Programmability of MCCFs

2PARMA project tackles the issue of programmability of MCCFs at both the programming language and OS level. At the programming language level, it is possible to identify two different types of parallelism that can be effectively exploited in MCCFs: task level and loop level parallelism.

Task parallelism represents the concurrent operation of several different routines, which may cooperate in a simple pipeline model, or with more complex interaction. The key aspect, however, is that each task represents a different type of computation, possibly with widely different degrees of complexity among the task set—in the pipeline example, different software stages can have different complexity, with some stages being further decomposed in parallel sub-tasks. Representing task level parallelism requires the encapsulation of the individual routines in task structures, and the deployment of connections to guarantee communication and synchronization, which should be explicitly represented by the application developer. In general, tasks are isolated from each other, communicating only through the exchange of messages on a set of explicitly defined channels.

Loop level parallelism (or *data parallelism*), on the other hand, represents a lower level of parallelism, dealing with massively replicated computations on large shared data structures. The key aspect in this case is that each parallel work item represents the same operation applied to one of a set of data items, as in a typical Single Instruction Multiple Data (SIMD) computation. The case of MCCFs differs from other architectures suited for execution of data parallel kernels (such as GPGPUs) in that MCCFs can easily support significant amount of control flow divergence—the possibility that the control flow of different work items diverges. This is generally a catastrophic event for a SIMD-oriented architecture, since different operations are now executed for each data item. However, it is well handled by MCCFs, which can in general follow a different control flow for each active thread without penalties.

The goal of the 2PARMA project is to allow the expression of both levels of parallelism, in order to support a wide range of application scenarios. To this end, the application developers employing the 2PARMA design flow will work in a top-down way, first decomposing the application in parallel tasks connected by communication and synchronization channels, then further parallelizing each task through the identification of parallel loops.

A distributed memory model will be therefore implemented to handle the communication between different tasks, while a shared memory model will be employed within the task to allow communication among the work items. This approach allows an easier deployment of tasks across clustered fabrics, while preserving the efficiency of a shared memory model at the lower level of parallelism.

5.3.1.1 Task Level Parallelism

In the 2PARMA compiler tool chain, tasks are represented through the increasingly popular Component-Based Software Engineering (CBSE), leveraging the concepts and tools developed in the cross-disciplinary "Nucleus" flagship project [4] of the UMIC Research Center at RWTH Aachen University. In this framework, each task, encapsulating a critical algorithmic kernel, is represented by a (possibly hierarchically decomposed) component, called a Nucleus.

In a later stage these Nuclei are assembled to construct the complete application, as shown in Fig. 5.3, by connecting them through FIFO channels. In contrast to existing CBSE tool-chains, requirements such as latency and throughput are integrated into the application description directly. Furthermore, the Nuclei are mapped to Flavors—efficient and optimised implementations for one Nucleus on a particular Processing Element (PE)—that are kept within the Board Support Package of a given HW platform. This allows the mapping tools to identify possible implementation options by performing interface and constraint checks. Among these different options, designers can select the final implementation that achieves the best performance.

5.3.1.2 Loop Level Parallelism

At this level, parallelism is expressed by means of parallel loops. These could be expressed explicitly, by tagging *for* loops with appropriate OpenMP directives, or

Fig. 5.3 Support of task-level parallelism through nucleus-based tool chain

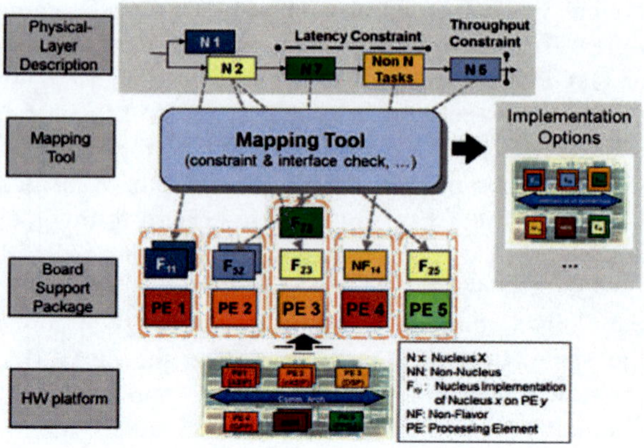

Fig. 5.4 Architecture of the compiler tool chain to support loop-level parallelism

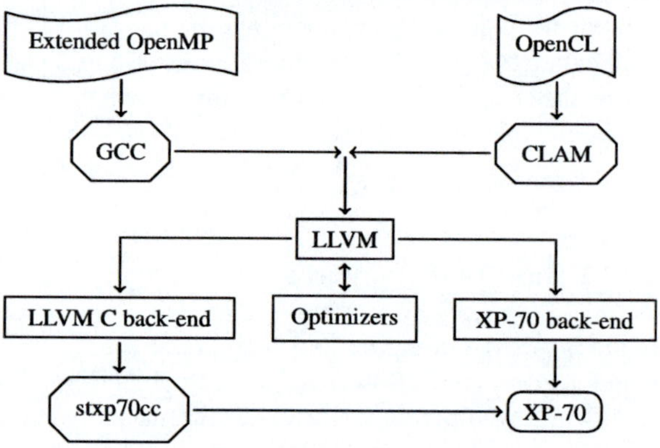

as parallel kernels defined by the OpenCL specification. In the 2PARMA project, the OpenCL kernel will be the default data parallel construct, though explicit representation of parallel loops will also be available to the programmer. Figure 5.4 describes the data parallel tool chain.

The 2PARMA compiler tool chain will benefit from techniques that automatically handle memories local to processor clusters usually available in GPGPU architectures, as well as automated lowering of higher-level code to OpenCL [5]. This will allow the programmer to first design the application under a shared memory paradigm, and then perform fine-tuning on a view of the application where the shared memory abstraction is removed.

5.3.2 Run-time Management

2PARMA project aims at improving energy efficiency with respect to conventional power management strategies, by supporting efficient and optimal task, data and resource management able to dynamically adapt to the changing context, taking into account the Quality-of-Service (QoS) requirements imposed by the user to each application. Multiple configurations of application parameters, dynamic workloads and time-dependent QoS impose strong requirements for application-specific and run-time adaptive dynamic memory management of running applications.

In 2PARMA, we push the boundary of this trade-off to reduce the design-time effort and employ the run-time resource manager, at different abstraction levels. We accomplish this task by providing a Run Time Manager (RTM) with metadata information, covering both design-time and run-time knowledge of both hardware and software.

The design-time knowledge of the system is generated by employing efficient design space exploration techniques. This reveals the operating points of the system/applications and their resource requirements. Then at run-time, the RTM exploits this knowledge and performs operating point switching, thus adapting to the applications' needs. More specifically, the run-time management performs adaptive task mapping and scheduling [6], dynamic data management [7] and system-wide power management [8].

The first role of the RTM is to monitor the dynamic applications behaviour to control the available and required platform resources at run-time while meeting user requirements. To handle multiple tasks competing at run-time for limited resources, a scenario-based mapping technique supporting inter-task scenarios will be developed. To perform this, a run-time monitoring technique will be developed to tune well-chosen parameters based on the input data to meet the requirements while maximizing the output quality.

The second role of the run-time manager is to handle the dynamism in the control flow and data usage (dynamic (de)allocation of data), by determining suitable data allocation strategies that meet the application needs.

Finally, the run-time manager is responsible for the adaptive power management of the MCCF architecture. This is achieved by identifying a suitable top-level modelling of the entities composing the overall systems, in terms of exchanged data, exposition of control settings and status information of the components/devices. The manager is responsible to combine the adaptive run-time task and data management schemes (component/device specific optimisations) with the adaptive power management policies, being aware of the presence of local optimisation strategies exposed by the rest of the system components (e.g., device drivers and in general any other resource manager).

The end result is a distributed run-time QoS Constrained Power Manager (CPM) working at the OS-level [8], based on the following concepts. *System-Wide Metrics* (SWMs) which are parameters describing behaviours of a running system and represent QoS requirements. They could be either "abstract" metrics (ASMs) or platform dependent (PSMs). The first ones are exposed to user-space and can be used by application to assert QoS requirements. The second ones instead are defined in the platform code and are used to keep track of hardware inter-dependencies. *Device Working Regions* (DWRs) define the mapping between the operating modes of a devices and the SWMs that define the QoS level supported by each operating mode. *Feasible System Configurations* (FSCs) are the n-dimensional intersections of at least a DWR for each device (where n is the number of SWMs defined). They identify the system-wide working points of the target platform where certain QoS levels are granted. *Constraints* on SWMs defined at run-time according to the QoS requirements of applications or drivers on these parameters. All the QoS requirements on the same SWM are translated to a constraint using an aggregation function which depends on the type of the parameter. *Multi-Objective optimization*, which could consider different performance parameters, by assigning a weight to each SWM, and energy consumptions, by assigning a power consumption measure to each FSC.

In practice, CPM-related activities inside the OS, can be grouped in three main phases:

- FSC Identification: at boot time all the device drivers registers to CPM by exposing their DWRs. All FSCs can be automatically identified by performing the intersection of DWRs.
- FSC ordering: every time the optimization goals change, the FSC are sorted according to the global optimization policy. This happens usually when the device usage scenarios change.
- FSC selection: at run-time applications can assert QoS requirements on a specific SWM. These requirements are aggregated to produce a new constraint for each SWM. These constraints could invalidate some FSC. If the current FSC is also invalidated then a new candidate is selected according to the ordering defined in the ordering phase.

Finally all the drivers are notified about the new FSC and required to update accordingly their operating mode. The CPM model has been preliminary implemented as a Linux kernel framework (version 2.6.30) and tested under some use-cases to evaluate its overhead, which is negligible (always less than 0.01%) [8].

5.3.3 Design Space Exploration

Design space exploration (DSE) plays a crucial role in designing many-core computing platforms [9–11]. Design alternatives may consist of the tuning of processor micro-architectural components, different mappings of software tasks to resources, different scheduling policies implemented on shared resources as well as lower level design parameters. In this context, the 2PARMA project provides to the designer DSE methodologies to trade-off the system-level metrics (such as energy and delay) by considering the dynamic evolution of the system. 2PARMA goes beyond traditional design space exploration by defining a methodology to provide synthetic information about the points of operation of each application with respect to the subsets of resources available to it.

Focusing on the combined optimisation of parallel programming models and architectural parameters for many-core platforms, it is expected that conventional or state-of-the-art profiling techniques cannot be used for the task of analysing and profiling. Profiling memory accesses on a cycle accurate basis [12] is not sufficiently supported by available profiling tools due to the fact that only shared memory architectures were modeled at the time. Moreover, many-core platforms will be built upon completely different interconnection networks requiring new profiling techniques taking into account connection topologies. The influence on the overall system performance of the implemented connection topology and the resulting fragmentation of memory accesses abroad the distributed memory will be evaluated by a set of tools. Based on the profiling methodologies developed in the project, it will be possible to get an in depth view of how parallel programming models behave on many-core platforms. The results will be used to co-optimise the programming model and the architecture of the target platform.

5.3.4 Preliminary Exploration of the OpenCL Programming Model

One of the key assumptions in the 2PARMA project is that the system performance is significantly impacted by software-level parameters such as the OpenCL work-group size. As we shall see, the tuning of such kind of parameters is of paramount importance to squeeze the performance out of parallel computing fabrics and the programmer should be helped in some way to tune his application towards the maximum exploitation of the available resources.

A preliminary investigation has been performed by using a recent NVIDIA GeForce GTX 260 (hardware revision of June 2008 in 65 nm technology). The parallel architecture provides 24 multi-processor cores (which correspond to OpenCL computing units), where each multi-processor core has eight processing elements. As stated by NVIDIA [13], these resources can be shared by 1024 actively running threads at most. A *warp* is a cluster of 32 threads which are likely to share the same execution path and thus is treated efficiently by the instruction scheduler of the multi-processor. Technically, the multi-processor can execute 8 threads at a time which correspond to a quarter of a warp.

To understand how the target architecture can perform sub-optimally whenever the work-group size is not specified as suggested by the manufacturer, we set up an experimental campaign in which we sampled various values of the work-group size and profiled the execution time of the *oclMatVecMul* OpenCL benchmark. The benchmark performs a matrix-vector multiplication by making the kernels within a work-group cooperate on a subset of the rows of the original matrix. Local memory is exploited only in the coalesced case to allow the kernels belonging to a work-group a faster access to shared data. Barriers are used to enable consistency in the *reduction* steps associated with the row-by-column multiplication.

Figure 5.5 shows the behaviour in terms of throughput (inverse of the time needed for elaborating the complete dataset for the specific benchmark (measured in bursts per seconds) for the "Coalesced-1" implementation. The occupancy stress highlighted two main characteristics:

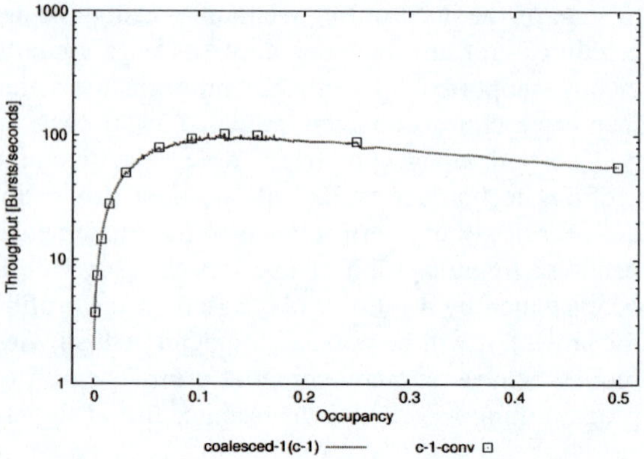

Fig. 5.5 oclMatVecMul performance (throughput) with mono-dimensional iteration space and varying occupancy

- For some manufacturer-suggested work-group size values (multiple of 32, indicated by the label '-conv' Fig. 5.5), the reduction algorithm is not applicable.
- The throughput tends to decrease after a sweet spot (around 10% of occupancy). Again, this can be due to several factors, the most important one being that there exists an increasing amount of synchronization between threads when work-groups are increased in size. Unfortunately, this is not masked by having multiple work-groups running on the multi-processor.

So far, we have thus seen that the determination of a sweet spot of the work-group size is not trivial to identify a-priori, even with the suggested values from the manufacturer. Thus, it is difficult to devise a work-group size which can be robust enough to be optimal on a wide range of architecture families. To this we may add the problem of software programmers who are relatively unaware of the architectural features of the accelerator they are using. Although some "best programming practices" do exist for some of the accelerators, we believe that the practice software engineering of OpenCL programs should be supported more pragmatically and efficiently by using automatic tools for tuning these parameters. We believe that conventional design space exploration (DSE) techniques play a crucial role in this field and we plan to apply those techniques to solve the above problems.

5.4 Applications

The MCCF template is designed as a coprocessor for computationally intensive applications in high-end embedded scenarios. To prove its effectiveness, and the effectiveness of the design flow and tools produced in the 2PARMA project, it is necessary to employ real world applications of considerable industrial impact. These applications will be engineered, optimised and specialised using the methodologies described in Sect. 5.3, and tested on the two target implementations of the MCCF template. In this section, we introduce the three applications chosen for the 2PARMA project: *Scalable Video Coding* (SVC), *Cognitive Radio*, and *Multi View Video* (MVV).

5.4.1 Scalable Video Coding

SVC [14], also known as layered video coding, has already been included in different video coding standards in the past. Scalability has always been a desirable feature of a media bit stream for different services and especially for best-effort networks that are not provisioned to provide suitable QoS and especially suffer from significantly varying throughput. Thus a service needs to dynamically adapt to the varying transmission conditions. E.g., a video encoder shall be capable of adapting the media rate of the video stream to the transmission conditions to provide at least acceptable quality at the clients, but shall also be able to explore the full benefits of available higher system resources. Within a typical multimedia session the video consumes the major part of the total available transmission rate compared to control and audio data. Therefore, an adaptation capability for the video bit rate is of primary interest in a multimedia session. Strong advantages of a video bit rate adaptation method relying on a scalable representation are drastically reduced processing requirements in network elements compared to approaches that require video re-encoding or transcoding. Thus, H.264/AVC-based SVC is of major practical interest and it is therefore highly important to investigate implementation aspects of SVC. SVC is an ideal application for demonstrating run-time resource management, including power management techniques.

5.4.2 Cognitive Radio

The Cognitive Radio application considered for the project includes both physical and MAC-layer processing. Especially, targetting low latency, high throughput and reconfiguration requirements of state-of-the-art wireless communication standards makes the cognitive radio application a highly appropriate use case for the 2PARMA project and its parallel programming models. Functional commonalities among MAC protocols are identified as fundamental building blocks so that a particular protocol can be realised by simply combining the required set of functionalities together. We have developed a tool (called the *wiring engine*) that combines the different components of the MAC together by coordinating the control and data flow among the blocks. The wiring engine is also able to exploit the parallelism in a particular MAC realization to achieve execution efficiency. Our approach of composing MAC protocols based on the same set of functional components using the wiring engine, leads to the realization of a wide range of protocols and allows run-time adaptation [15]. By implementing the MAC modules demanding high degree of computations and communication in the silicon as kernel functionalities, our approach allows meeting the strict timing deadlines thereby giving high degree of performance gains and flexibility. Furthermore, our methodology also facilitates much deeper cross-layer designs between MAC and physical layer kernels, which are demanded by cognitive and spectrum agile MACs [16].

5.4.3 Multi-View Video

With the current development of electronic, network, and computing technology, Multi-View Video (MVV) becomes a reality and allows countering the limitations of conventional single video. MVV refers to a set of N temporal synchronised video streams coming from cameras that capture the same scene from different viewpoints.

In particular, within the context of the 2PARMA project, we consider a cross-based stereo matching algorithm [17], assuming two aligned left and right cameras. The algorithm compute stereo pixel depth by means of their disparity (difference on the x coordinate), which can then be visualised in grayscale encoding. The cross-based stereo matching algorithm is an area-based local method, where the disparity computation at a given pixel only depends on intensity values within a finite window. The challenge of this method is that the support window should be large enough to include enough intensity variation for reliable matching, while it should be small enough to avoid disparity variation inside the window. Therefore, to obtain accurate disparity results at reasonable costs, an appropriate support window should be selected adaptively.

5.5 Conclusions

The 2PARMA project tackles the issue of programming and managing a MCCF—a novel architectural template represented within the project by STM Platform 2012 and IMEC Cobra architectures.

A design flow has been defined, starting with the high-level implementation of the application and leading to run-time management of the application execution, in a highly-variable context where multiple applications compete for resources. Design space exploration and profiling techniques close the feedback loop, helping the designer in refining the application for each target platform.

Finally, a set of high-impact applications has been selected to demonstrate and validate the effectiveness of the proposed methodologies.

References

1. Derudder V, Bougard B, Couvreur A, Dewilde A, Dupont S, Folens L, Hollevoet L, Naessens F, Novo D, Raghavan P, Schuster T, Stinkens K, Weijers JW, der Perre LV (2009) A 200 mbps+ 2.14nj/b digital baseband multi processor system-on-chip for sdrs. In: 2009 Symposium on VLSI Circuits, pp 292–293

2. Mei B, Vernalde S, Verkest D, Man HD, Lauwereins R (2003) Adres: an architecture with tightly coupled vliw processor and coarse-grained reconfigurable matrix. In: Cheung PYK, Constantinides GA, de Sousa JT (eds) FPL. Lecture Notes in Computer Science, vol 2778. Springer, pp 61–70
3. Naessens F, Derudder V, Cappelle H, Hollevoet L, Raghavan P, Desmet M, Abdel-Hamid A, Vos I, Folens L, O'Loughlin S, Singirikonda S, Dupont S, Weijers JW, Dejonghe A, der Perre LV (2010) A 10.37 mm^2 675 mw reconfigurable ldpc and turbo encoder and decoder for 802.11n, 802.16e and 3gpp-lte. In: 2010 Symposium on VLSI Circuits, pp 292–293
4. Ramakrishnan V, Witte EM, Kempf T, Kammler D, Ascheid G, Meyr H, Adrat M, Antweiler M (2009) Efficient and portable SDR waveform development: the nucleus concept. In: IEEE military communications conference (MILCOM 2009)
5. Andrea DB, Giovanni A (2010) Improved programming of GPU architectures through automated data allocation and loop restructuring. In: Proceedings of the 2PARMA Workshop (ARCS2010 Workshop)
6. Ma Z, Marchal P, Scarpazza DP, Yang P, Wong C, Gmez JI, Himpe S, Ykman-Couvreur C, Catthoor F (2007) Systematic methodology for real-time cost-effective mapping of dynamic concurrent task-based systems on heterogenous platforms. Springer Publishing Company, Incorporated
7. Bartzas A, Peon-Quiros M, Poucet C, Baloukas C, Mamagkakis S, Catthoor F, Soudris D, Mendias JM (2010) Software metadata: systematic characterization of the memory behaviour of dynamic applications. J Syst Software 83(6):1051–1075
8. Bellasi P, Fornaciari W, Siorpaes D (2010) A hierarchical distributed control for power and performances optimization of embedded systems. In: Müller-Schloer C, Karl W, Yehia S (eds) ARCS. Lecture Notes in Computer Science, vol 5974, Springer, pp 37–48
9. Chang H, Cooke L, Hunt M, Martin G, McNelly AJ, Todd L (1999) Surviving the SOC revolution: a guide to platform-based design. Kluwer Academic Publishers, Norwell, MA, USA
10. ARTEMIS Strategic Research Agenda Working Group (2006) Strategic research agenda: design methods and tools. Technical report, ARTEMIS
11. Duranton M, Yehia S, Sutter BD, Bosschere KD, Cohen A, Falsafi B, Gaydadjiev G, Katevenis M, Maebe J, Munk H, Navarro N, Ramirez A, Temam O, Valero M (2009) The HiPEAC 2012-2020 vision. Technical report, HiPEAC
12. Hübert H, Stabernack B (2009) Profiling-based hardware/software co-exploration for the design of video coding architectures. IEEE Trans. Circuits Sys Video Technol 19(11):1680–1691
13. Corp N (2008) Nvidia geforce gtx 200 gpu. Architectural overview
14. Schwarz H, Marpe D, Wiegand T (2007) Overview of the scalable video coding extension of the h.264/avc standard. IEEE Trans Circuits Syst for Video Technol 17(9):1103–1120
15. Ansari J, Zhang X, Achtzehn A, Petrova M, Mähönen P (2011) A flexible MAC development framework for congnitive radio systems. In: Procedings of the IEEE WCNC, Cancun, Mexico
16. Claudia C, Kaushik RC (2009) A survey on MAC protocols for cognitive radio networks. Ad Hoc Networks 7(7):1315–1329
17. Zhang K, Lu J, Lafruit G (2009) Cross-based local stereo matching using orthogonal integral images. IEEE Trans Cir and Sys for Video Technol 19(7):1073–1079

Part II
Embedded System Design

Chapter 6
Adaptive Task Migration Policies for Thermal Control in MPSoCs

David Cuesta, Jose Ayala, Jose Hidalgo, David Atienza, Andrea Acquaviva and Enrico Macii

Abstract In deep submicron circuits, high temperatures have created critical issues in reliability, timing, performance, coolings costs and leakage power. Task migration techniques have been proposed to manage efficiently the thermal distribution in multi-processor systems but at the cost of important performance penalties. While traditional techniques have focused on reducing the average temperature of the chip, they have not considered the effect that temperature gradients have in system reliability. In this work, we explore the benefits of thermal-aware task migration techniques for embedded multi-processor systems. We show the implementation issues of task migration policies on next generation architectural template of distributed memory multicore systems and we discuss the programmer's implications. Built on top of this programming model, we propose several policies that are able to reduce the average temperature of the chip and the thermal gradients with a negligible performance overhead. With our techniques,

D. Cuesta · J. Ayala · J. Hidalgo
Complutense University, Madrid, Spain
e-mail: dcuestag@pdi.ucm.es

J. Ayala
e-mail: jayala@fdi.ucm.es

J. Hidalgo
e-mail: hidalgo@fis.ucm.es

D. Atienza
Ecole Polytechnique Fédérale de Lausanne, Lausanne, Switzerland
e-mail: david.atienza@epfl.ch

A. Acquaviva (✉) · E. Macii
Politecnico di Torino, Turin, Italy
e-mail: andrea.acquaviva@polito.it

E. Macii
e-mail: enrico.macii@polito.it

N. Voros et al. (eds.), *VLSI 2010 Annual Symposium*,
Lecture Notes in Electrical Engineering 105, DOI: 10.1007/978-94-007-1488-5_6,

hot spots and temperature gradients are decreased up to 30% with respect to state-of-the-art thermal management approaches.

6.1 Introduction

Recent works have demonstrated that large temperature variations cause low reliability and they also impact leakage current. Temperatures over a threshold in localized areas of the chip (hot spots) can produce timing delay variations, transient reduction in overall system performance or permanent damages in the devices [1].

The reliable and efficient functioning of MPSoCs can be satisfied by guaranteeing the operation below a temperature threshold and power budget. It is in this control problem where thermal management and balancing policies come into play. Task and thread migration policies can be proposed to manage the thermal profile in embedded multi-processor systems [2]. While traditional dynamic thermal management (DTM) techniques have already been applied, they have not considered the spatial and temporal gradients that determine the mean-time-to-failure of the devices.

Thermal simulation of MPSoCs, where the exploration of the interaction between the hardware architecture and the software layer that performs the task migration is crucial, can take an unaffordable time. Thus, in order to explore the HW/SW interaction, FPGA-based thermal emulators have been developed [3, 4]. The experimental work carried out in this work is also developed for an FPGA-based MPSoC emulation platform [5] that speeds up the simulation time and provides high flexibility in the thermal analysis.

Thus, this chapter focuses on the design an implementation of three different task migration policies that are able to minimize the average temperature in MPSoCs as well as the spatial and temporal variations of the thermal profile. Our results show that they reduce the impact on the system performance to a minimum as compared to previous published approaches [2, 5, 6]. The specific contributions of our work are the following:

- three novel task migration policies based on adaptable weighted functions of three different factors: average thermal deviation between processors, maximum temperature of the overall chip and thermal gradient between cores.
- the proposed policies minimize the peak temperature and thermal gradients by considering a floorplan-aware task migration approach, at the same time as the time history of thermal gradients and thermal deviation of the different processors.
- the reliability of the system is improved by a combined minimization of time-based thermal unbalance (thermal cycles) and space-based thermal variations (hot spots).
- the experiments has been developed on a realistic MPSoC emulation platform [5], and the policies have been embedded in a multi-processor OS to assess its real-life task migration overheads in performance and temperature profile.

6.2 Background

6.2.1 Multiprocessor Architecture Organization

In this subsection we first describe two different kind of multiprocessor architectures, depending on the coupling of processors and memory, then we describe distinctive features of embedded multiprocessors systems-on-chip.

6.2.1.1 Shared Memory Multiprocessors

In a shared memory multiprocessor, all main memory is accessible to and shared by all processors. The cost of accessing shared memory is the same for all processors. In this case, from a memory access viewpoint they are called UMA (Uniform Memory Access) systems. A common communication medium links several memory modules to computing tiles consisting of a cache and one or more processor elements. Also I/O devices are attached directly to it. In tightly coupled shared memory Symmetric Multi-Processor (SMP) systems, which belong to this category, all the processors run a single copy of an operating system that coordinates global activities [7]. Synchronization is maintained through a cache-coherent low-latency shared memory. A particular category of shared memory multiprocessor is Non-Uniform Memory Access (NUMA). In a NUMA architecture, all physical memory in the system is partitioned into modules, each of which is local to and associated with a specific processor. As a result, access time to local memory is less than that to non local memory. NUMA machines may use an interconnection network to connect all processors to memory units, or use cache-based algorithms and a hierarchical set of buses for connecting processors to memory units. In both machines, I/O devices can be attached to individual processor modules or can be shared.

6.2.1.2 Distributed Multi-Processors

The individual processing units reside as separate nodes. Each processor runs its own operating system and synchronizes with other processors using messages or semaphores over an interconnect. From a memory access viewpoint, each processor has its own local memory that is not shared by other processors in the system (NO Remote Memory Access - NORMA - Multiprocessors). Computer clusters (CC) are examples of non-shared memory multiprocessors. Workstation clusters usually do not offer specialized hardware for low-latency inter-machine communication and also for implementation of selected global operations like global synchronization or broadcast. In general, NORMA machines do not support cache or main memory consistency on different processors memory modules. Such consistency is guaranteed only for local memory and caches (i.e., for non-shared memory), or it must be explicitly enforced for shared memory by user- or compiler-generated code.

6.2.1.3 Embedded MPSoCs

Modern Multiprocessor Systems-on-Chip are characterized by a combination of constraints. They must satisfy high computational rates but also real-time requirements, low power and cost [8]. Processing elements can be heterogeneous, homogeneous or even configurable, depending on the application requirements. On the memory viewpoint, heterogeneous hierarchical organizations are useful to increase predictability. In fact, in presence of a shared bus, memory access timings are difficult to estimate, leading to real-time constraint violations. Some embedded MPSoCs follows the SMP model, implementing a cache-coherent shared memory where the operating system and user-level code and data reside [9]. A more scalable organization is distributed, where local and shared memories are present, so the access is non uniform (NUMA). Communication happens through explicit message passing and each processor has its local code and data stored in the local memory. This is the architecture we consider in this work.

6.2.2 Communication and Synchronization in Embedded MPSoCs

Considering a distributed memory architecture, communication and synchronization must be explicit. Concerning communication, a hardware/software support for message passing on this kind of architecture is presented in [10]. Communication is based on DMA engine and support both shared memory and scratch-pad based communication. An hardware support is also proposed in [11] through coprocessors attached to each processor, to link it to local and remote memories. The cost of message passing primitives in such an architectural environment has been analyzed in terms of power and performance in [12]. In all of the previous approach, explicit synchronization between tasks can be performed through message passing. However, the usage of global variables in shared memory is also possible. However, the continuous polling of a shared variable leads to large contention on the system interconnect. In [13] a hardware module, called Synchronization Buffer (SB) is proposed which performs lock and unlock operations reducing contention in memory and interconnect.

In this work we will describe a message passing and synchronization support which is not based on additional hardware modules. We only assume the presence of hardware semaphores and interprocessor interrupt, that are commonly supported in MPSoCs.

6.2.3 Resource Management in Embedded MPSoCs

Due to the increasing complexity of these processing platforms, there is a large quantity and variety of resources that the software running on top of them has to manage. This may become a critical issue for embedded application developers, because resource allocation may strongly affect performance, energy efficiency and reliability. As a consequence, from one side there is need of efficiently exploit system resources, on the other side, being in an embedded market, fast and easy development of applications is a critical issue. For example, since multimedia applications are often made of several tasks, their mapping into processing elements has to be performed in a efficient way to exploit the available computational power and reducing energy consumption of the platform.

The problem of resource management in MPSoCs can be tackled from different perspectives. To the purpose of this chapter, it is useful to distinguish between static and dynamic resource management. Static resource managers are based on the a-priori knowledge of application workload. For instance, in [14] a static scheduling and allocation policy is presented for real-time applications, aimed at minimizing overall chip power consumption taking also into account interprocessor communication costs. Both worst case execution time and communication needs of each tasks are used as input of the minimization problem solved using Integer Linear Programming (ILP) techniques. In this approach authors first perform allocation of tasks to processors and memory requirement to storage devices, trying to minimize the communication cost. Then scheduling problem is solved, using the minimization of execution time as design objective.

Static resource allocation can have a large cost, especially when considering that each possible set applications may lead to a different use case. The cost is due to run-time analysis of all use cases in isolation. In [15] a composition method is proposed to reduce the complexity of this analysis. An interesting semi-static approach that deals with scheduling in multiprocessor SoC environments for real-time systems is presented in [16]. The authors present a task decomposition/clustering method to design a scalable scheduling strategy. Both static and semistatic approaches have limitations in handling varying workload conditions due to data dependency or to changing application scenarios. As a consequence, dynamic resource management came into play.

Even if scheduling can be considered a dynamic resource allocation mechanism, in this chapter we assume that a main feature of a dynamic resource manager in a multiprocessor system is the capability of moving tasks from processing elements at run-time. This is referred as to task migration. As such, dynamic resource management resorts to mainly two techniques, as far as processing elements are concerned: task migration and voltage/frequency selection, that we describe in the rest of this section.

6.2.4 Task Migration in Embedded MPSoCs

In the field of multiprocessor systems-on-chip, process migration can be effectively exploited to facilitate thermal chip management by moving tasks away from hot processing elements, to balance the workload of parallel processing elements and reduce power consumption by coupling dynamic voltage and frequency scaling [17–19]. However, the implementation of task migration, traditionally developed for computer clusters or symmetric multiprocessor, cache-coherent machines, poses new challenges [20]. This is specially true for non-cache-coherent MPSoCs, where each core runs its own local copy of the operating system in private memory. A migration paradigm similar to the one implemented in computer clusters should be considered, with the addition of a shared memory support for interprocessor communication.

For instance, many embedded system architectures do not even provide support for virtual memory, therefore many task migration optimization techniques applied to systems with remote paging support cannot be directly deployed, such as the eager dirty [21] or the copy-on-reference [22] strategies.

In general, migrating a task in a fully distributed system involves the transfer of processor state (registers), user level and kernel level context and address space. A process address space usually accounts for a large fraction of the process state, therefore process migration performance largely depends on the transfer efficiency of the address space. Although a number of techniques have been devised to alleviate this migration cost (e.g., lazy state transfer, pre-copying, residual dependencies [23]), a frequent number of migrations might seriously degrade application performance in an MPSoC scenario. As a consequence, assessing the impact of migration overhead is critical.

In the context of MPSoCs, in [24] a selective code/data migration strategy is proposed. Here authors use a compilation-level code profiling technique to evaluate the communication energy cost of transferring each function and procedure over the on-chip network. This information is used to decide whether it is worth migrating tasks on the same processor to reduce communication overhead or transferring data between them.

In [20], a feasibility study for the implementation of a lightweight migration mechanism is proposed. The user-managed migration scheme is based on code checkpointing and user-level middleware support. To evaluate the practical viability of this scheme, authors propose a characterization methodology for task migration overhead, which is the minimum execution time following a task migration event during which the system configuration should be frozen to make up for the migration cost. We take a similar approach in the experimental result section of this chapter to evaluate migration costs.

6.2.5 Voltage/Frequency Management

Run-time voltage and frequency scaling techniques have been extensively studied for single processor systems with soft real-time requirements. See [25] for an overview. Some of them are application dependent or exploit application information to make speed setting decisions [26]. Application independent policies are mostly based on overall processor utilization [27–29]. Among them, [29] proposes a per-task utilization based algorithm, which has been adopted into the ARM Intelligent Energy Manager standard [30].

In multiprocessor systems, approaches for combined DVS and adaptive body biasing in distributed time-constrained systems have been reported in [31]. A technique for combined voltage scaling of processors and communication links is proposed in [32]. The effect of discrete voltage/speed levels on the energy savings for multi-processor systems is investigated in [33], and a new scheme of slack reservation to incorporate voltage/speed adjustment overhead in the scheduling algorithm is also proposed.

Compared to previous techniques proposed for multiprocessor environments, in this chapter we do not focus on scheduling but just on scaling. Scheduling is assumed to be given inside each processing element. Moreover, we consider soft real-time systems rather than hard real-time systems, where deadlines are guarantee on the basis of worst case execution time (WCET) analysis. In this work we focus on techniques that do not require the knowledge of WCET or other information about task execution time, being application independent.

In [34] a semi-static solution of processor allocation and frequency/voltage selection is presented. The algorithm is based on the construction of Pareto curves for selecting the optimal speed based on throughput and utilization constraints. Authors use a fully functional system to test their approach. Compared to this work, we do not consider processor allocation, and we present a full on-line frequency selection policy.

Control theoretic approaches to DVS have been also proposed. A feedback control technique is presented in [35] focusing on average system performance. In [36, 37] buffers are used to save power and provide quality of service (QoS) guarantees. Feedback control on data buffers for DVS is used in [38] for a MPEG application. Here a proportional integrator (PI) controller is proposed that adjusts the decoder speed to keep constant the occupancy of the buffer between the decoder and the display. In [39] a non-linear feedback technique is proposed. This is the technique we used in this chapter. It outperforms traditional linear regulators in terms of voltage switching rate and ease of tuning.

6.2.6 Task Migration for Thermal Optimization

Load balancing techniques have been studied for general purpose parallel computers in the last decade [40, 41]. However, embedded systems and MPSoCs impose constraints, as the low-cost packaging and the portability, that make necessary to develop new techniques.

Nollet et al. [42] proposed a reuse technique that uses the debug registers of the processor to get the system workload information. Therefore, the initial overhead of a heterogeneous MPSoC task migration is diminished by considering these hardware devices which are not always available in current architectures.

Bertozzi et al. [43] presented an approach that dealt with MPSoCs task migration. They proposed a strategy where the user is responsible for setting the possible migration points in the application code. The architecture used in this work was composed by one master and an arbitrary number of slaves cores. Even though this chapter shows interesting results for such specific architecture, our work deals with a more general system where the task migration is dynamically performed.

Barcelos et al. [44] proposed a hybrid memory organization approach which supports the task migration algorithms with low-energy consumption constraints. In this approach, the data to be migrated can be provided either by the source node or from the shared memory. Barcelos' work is extended by Brião et al. [45] who takes into account the task migration overhead in a dynamic environment and discusses its impacts in terms of energy, performance and real-time constraints for MPSoCs based on Network on Chips (NoCs). In the context of streaming applications, the impact of task migration has been quantified by Pittau et al. [46] and Acquaviva et al. [47]. Following this line, our work considers the impact of task migration and minimizes this factor to optimize both performance and energy dissipation.

In the area of temperature optimization, several approaches have been proposed. Donald et al. [48] introduced several thermal management policies such as dynamic voltage and frequency scaling (DVFS) and thread migration based on current temperature, but their work do not consider the thermal history of the cores. This information gives a meaningful information about the future behavior of the system and can be exploited to improve the results of the migration.

The work by Puschini et al. [49] also manages dynamically the voltage and frequency assignment of each core based on game theory. However, the DVFS as a thermal optimization technique is limited by its implementation and its impact on performance.

On the other hand, Gomaa et al. [50] described techniques that, using the information provided by performance hardware counters, tried to balance the temperature by thread migration. However, it is considered that performance counters do not represent accurately the thermal profile.

In [51], Yang et al. showed an execution ordering approach that swaps hot and cool threads in cores to control the temperature. This can only be applied once the application has been profiled.

A recent work by Yeo et al. [52] presented a temperature-aware scheduler based on thermal grouping of the applications using a K-means clustering. This work provided interesting results but requires a very complex analysis phase, which grows largely in complexity with the number of considered cores.

Finally, in [5] it is proposed a heuristic optimization for thermal balancing in MPSoCs that adapts the current workload of the cores using DVFS and task migration, according to the standard deviation of the hottest and coldest cores at each moment in time during the execution. Although it shows clear benefits for thermal balancing with respect to previous thermal runaway approaches [48], it can still produce significant thermal unbalance in non-stable working conditions.

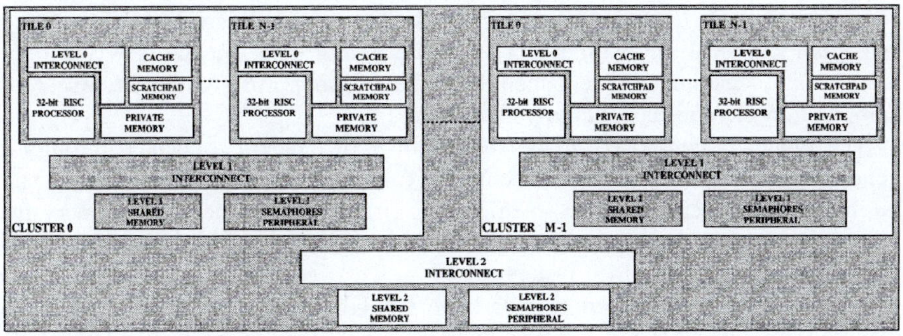

Fig. 6.1 Target processors-memory architectural template

(i.e., periods of small tasks being executed in the MPSoC or tasks being stop due to I/O processes) as we show in Sect. 6.7), because it does not take into account the recent thermal history of the system but just the instant thermal unbalance.

Our work outperforms previous approaches with the provision of three task migration policies that optimize the thermal profile of MPSoCs by balancing dynamically the weight of the on-chip thermal gradients, maximum temperature and effect of underlying floorplan on heat dissipation properties of each core. Moreover, the proposed policies are able to minimize the risk of system failure by the minimization of temperature-driven reliability factors, as it considers thermal unbalance in time and space, as they keep a history of the thermal profile of the target MPSoC, which minimizes the number of task migrations.

6.3 Target Architecture Template and Programmer's View

In this section we first describe a general architectural template targeted in this work. Next generation high performance computing systems are going toward a composable architectural template made of single processing forming a cluster of computational tiles with a distributed memory organization. Typical examples are ST P2012 [53] and Intel [54]. In this work we consider such an architectural template where each node runs its own local operating system instance.

6.3.1 General Architectural Template

Figure 6.1 describes the target architectural template. We focus in particular on processor-memory communication infrastructure. From this perspective, it appears as a generic template allowing a mix of shared memory and message passing communication. In this way it is possible to merge concept from cluster computing and from classical shared-memory multiprocessing. The model is based in a hierarchical, customizable connectivity scheme, supporting different levels of computation/memory/connectivity resources.

The model is composed of variable number of clusters connected through a generic interconnect (level 2 interconnect). Communication between clusters can be implemented using a shared-memory (level 2 shared memory), a memory for semaphores (level 2 semaphores memory), and using direct writes to private or shared memories inside the clusters.

Each cluster is composed of a variable number of tiles, connected through a generic interconnect (level 1 interconnect). Communication between tiles can be implemented using a shared-memory (level 1 shared memory), a memory for semaphores (level 1 semaphores memory), or using direct writes to private or shared memories inside the tiles. Each tile is composed of 32-bit RISC processor, a scratch-pad memory, a cache memory, and a private memory, connected through a generic interconnect (level 0 interconnect). The three levels of interconnect can be implemented using one or more of the available technologies (shared bus, hierarchical bus or NoC). A high generic target architecture allows the development of a software abstraction layer able to support both symmetric and heterogeneous multiprocessing.

6.3.2 Application Modeling

To exploit the potential of MPSoCs the application must be modeled and coded as a parallel program. In a parallel program, the functionalities are partitioned in chunks of code, and to each chunk of code a task is associated, with the assumption that tasks will be executed in parallel.

When mapping applications to MPSoCs a parallel representation often ease the work of the programmer, being typical MPSoC applications inherently parallel. A common representation of a parallel application is the task graph, i.e. a graph where each node represents a task, while each arc represents communication between tasks.

The possible parallel programming models approaches vary in the way the different parallels chunks of code are encapsulated and in the way synchronization and specially communication are modeled and implemented, and in the way the programmer cooperates with the underlying OS/middleware specializing its code.

The main feature of the proposed framework is that task migration is supported. In this way, dynamic resource management policies can take care run time mapping of task to processors, to improve performance, power dissipation, thermal management, reliability and/or other metrics. The programmer can avoid to take care of task mapping, it has only to manually insert checkpoints in the code, because migration is enabled only corresponding to checkpoints.

6.3.2.1 Task Modeling

In our approach, each task is represented using the process abstraction. In practice this means that task communication has to be explicit because shared variables between tasks are not allowed, each task has its own private address space. This is the main difference with respect to multi-threaded programming. Data can be shared between tasks using explicit services given by the operating system, using one or both of the available communication models: message passing and shared memory. Moreover dedicated services provided by the underlying middleware/OS are needed to enable tasks synchronization.

6.3.2.2 Task Communication and Synchronization

Both shared memory and message passing programming paradigms are supported by the proposed framework. Using message passing paradigm, when a process requests a service from another process (which is in a different address space) it creates a message describing its requirements, and sends it to the target address space. A process in the target address space receives the message, interprets it and services the request. Send and receive message functions can be either blocking or non-blocking.

To use shared memory paradigm, two or more tasks are enabled to access the same memory segment after they called shared malloc and were returned pointers to the same actual memory. When one task changes a shared memory location, all the other tasks see the modification. To use a shared memory segment, one process must allocate the segment using the dynamic shared memory allocation function. Then, using message passing, it communicates to the other tasks the starting address of the segment to share. After the communication is finished, one of the tasks has to deallocate the segment using a proper deallocation function.

Synchronization is supported providing basic primitives like mutexes and semaphores. Both spinlock and blocking mutexes and semaphores are implemented. Implementation of all these features is described in Sect. 6.4.

6.3.3 Checkpointing

The task migration support is not completely transparent to the programmer. Indeed, task migration can occur only corresponding to special function calls (checkpoints) manually inserted in the code by the programmer. Migration techniques involve saving and restoring the context of a process so it can be safely executed on a new core. Both in computer cluster and shared memory environments only the user context is migrated. System context is kept either on the home node or in shared memory. In our migration framework, all the data structure describing the task in memory is migrated. The use of suitable checkpointing

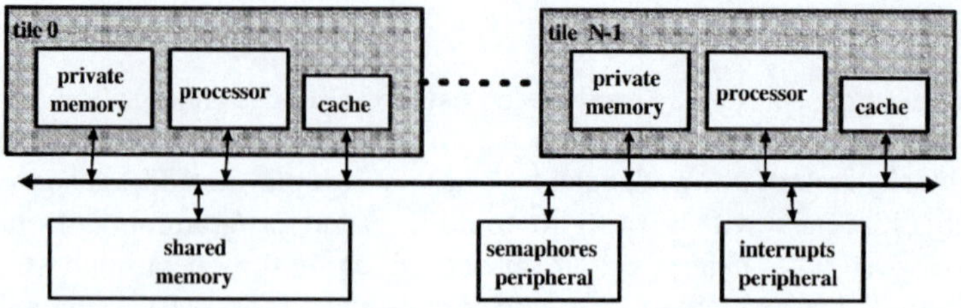

Fig. 6.2 Homogeneous Cluster-on-Chip architecture

strategy avoids the implementation of a link layer (like in Mosix) that impacts predictability and performance of the migrated process, which in our system does not have the notion of home node.

The programmer must take care of this by carefully selecting migration points or eventually re-opening resources left open in the previous task life. In fact, the information concerning opened resources (such as I/O peripherals) cannot be migrated, so the programmer should take into account this when placing checkpoints. In this case, a more complex programming paradigm is traded-off with efficiency and predictability of migration process. This approach is much more suitable to an embedded context, where controllability and predictability are key issues.

Checkpointing-based migration technique relies upon modifications of the user program to explicitly define migration and restore points, the advantage being predictability and controllability of the migration process. User level checkpointing and restore for migration has been studied in the past for computer clusters.

6.3.4 Homogeneous Cluster-on-a-Chip: A Case study

Given the generic Cluster-on-a-Chip architectural template and its programming model previously described, we now focus on a case study that has been used to implement and test the middle-ware layer presented in this chapter. Many new generation MPSoCs are designed based on the symmetric multi-processing paradigm [9, 55] exploiting homogeneous hardware architecture with general purpose processing elements, with no hierarchy levels. For these kind of architecture we can currently build realistic simulation models, that allow us to quantitatively assess the effectiveness of middleware level solution. That is why we resort to the implementation of our middleware on top of this architecture, that we called "homogeneous cluster" (shown in Fig. 6.2).

It is based on the following components:

- a configurable number of 32-bit RISC processors, without memory management unit (MMU),

- their caches,
- their private memories,
- a non-cached shared memory,
- a hardware interrupt module,
- a hardware *test-and-set* semaphores module,
- a 32-bit bus among them all.

The processor cores store private or local data in their private memory. In order to communicate each others, they use the non-coherent shared memory. For the synchronization among the processors, semaphore and interrupt facilities are also needed: (i) a core can send interrupt signals to each other using the hardware interrupt module mapped on in the global addressing space; (ii) several cores can synchronize themselves using the semaphore module that implements test-and-set operations. This is a basic hardware requirement to implement synchronization in a multiprocessor environment.

6.4 Operating System/Middleware Infrastructure

Modern Multiprocessor Systems-on-Chip are usually equipped with local and shared memory, so the access is non-uniform (NUMA). In this work, we target MPSoCs where each processor accesses a local private memory. Cache-coherency is guaranteed on private memories where a local copy of uClinux operating system [56] runs for each core (distributed operating system). This is follows the structure envisioned for non-cache-coherent MPSoCs [57, 58]. The uClinux OS is a derivative of Linux 2.4 kernel intended for microcontrollers without Memory Management Units (MMUs).

On the same on-chip bus, a non-coherent shared memory can be accessed, thus providing support for interprocessor communication. Hardware semaphores and inter-processor interrupts are needed to enable synchronization between the different OSes.

Each system image (one for each processor) is loaded in the private memory of each processor. This way each processor has its own filesystem, in the form of ROM filesystem. Although the ROM filesystem should be installed in a nonvolatile shared memory, we decided to have a distributed file system for simplicity of porting. Since uClinux is natively designed to run in a single-processor environment, we added the support for interprocessor communication at the middleware level.

This organization is a natural choice for a loosely coupled distributed systems with no cache coherency, to enhance efficiency of parallel application without the need of a global synchronization, that would be required by a centralized OS.

On top of local OSes we developed a layered software infrastructure to:

- provide and efficient parallel programming model for MPSoC software developers

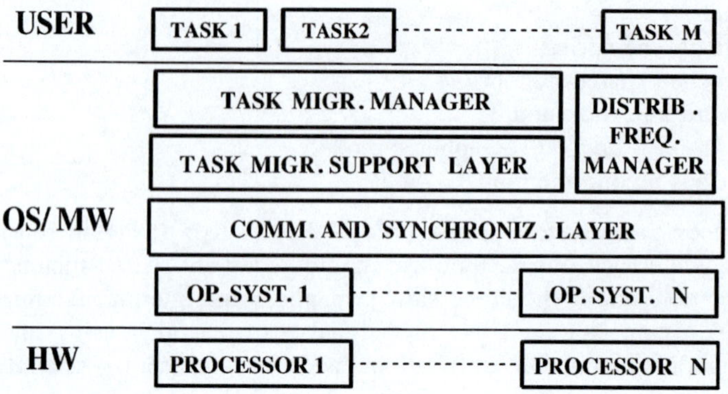

Fig. 6.3 Illustration of the software abstraction layer

- provide global synchronization
- support efficient runtime resource allocation and management

The proposed software abstraction layer, described in Fig. 6.3 is based on four main components: (1) Stand alone OS/schedulers for each processor running in private memory; (2) lightweight middle-ware layer providing synchronization and communication services (3) task migration support and dynamic resource management layer.

6.4.1 Basic Services: Communication and Synchronization Support

In our distributed system architecture, a process always remains within its address space; communication among processes happens using two mechanisms: shared memory and message passing.

Using message passing paradigm, when a process requests a service from another process (which is in a different address space) it creates a message describing its requirements, and sends it to the target address space. A process in the target address space receives the message, interprets it and services the request. We implemented a lightweight message passing scheme able to exploits scratch-pad memories or physical shared memory to implement ingoing mailboxes for each processor core. We defined a library of mixed user-level functions and system calls that each process can use to perform blocking write and read messages of data buffers. We defined a mailbox for each core and not for each task to avoid allocation/deallocation of mailboxes depending on process lifetime.

To use shared memory paradigm, two or more tasks are enabled to access the same memory segment after they called shared malloc and were returned pointers to the same actual memory. When one task changes a shared memory location, all the other tasks see the modification. Allocation in shared memory is implemented using a parallel version of the Kingsley allocator, commonly used in linux kernels.

Task and OS synchronization is supported providing basic primitives like binary or general semaphores. Both spinlock and blocking versions of semaphores are provided. Spinlock semaphores are based on hardware test-and-set memory-mapped peripherals, while non-blocking semaphores also exploit hardware inter-processor interrupts.

6.4.2 Advanced Services for Dynamic Resource Management: Task Migration Support

We explain here the general features of task migration support, outlining two possible implementations of task respawning, one with task replication and one based on task-recreation. The former, being based on task replicas on the various processors, has a lower cost in terms of migration delay, but imposes a larger memory occupation. The latter completely deallocates memory assigned to tasks that are not running on a given processor, reducing memory requirements but imposing a larger overhead for memory re-allocation in case of migration. One of the two mechanisms can be selected at configuration time depending on application and architectural constraints.

6.4.2.1 Task Respawning with Task Replication

The first method described is task replication. In this case the data structure used by the OS to manage an application is replicated in each private OS. When an application is launched, a fork is launched for each task of the application in each private OS. Only one processor at a time however is enabled to run a task. In this processor the task is managed as a standard task, while in the other processors the task is in the suspended tasks queue. In this way tasks which can be migrated and task which are not enabled for migration can coexist transparently for the private OS. Not all the data structure of a task is replicated, just the Process Control Block (PCB), which is an array of pointers to the resources of the task, and the local resources.

The migration process is managed using two kinds of kernel daemons, a master daemon running in a single processor, and slave daemons running in each processor. The master daemon is directly interfaced to the decision engine providing the selected policy for run time resource allocation. The processor where master daemon run is supposed to be the processor where task are launched or terminated by the user.

The master daemon performs four activities:

- The master periodically reads a data structure in shared memory where each slave daemon writes the statistics related to the own processor, and provides it to the module implementing the dynamic task allocation policy (decision engine).

The decision engine processes this data and run-time decides task allocation, eventually issuing task migrations.

- When a new application is launched by the user the master daemon communicates this information to the decision engine and sends a message to each slave communicating that the application should be initialized. The decision engine communicates to the daemon the identifier of the processor where the task is designed to start. The daemon forwards this communication to the slave daemon of that processor, so the task will be started. All the communications between master and slave daemons are implemented using dedicated, interrupt-triggered mailboxes in shared memory.
- When the decision engine decides a task migration, it triggers the master daemon, which signals to the slave daemon of the processor source of the migration that the task X has to be migrated to the processor Y.
- When an application ends or when it is stopped by the user, the master intercepts the information and forwards it to the slave daemons and to the decision engine. So the former can deallocate the task while the latter can update its data structures.

The slave daemon performs four activities:

- When the communication that a new application was launched arrives from the master daemon, it forks an instance task for each task of the application. Each task is stopped at its initial checkpoint and it is put in the suspended tasks kernel queue. The memory for the process is not allocated.
- Periodically it writes in the dedicated data structure in shared memory the statistics related to its processor which will be the base for the actions of the decision engine.
- When the master signals that a task has to be migrated from its own processor to a destination processor:
  - waits until the task to be migrated reaches a checkpoint, and puts it in the queue of the suspended tasks.
  - copy the block of data of the task to the scratch-pad memory of the destination process (if it is available and if there is space enough) or to the shared memory.
  - communicates to the slave daemon of the processor where the task must be moved that the data of the task is available in the scratch-pad or in the shared. A dedicated interrupt-based mailbox is used.
  - deallocates the memory dedicated to the block of the migrated task, making it available for new tasks or for the kernel.
  - put the migrated task PCB in the suspended tasks queue.
- when the salve daemon of the processor source of the migration communicates an incoming task:
  - allocates the memory for the data of incoming task and copies the data from the scratch-pad or from the shared memory to its private memory
  - puts the PCB of the incoming task in the ready queue

6.4.2.2 Task Respawning with Task Re-creation

This method is based on the run-time destruction of a task in the source processor, copy of task data to the destination processor, and task recreation in the destination processor. Also in this case the migration process is managed using a master daemon running in a single processor, and slave daemons running in each processor.

The master daemon performs four activities:

- It periodically reads a data structure in shared memory where each slave daemon writes the statistics related to the own processor, and provides it to the decision engine. The decision engine processes this data and run-time decides task allocation, eventually issuing task migrations.
- When a new application is launched by the user the master communicates this information to the decision engine which decides the processor where to start the application and communicates it to the master. The master sends a message to the slave daemon issuing the fork of a new task.
- When the decision engine decides a task migration, it triggers the master which signals to the slave daemon of the processor source of the migration that the task X has to be migrated to the processor Y.
- When an application ends or when it is stopped by the user the master intercepts the information and forwards it to the slave daemons and to the decision engine.

The slave daemon performs four activities:

- Periodically it writes in the dedicated data structure in main memory the statistic related to its processor which will be the provided to the decision engine.
- When the communication that a new application has to be launched arrives form the master daemon, it forks the new task.
- When the master daemon signals that a task has to be migrated from its own processor to a destination processor:
  - waits until the task to be migrated reaches a checkpoint, suspends it, and copy its data to the scratch-pad of the destination processor or to the shared memory.
  - communicates to the slave daemon of the processor destination of the migration that the data of the task is available in the scratch-pad or in the shared. A dedicated interrupt-based mailbox is used.
  - kills the task
- when the slave daemon of the processor source of the migration communicates an incoming task:
  - fork a new task
  - copies the incoming task data from its scratch-pad or from the shared memory to its private memory
  - performs an exec overwriting the data of the new task with the data copied from the scratch-pad or from the shared

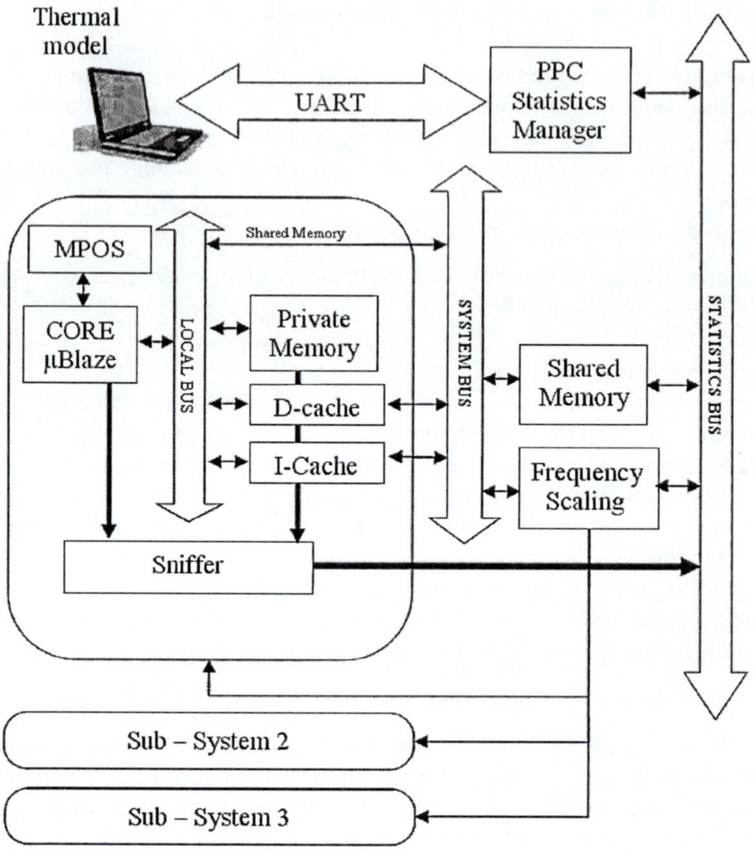

Fig. 6.4 Schematic view of the emulation platform

6.5 Emulation Platform

The thermal analysis conducted in this work requires an efficient mechanism to evaluate the performance and thermal statistics of the multi-processor system. The accuracy and the fast emulation of the system are the main constraints for the platform. Also, it is needed an MPOS that implements and manages the task migration policies.

In this work, we have used a complete FPGA-based estimation framework, implemented in a Virtex II pro VP30 and based on [4]. Figure 6.4 shows an schematic view of this emulation platform detailing a single core system. Using this framework we can retrieve the memory and processor statistics required by the thermal model and the migration policies (power consumption, memory misses and memory hits) by mean of hardware sniffers. This platform also includes a complete MPOS and task migration support library between the three cores of the emulated MPSoC (see Fig. 6.8).

In this emulation platform, the collected statistical data are sent to the host PC through the serial port. In the multiprocessor system, a dedicated PowerPc is the one in charge of processing and sending the statistics to the host PC. The host translates the received information into temperature values by means of a thermal library. This thermal library splits the floorplan of the emulated system in unitary cells, which are modeled as simple $R_{\text{thermal}}C_{\text{thermal}}$ circuits. The resolution of the linear equations created by an RC grid provides the evolution in time of the temperature of the system [59].

The emulated architecture is an homogeneous multi-processor system with three 32-bit RISC cores and the PowerPC. These processor do not include a memory management unit (MMU) and the access to the cacheable private memories and to a non-cacheable shared memory is managed by the OS.

Each core runs a uClinux OS. This is based on a Linux 2.4 kernel for microprocessors without an MMU, but upgraded to support the interprocessor communication found in our target system. The OS implements the task migration policies based on *task-replication*. Thus, there is a replica of each task in every local OS, but only one processor at a time can execute it. This method requires a slightly larger private memory to hold the tasks and task intermediate states/data before migrations, but it speeds up the task migration phase because the memory allocation required by the replication of tasks is avoided. Then, the task migration takes place only at predefined checkpoints chosen by the programmer between phases of the streaming execution (e.g., between processing different frames).

Several modifications have been done in the OS kernel to support the floorplan-aware policy. First, the identifier and weight of the cores (used by the policies to select the candidate in the task migration, as it will be presented later) are allocated in the shared memory. Second, the OS can then access this information to apply the task migration algorithm and achieve the thermal optimization.

In summary, the emulation platform is composed of these layers:

- *Application layer*: built as a set of independent tasks found in every processor of the system. The tasks are executed under the OS demand:
- *OS/Middleware layer*: controls the task migration and the communication and synchronization of the cores through the shared memory.
- *HW layer*: composed of three core-subsystems and a shared memory.

Finally, the emulation system has also been upgraded with a floorplan-temperature visualization tool. This tool communicates with the thermal library and, in real-time, provides a colored floorplan thermal map of the emulated MPSoC (see Fig. 6.10). The developed tool enables a rapid inspection of the hot spots, the evolution in time of the temperature and the spatial and temporal heat spread.

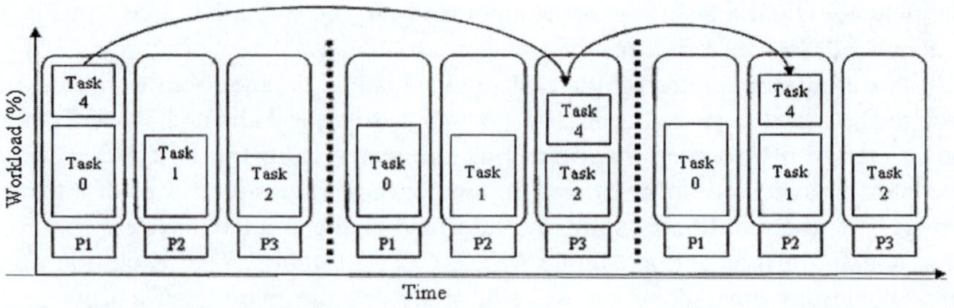

Fig. 6.5 Migration example between three cores

6.6 Adaptive and Floorplan Aware Policies for Thermal Balancing

As previously mentioned, the task migration policies that we present in this chapter are devoted to reduce the thermal gradients and mean temperature in a multi-processor system, because both facts affect negatively the reliability and the leakage of the chip [1]. This assumption is even more critical for embedded systems, where the power and temperature constraints must be satisfied in parallel with requirements of high-performance execution.

The FPGA-based multi-processor platform used in our experiments has been extended with a DVFS policy as an effective way to manage the voltage and frequency settings of the cores depending on the working load. The DVFS technique implemented follows the *vertigo* policy [60]. To apply the vertigo policy a previous characterization of the tasks is needed attending to their full-speed-equivalent (FSE), defined as the load that a task imposes when it is run at full speed in a core. Therefore, if one core is running a task that loads it, e.g. 45%, the core can adapt its frequency to 45% of its maximum.

Task migration policies are proposed to balance the load in the processors and, consequently, obtain a homogeneous distribution of temperature. Figure 6.5 presents an example. Three cores are running four tasks with different workload. This workload in the processors is translated into temperature due to the relation with the electric activity and dynamic energy; hence, this situation will create a thermal gradient due to the unbalanced distribution of the load, being **core 1** the hottest one. Thermal balance will be achieved migrating one task from this core to one of the colder processors.

If the temperature of the chip varies slower than the rate of task migration,[1] thermal balance will be achieved. In this case, we can assume that the real workload of each processor is the average of the total, in the example, around 55%. However, task migration must be applied carefully because it affects the performance of the system due to the overhead introduced by data transfers.

[1] This is a common assumption because the thermal evolution is a slow diffusion process.

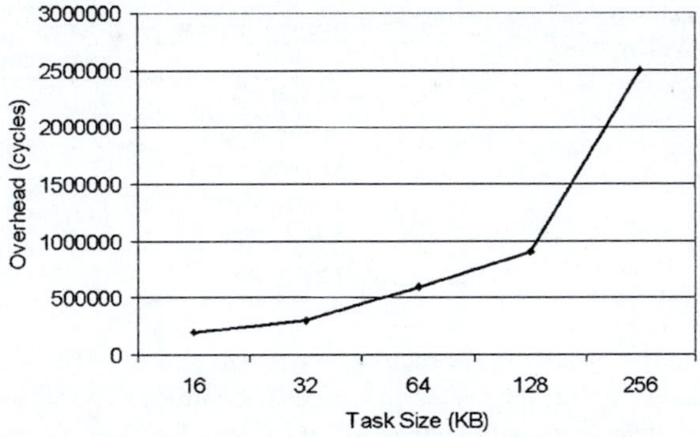

Fig. 6.6 Overhead of the task migration mechanism

The following paragraphs analyze the state-of-the-art task migration techniques, and the policies that we propose to specifically adapt the workload of the system depending on the state of the processors.

6.6.1 Compared State-of-the-Art Thermal Control Policies

- *Enhanced Migration* (*Mgr*) [6]: moves the task that is running in a hot core when it exceeds a threshold temperature to the coolest core. This policy could be considered as an even improved solution of the original policy of Heat & Run, because it adds task migration at run-time, as proposed in [5], not just between stopped or starting tasks.
- *Task rotation* (*Rot*) [2]: inspired by a Round Robin mechanism, migrates a task between processors every time slot. This policy achieves the thermal balance in the system at the cost of an important overhead due to the frequent migrations.
- *Thermal Thresholds* (*Thres*), presented in [5], moves the task running in the processor that exceeds an upper or lower threshold to a destination core. This is chosen considering the weight of the task that is going to be migrated and its impact on the workload of the processor.

6.6.2 Atomic Policies Pre-Characterization

The definition of our new task migration policies begins with the characterization of atomic policies in the multi-processor system. These atomic policies perform simple migrations only according to the temperature and the workload of the cores. The migration of the task is executed from one processor to another one with a negligible computation cost. Figure 6.6 shows the overhead introduced by the task replication mechanism for different sizes of the migrated task. As can be seen, the impact of migrating a 64 KB task (the one considered in our experimental work) is of 6E5 cycles, which translates into a delay of 6 ms for the worst case, depending

Table 6.1 Characterization of atomic policies

Atomic policy	Mean temperature	Max. temperature	Thermal gradient
Hot–Cold	4	5	4
Warm–Cold	2	2	1
Hot–Warm	5	4	4
Cold–Warm	1	1	1
Warm–Hot	3	3	1
Cold–Hot	1	1	2

on the operating frequency (from 100 to 500 MHz) of our system. This delay could have important issues in process' deadlines for real-time tasks.

The results of the analysis of these policies are classified in several sets depending on their response to pre-defined metrics. These metrics evaluate the capability of the atomic task to reduce the thermal gradient, the maximum temperature or the mean temperature in the chip. We also performed a statistic study to classify the policies in these groups and assign a quality mark that goes from 1 (very bad response) to 5 (very good response). The granularity of the classification is enough to represent the variability expected in the results and to reflect the variations found in the metrics.

Table 6.1 shows a reduced sub-set of the atomic policies that have been considered and their classification after the statistic analysis. In this table, the first column is the name of the atomic policy (it designs the origin and destination cores in the migration), being *hot* the reference for the hottest processor, *cold* for the coldest one and *warm* is the name given for those cores whose temperature is in between both hottest and coldest ones. As the goal of the analysis is the characterization of the policies, these will be always activated and the migrations will take place continuously. Finally, the initial workloads in the cores of the system are deliberately unbalanced to force the execution of the atomic policies. Next columns show the assigned quality mark for every metric.

The pre-characterization study also considered the thermal history of the cores (cores that have been cold or hot during a certain period in the past), which brought out the possibility to minimize the overhead in terms of number of migrations and amount of data transferred due to migrations.

The time window has been selected as the largest with the minimum impact on the temperature gradient after a detailed experimental study [5]. This selection of 300 ms for the time window is independent of the application run by the processors and only should be revisited in case of a new package.

6.6.3 Proposed Policies

6.6.3.1 Heuristic Algorithm (Heu)

This algorithm is able to select efficiently among the atomic policies and achieve the thermal optimization with a minimum performance impact. The implementation of this heuristic is based on the information retrieved by the characterization

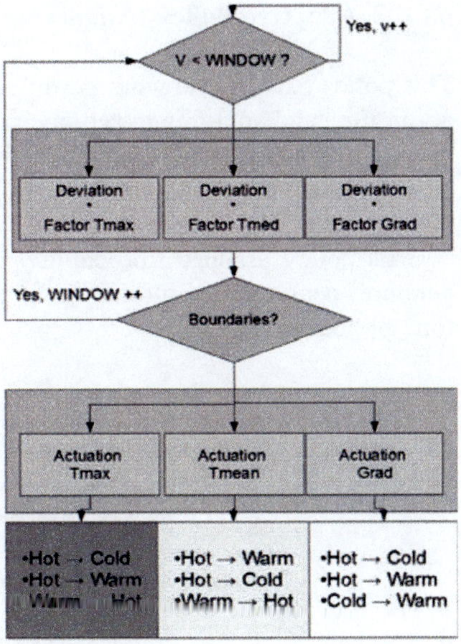

Fig. 6.7 Heuristic algorithm decision chart

phase, which provides the information about the thermal profile under the execution of the different atomic policies.

The algorithm works as follows: A time window is set and the workload and thermal information of the processors are collected at run-time during this time slot. At the end of the time window, we evaluate the data and compare them with the preferred working parameters (in terms of mean temperature, gradient and peak temperature). The atomic policy to apply is selected in order to solve the divergence of metrics between the current state and the desired one. Figure 6.7 shows the decision chart that explains the functioning of this heuristic.

In this figure the *Deviation* is the difference between the preferred working value (which is 50°C for the mean temperature, 75°C for the peak temperature and 6°C difference for the thermal gradient) and the current state value. These values have been selected to assure a proper operation of the system. *Factor* has been tuned experimentally to balance the importance of the different decision sets, namely, giving twice more weight to the mean temperature with respect to the gradient and 1.5 more than the maximum temperature.

The proposed heuristic defines a multi-objective optimization problem. The implementation of the heuristic applies sequentially the atomic policies in case of identical unbalance in the three metrics. In this way, the complexity in the decision process is minimized to simplify the heuristic. In order to alleviate the constraint imposed by this simplified thermal controller, an adaptive policy is introduced.

6.6.3.2 Adaptive Policy (Adapt)

This policy extends the work performed by the previous approach, collecting data at run-time and applying the atomic policies to achieve the optimum thermal state. This policy adapts the selection of the atomic policy by means of the statistical information of the cores, which predicts the behavior of the processors attending to the information about the past time.

This policy assigns a probability to every set of atomic policies (mean temperature, peak temperature, thermal gradient) and updates this probability every time period as follows:

$$P_t = P_{t-1} + W \tag{6.1}$$

$$W_{\text{init}} = M_{\text{pref}} - M_{\text{avg}} \tag{6.2}$$

$$W = \begin{cases} \alpha_{\text{inc}}(T_{\text{mean}}, T_{\text{peak}}, T_{\text{gradient}}) \cdot W_{\text{init}} & W_{\text{init}} > 0 \\ \alpha_{\text{dec}}(T_{\text{mean}}, T_{\text{peak}}, T_{\text{gradient}}) \cdot W_{\text{init}} & W_{\text{init}} < 0, \end{cases} \tag{6.3}$$

where W is the weight assigned to the sets every time period; M represents the different sets of atomic policies, as explained before; M_{pref} is the preferred working state and M_{avg} is the current state. The expressions for the increase and decrease of the probabilities are parametrized for every set of atomic policies, and the obtained probabilities are normalized in order to maintain math consistency. M_{pref} is the safe operating state already defined.

Using the previous equations, our extended OS updates the probabilities of selecting atomic policies every time window, and decides the working state by the execution of these policies. The design of the Adaptive Policy is supported by the pre-characterization of atomic policies. This initial study gives us the information of the best candidates (those atomic policies that obtain the maximum minimization of the metrics) for a task migration or task swapping in order to achieve a desired working state.

The atomic policies implemented in this adaptive technique always migrate a task from a source core to a destination core. As the temperature of the destination core is the only variable considered in the decision, more than one processor can satisfy the requirements. The last proposed policy extends the variables with the placement of the core for a more accurate selection of the destination core.

6.6.3.3 Floorplan-Aware Policy (FloorAdapt)

This policy considers the information about the floorplan. In this way, the OS is aware of the location of the cores and accordingly selects the destination processor in a task migration. This is implemented in the kernel of the OS with the assignment of different weights to each core. The smaller this weight is, the better candidate the core is to receive tasks. This factor is calculated with the following equation:

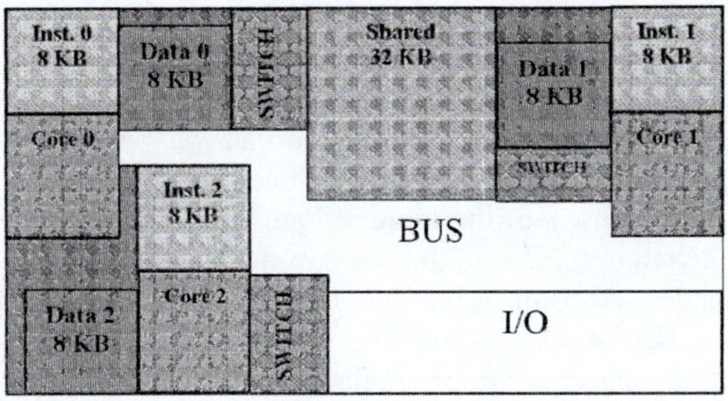

Fig. 6.8 Floorplan design

$$G = d_{\text{edge}}^3 + \frac{1}{d_{\text{core}}^2} + d_{\text{shared}} \tag{6.4}$$

where d_{edge} is the distance to the edge of the chip, d_{core} is the distance to another core (which is a heat source), and d_{shared} is the distance to the shared memory (which is a heat sink [61]). This expression has been created to resemble the strong influence of the ambient as a heat sink (cubic factor), the medium influence of the near cores as heat sources (quadratic factor) and the light influence of the shared memory as a heat sink (linear factor). The strength of the factors considers the proximity of the heat/sink and the thermal resistance of the joint.

Every time window, the thermal history of the processors is analyzed to solve possible hot spots, critical thermal gradients, or values over the safe peak temperature (75°C). However, if the system is still working in a safe state, the task migrations will not occur and the overhead of the policies will be avoided.

The knowledge of the thermal characteristics of the cores depending on the placement is a precious information for the task migration policies. The location of the cores in the chip surface produces very different thermal behavior due to the proximity to heat sinks or heat sources which dissipate the temperature. In our floorplan design shown in Fig. 6.8, **core 0** is close to **core 2** and both processors are prone to heat up due to the thermal diffusion from one to the other. On the other hand, **core 1** is far from the other processors but close to the edge of the chip, which increases the possibility to cool easily. Therefore, **core 1** would be selected to receive a heavy workload in case of a task migration.

The floorplan-aware policy incorporates this information about the core placement to adapt and select the probabilities of migrating or receiving a task.

6.7 Experimental Work

The experimental work has been conducted with the emulation platform described in Sect. 6.5, which has been used to model a multi-processor system with three working processors (μBlaze) and a PowerPC serving as the arbiter of the communication. The benchmark selected for the analysis is a real-life streaming application that loads the cores. The experiments have been run considering a special package derived from real-life streaming SoCs [62] for mobile embedded devices where the temperature can vary as much as 10 degrees in less than a second. The chip package has been selected to stress the number of required task migrations and, therefore, create a worst-case scenario for the validation of our techniques. Finally, the cores in the system can work at different clock frequencies under selection of the OS: 100, 200, 300, 400 and 500 MHz.

The validation of the task migration techniques has been accomplished attending to some pre-defined metrics that cover the spectrum of thermal aware optimization:

1. Spatial variation of the temperature of the processors: measured as the linear distance per area unit between cores at a different temperature. This metric quantifies the heat spread on the chip surface and the probability of thermal gradients.
2. Mean temperature of the chip: calculated as the arithmetic mean of the processor and memory temperatures in the chip. This metric relates the temperature of the devices to the energy consumption and cooling necessities.
3. Maximum temperature of the chip: measured as the maximum temperature value on the chip surface. It is related with the susceptibility to temperature-driven reliability factors.

The results obtained during the validation phase have been also compared with the results provided by the policies described in Sect. 6.6.

6.7.1 Description of the Application

The software that is executed by the platform is a Software FM Defined Radio [5] application, which is a typical example in multimedia streaming. This application is composed of several tasks that can be assigned to the different processors in the system. The input data is a digitalized PCM radio signal which has to be processed in several steps to obtain an equalized base-band audio signal.

The first step in the processing phase is a low pass filter (`LPF`), and the resulting signal is demodulated (`DEMOD`) and shifted to the baseband. After that, the signal is forked in three branches to be equalized by three different band pass filters (`BPF`). Finally a consumer (`SUM`) collects the data from every `BPF`. The communication between tasks is done using FIFO queues that transfer the data. Each task is allocated in a different processor during the load of the application. Then, the policy implemented in the OS migrates the tasks depending on the temperatures of the cores. Figure 6.9 shows an schematic view of this application an the relations among the processing steps.

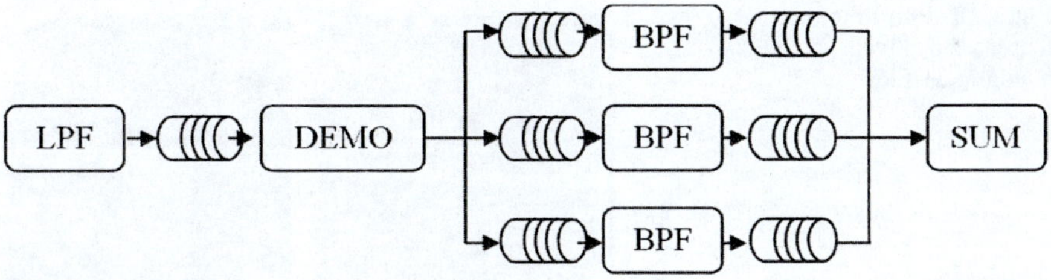

Fig. 6.9 Schematic view of the *SDR* application

Table 6.2 Initial working state

Core (Frequency)	Load (%)	Temperature (K)
Core 0 (533 MHz)	44	340
Core 1 (533 MHz)	83	339.5
Core 2 (266 MHz)	29	328.5

6.7.2 Evaluation of the Policies

The task migration policies implemented in the OS kernel were applied to the benchmark and the pre-defined metrics were collected to perform the evaluation.

The execution of the application in the emulation platform consists of two phases. The first one is the initialization of the OS and the tasks. As this phase does not exhibit a critical thermal state and it occurs just once during the system boot-up, the task migration policies are deactivated at this time. When this initial phase finishes, the thermal and workload state of the system is the one described in Table 6.2. Our experimental work starts at this point setting a thermal unbalance that motivates the activation of the migration policies. In the second phase, when the execution of the application starts, all the policies described in this chapter are evaluated separately.

The analysis performed for the task migration policies is two fold. Firstly, a visual inspection of the thermal distribution in the chip surface is done using the developed graphical tool. With this analysis, the evolution of temperature in real-time is obtained, as shown in Fig. 6.10. This figure shows an example of the run-time behavior for the (a) proposed `adaptive` and the (b) `migration` [5] policies.

As shown, both policies start similarly, decreasing rapidly the presence of hot spots. However, as time evolves, the adaptive policy obtains lower temperature values and a more homogeneous thermal distribution due to the presence of short-time execution tasks. In fact, for the SDR benchmark, all the cells in the floorplan are within a range of temperature of 5 degrees when the adaptive policy is applied, while differences of more than 15 degrees can be found in certain periods for the migration policy. Similar results occur with the other task migration techniques.

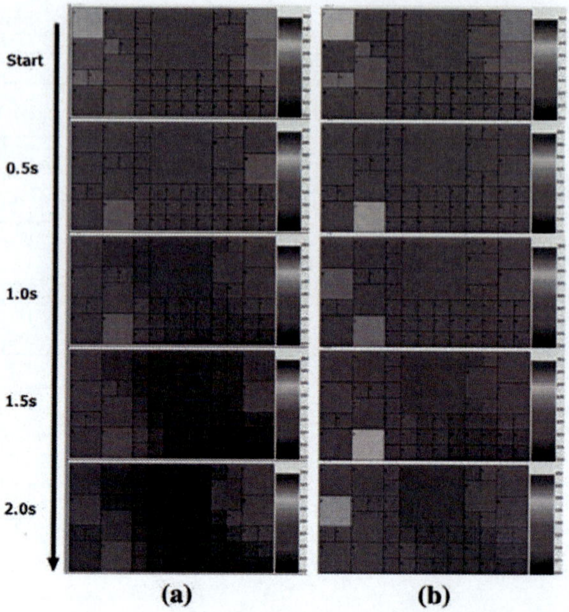

Fig. 6.10 Run-time thermal maps. **a** adaptive, **b** migration [26]

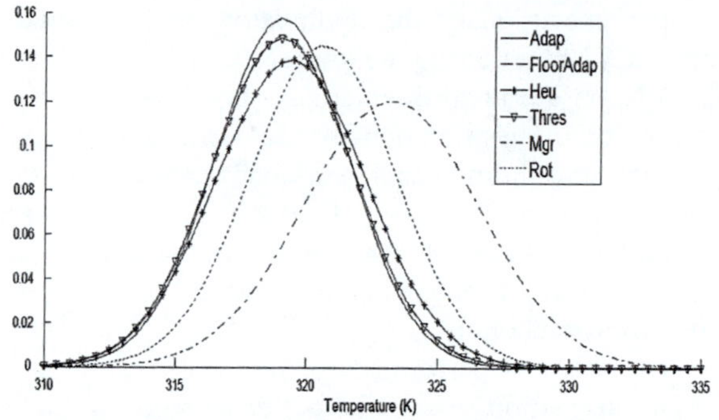

Fig. 6.11 Normalized statistical distributions

Secondly, a statistical study of the distribution of temperatures in the chip under the execution of the task migration policies is accomplished. This analysis evaluates which policies have better results when applied in the multi-processor system. The mean and sigma values of the temperature for every policy are calculated in the statistic analysis and fit to a normal distribution (see Fig. 6.11).

As can be derived from the values in the Figure, the best results in terms of thermal distribution and absolute values are achieved with the three policies specifically proposed in this chapter. In particular, the adaptive algorithm concentrates the temperature of the cells within a small range of temperatures centered in the mean temperature (mean temperature 319.038 with a σ of only 2.53). The curves for the three proposed policies present: lower mean value (translated into a

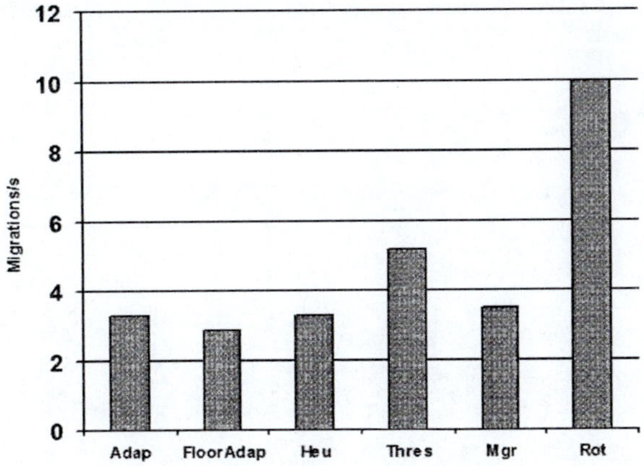

Fig. 6.12 Number of migrations per time unit

Table 6.3 Performance overhead

	Adap	FloorAdapt	Heu	Thres	Mgr	Rot
Overhead (%)	0.85	0.52	0.85	1.2	0.93	2.4

decrease in the average temperature of the chip) and narrower shape of the curve (translated in a smaller sigma and, therefore, a decrease in the thermal gradient of up to 30% with respect to state-of-the-art techniques [2, 6, 63]).

Another interesting quality factor in the development of task migration techniques is the number of migrations per unit. As has been previously discussed, task migration policies introduce a performance overhead due to the time required for the memory allocation, as well as an energy waste. This impact can be characterized by means of the number of effective migrations per time unit. Figure 6.12 shows the number of migrations per time unit for all the policies considered in our study. As can be seen, our proposed policies not only achieve similar results to the threshold technique [5] in terms of mean temperature and sigma of the thermal distribution, but they also decrease the impact on performance by a 40% because less task migrations are required. Table 6.3 summarizes the performance overhead imposed by every task migration technique, where the minimum impact of our proposed policies can be observed.

Finally, two factors with a very strong impact on the reliability of the system have been evaluated: the percentage of hot spots in the chip area, and the thermal cycles. Both metrics have been calculated assuming that a hot spot in our set-up is represented by a temperature value over 338 K. Figure 6.13 shows the percentage of hot spots in the chip area, averaged along the execution of the benchmark, and for every migration policy. As can be seen, our Adaptive policy behaves better than the traditional approaches, only outperformed by the Rotation policy which, on the contrary, has a strong impact on performance. The percentage of hot-spots is reduced to 1% and, therefore, the probability of system failure is minimized.

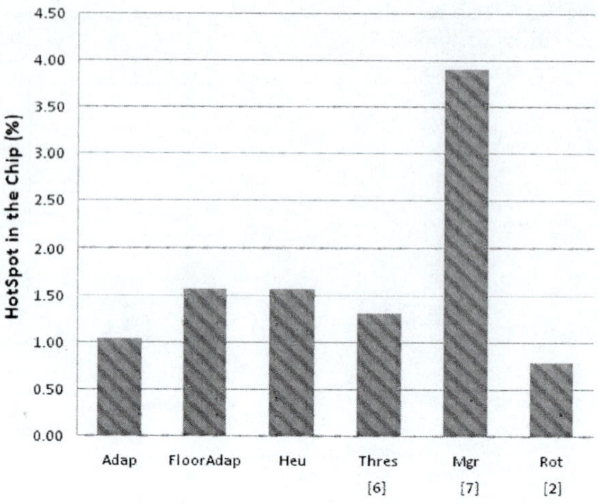

Fig. 6.13 Percentage of hot-spots

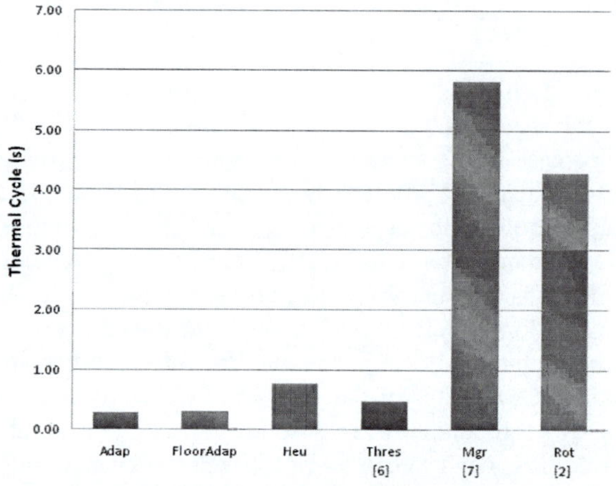

Fig. 6.14 Thermal cycles

Figure 6.14 shows the thermal cycles for the same system configuration and task migration policies. As can be seen, our proposed approaches are able to reduce the thermal cycles to a minimum, showing better results than the traditional approaches (25% better than [5] and up to 4× less thermal cycles than [2] and [6]); and, moreover, with the smallest performance overhead (less than 0.9% impact on execution time).

6.8 Conclusions

In this chapter, we have investigated and proposed OS-level task migration policies for thermal management in embedded multi-processor systems. We have showed that the proposed techniques achieve low and balanced temperatures profiles, diminishing the percentage of hot spots, thermal cycles, and thermal gradients. As compared with traditional techniques, our policies incorporate the floorplan information in the OS, dynamically adapt the migration to the thermal profile of the application, and improve the thermal behavior of the chip with a negligible performance overhead.

References

1. Semenov OeA (2006) Impact of self-heating effect on long-term reliability and performance degradation in CMOS circuits. IEEE Trans Device Mater Reliab 6(1):17–27
2. Chaparro PeA (2007) Understanding the thermal implications of multi-core architectures. IEEE Trans Parallel Distrib Syst 18(8):1055–1065
3. Carta S, Acquaviva A, Del Valle PG, Atienza D, De Micheli G, Rincon F, Benini L, Mendias JM. (2007) Multi-processor operating system emulation framework with thermal feedback for systems-on-chip. In: Proceedings of the 17th ACM GLS on VLSI, pp 311–316
4. Atienza D, Del Valle PG, Paci G, Poletti F, Benini L, Micheli GD, Mendias JM, Hermida R (2007) HW-SW emulation framework for temperature-aware design in MPSoCs. ACM Trans Des Autom Electron Syst 12(3):1–26
5. Mulas F, Pittau M, Buttu M, Carta S, Acquaviva A, Benini L, Atienza D (2008) Thermal balancing policy for streaming computing on multiprocessor architectures. In: Proceedings on DATE, pp 734–739
6. Gomaa M, Powell MD, Vijaykumar TN (2004) Heat-and-run: leveraging SMT and CMP to manage power density through the operating system. SIGOPS Oper Syst Rev 38(5):260–270
7. Dharmasanam S. Multiprocessing with real-time operating systems. http://www.embedded.com/story/OEG20030512S0080
8. Jerraya AA, Tenhunen H, Wolf W (2005) Guest editors introduction: multiprocessor systems-on-chips, IEEE Computer. pp 36–40
9. ARM Ltd, ARM11 MPCore. http://www.arm.com/products/CPUs/ARM11MPCoreMultiprocessor.html
10. Poletti F, Poggiali A, Marchal P (2005) Flexible hardware/software support for message passing on a distributed shared memory architecture. In: Proceedings of DATE, pp 736–741
11. Han S-I, Baghdadi A, Bonaciu M, Chae S-I, Jerraya AA (2004) An efficient scalable and flexible data transfer architecture for multiprocessor SoC with massive distributed memory. DAC, pp 250–255
12. Loghi M, Benini L, Poncino M (2004) Analyzing power consumption of message passing primitives in a single-chip multiprocessor. In: Proceedings of DATE, 2004
13. Monchiero M, PALERMO G, Silvano C, Villa O (2006) Power/Performance hardware optimization for synchronization intensive applications in MPSoCs. In: Proceedings of DATE, 2006
14. Ruggiero M, Acquaviva A, Bertozzi D, Benini L (2005) Application-specific power-aware workload allocation for voltage scalable MPSoC platforms. ICCD05, pp 87–93
15. Kumar A, Mesman B, Corporaal H, van Meerbergen J, Yajun H (2006) Global analysis of resource arbitration for MPSoC, In: Proceedings of digital system design, 9th Euromicro conference, DSD 06
16. Ma Z, Catthoor F (2006) Scalable performance-energy trade-off exploration of embedded real-time systems on multiprocessor platforms. In: Proceedings of DATE, 2006
17. Hung W-L, Xie Y, Vijaykrishnan N, Kandemir M, Irwin MJ (2005) Thermal-aware allocation and scheduling for systems-on-a-chip design. In: Proceedings of DATE, 2005
18. Li F, Kandemir M (2005) Locality-conscious workload assignment for array-based computations in MPSOC architectures, In: Proceedings of the 42nd annual conference on design automation, pp 95–100

19. Kandemir MT, Chen G (2005) Locality-aware process scheduling for embedded MPSoCs, In: Proceedings of DATE, pp 870–875
20. Bertozzi S, Acquaviva A, Poggiali A, Bertozzi D (2006) Supporting task migration in MPSoCs: a feasibility study. In: Proceedings of design, automation and test in Europe (DATE)
21. Barak A, La'adan O, Shiloh A (1999) Scalable cluster computing with MOSIX for Linux. In: Proceedings Linux expo '99, pp 95–100
22. Zayas E (1987) Attacking the process migration bottleneck. In: Proceedings of the eleventh ACM symposium on operating systems principles, pp 13–24
23. Milojicic D, Douglis F, Paindaveine Y, Wheeler R, Zhou S (2000) Process migration survey, ACM computing surveys
24. Ozturk O, Kandemir M, Son SW, Karakoy M (2006) Selective code/data migration for reducing communication energy in embedded MpSoC architectures. GLSVLSI 2006
25. Benini L, Bogliolo A, De Micheli G (2000) A survey of design techniques for system-level dynamic power management. IEEE Trans VLSI Systems 8(3):299–316
26. Pouwelse JA, Langendoen K, Sips H (2001) Voltage scaling on a low-power microprocessor. Mobile computing conference (MOBICOM)
27. Kwon W, Kim T (2003) Optimal voltage allocation techniques for dynamically variable voltage processors. IEEE Trans VLSI Systems, pp 125–130, June 2003
28. Pillai P, Shin K (2001) Real-time dynamic voltage scaling for low-power embedded operating systems. ACM SIGOPS 01, pp 89–102, October 2001
29. Flautner K, Mudge TN (2002) Vertigo: Automatic performance-setting for Linux. OSDI 2002
30. ARM Intelligent Energy Manager (2005) Dynamic power control for portable devices. http://www.arm.com/products/CPUs/cpu-arch-IEM.html
31. Andrei A, Schmitz M, Eles P, Peng Z, Al-Hashimi BM (2004) Overhead-conscious voltage selection for dynamic and leakage energy reduction of time-constrained systems. DATE04, pp 518–523
32. Andrei A, Schmitz M, Eles P, Peng Z, Al-Hashimi BM (2004) Simultaneous communication and processor voltage scaling for dynamic and leakage energy reduction in time-constrained systems. ICCAD04, pp 362–369
33. Zhu D, Melhem R, Childers B (2003) Scheduling with dynamic voltage/speed adjustment using slack reclamation in multi-processor real-time systems. IEEE Trans Parallel Distrib Syst 14:686–700
34. Ruggiero M, Acquaviva A, Bertozzi D, Benini L (2005) Application-specific power-aware workload allocation for voltage scalable MPSoC platforms. ICCD05
35. Lu Z, Hein J, Humphrey M, Stan M, Lach J, Skadron K (2002) Control theoretic dynamic frequency and voltage scaling for multimedia workloads. CASES02, pp 156–163
36. Lu Y, Benini L, De Micheli G (2002) Dynamic Frequency scaling with buffer insertion for mixed workloads. IEEE Trans Comput Aided Des Integr Circuits Syst 21(11):1284–1305
37. Im C, Kim H, Ha S (2001) Dynamic voltage scaling technique for low-power multimedia applications using buffers. ISLPED01, pp 34–39
38. Lu Z, Lach J, Stan M (2003) Reducing Multimedia Decode Power using Feedback Control. ICCD03
39. Carta S, Alimonda A, Acquaviva A, Pisano A, Benini L (2006) A control theoretic approach to energy efficient pipelined computation in MPSoCs. To appear on transaction on embedded computing systems (TECS), 2006
40. Suen TTY, Wong JSK (1992) Efficient task migration algorithm for distributed systems. IEEE Trans Parallel Distrib Syst 3(4):488–499
41. Chang HWD, Oldham WJB (1995) Dynamic task allocation models for large distributed computing systems. IEEE Trans Parallel Distrib Comput Syst 6:1301–1315
42. Nollet V, Avasare P, Mignolet JY, Verkest D (2005) Low cost task migration initiation in a heterogeneous MP-SoC. In: Proceedings of the conference on DATE, pp 252–253
43. Bertozzi S, Acquaviva A, Bertozzi D, Poggiali A (2006) Supporting task migration in multi-processor systems-on-chip: a feasibility study. In: Proceedings of the conference on DATE, pp 15–20

44. Barcelos D, Brião EW, Wagner FR (2007) A hybrid memory organization to enhance task migration and dynamic task allocation in NoC-based MPSoCs. In: Proceedings of the 20th annual conference on Integrated circuits and systems design, pp 282–287
45. Brião EW, Barcelos D, Wronski F, Wagner FR (2007) Impact of task migration in NoC-based MPSoCs for soft real-time applications. In: Proceedings of the international conference on VLSI, pp 296–299
46. Pittau M, Alimonda A, Carta S, Acquaviva A (2007) Impact of task migration on streaming multimedia for embedded multiprocessors: A quantitative evaluation. In: Embedded systems for real-time multimedia, 2007. ESTIMedia 2007. IEEE/ACM/IFIP Workshop on, pp 59–64
47. Acquaviva A, Alimonda A, Carta S, Pittau M (2008) Assessing task migration impact on embedded soft real-time streaming multimedia applications. In: EURASIP journal on embedded systems, Vol. 2008, Article ID 518904
48. Donald J, Martonosi M (2006) Techniques for multicore thermal management: Classification and new exploration. In: Proceedings of the 33rd international symposium on computer architecture, pp 78–88
49. Puschini D, Clermidy F, Benoit P, Sassatelli G, Torres L (2008) Temperature-aware distributed run-time optimization on MP-SoC using game theory. In: IEEE computer society annual symposium on VLSI
50. Gomaa M, Powell MD, Vijaykumar TN (2004) Heat-and-run: leveraging SMT and CMP to manage power density through the operating system. In: Proceedings of the 11th international conference on architectural support for programming languages and operating systems, pp 260–270
51. Yang J, Zhou X, Chrobak M, Zhang Y, Jin L (2008) Dynamic thermal management through task scheduling. In: Proceedings of the IEEE international symposium on performance analysis of systems and software, pp 191–201
52. Yeo I, Kim EJ (2009) Temperature-aware scheduler based on thermal behavior grouping in multicore systems. In: Proceedings of the conference on DATE 2009
53. ST Microelectronics and CEA, Platform 2012: A Many-core programmable accelerator for ultra-efficient embedded computing in nanometer technology, ST Whitepaper, 2009, http://www.cmc.ca/en/NewsAndEvents//media/English/Files/Events/STP2012_20101102_Whitepaper.pdf
54. Intel, Single-Chip Cloud Computer. http://techresearch.intel.com/ProjectDetails.aspx?Id=1
55. Pham D, et al. (2003) The design and implementation of a first generation CELL processor. IEEE/ACM ISSCC, pp 184–186, 2005. July 2003
56. uClinux, Embedded Linux Microcontroller Project. http://www.uclinux.org/
57. Friebe L, Stolberg H-J, Berekovic M, Moch S, Kulaczewski MB, Dehnhardt A, Pirsch P (2003) HiBRID-SoC. A system on chip architecture with two multimedia DSPs and a RISC core. IEEE international SOC conference, September 2003, pp 85–88
58. van der Wolf P, de Kock E, Henriksson T, Kruijtzer W, Essink G (2004) Design and programming of embedded multiprocessors: an interface-centric approach, CODES+ISSS, pp 206–217
59. Paci G, Marchal P, Poletti F, Benini L (2006) Exploring temperature-aware design in low-power MPSoCs. In: Proceedings of the DATE, vol 1. pp 1–6
60. Flautner K, Mudge T (2002) Vertigo: automatic performance-setting for Linux. SIGOPS Oper Syst Rev 36(SI):105–116
61. Huang W, Stant MR, Sankaranarayanan K, Ribando RJ, Skadron K (2008) Many-core design from a thermal perspective. In: Proceedings of the 45th annual DAC, pp 746–749
62. Skadron K, Stan MR, Sankaranarayanan K, Huang W, Velusamy S, Tarjan D (2004) Temperature-aware microarchitecture: Modeling and implementation. ACM Trans Archit Code Optim 1(1):94–125
63. Mulas F, Atienza D, Acquaviva A, Carta S, Benini L, De Micheli G (2009) Thermal balancing policy for multiprocessor stream computing platforms. IEEE transactions on computer-aided desing of integrated circuits and systems, Vol 28(12):1870–1882

Chapter 7
A High Level Synthesis Exploration Framework with Iterative Design Space Partitioning

Sotirios Xydis, Kiamal Pekmestzi, Dimitrios Soudris and George Economakos

Abstract This chapter introduces a methodology for fast and efficient Design Space Exploration during High Level Synthesis. Motivated by the fact that higher quality design solutions are delivered when a larger number of parameters are explored, we study an augmented instance of the design space considering the combined impact of loop-unrolling, operation chaining and resource allocation onto the final datapath. We propose an iterative design space partitioning exploration strategy based on the synergy of an exhaustive traversal together with an introduced heuristic one. The introduced heuristic is based on a gradient-based pruning technique which efficiently evaluates large portions of the solution space in a quick manner. We show that the proposed exploration approach delivers high quality results, with considerable reductions of the exploration's runtime in respect to the fully exhaustive approach.

This research is partially supported by the E.C funded program MOSART IST-215244,Website: http://www.mosart-project.org/.

S. Xydis (✉) · K. Pekmestzi · D. Soudris · G. Economakos
National Technical University of Athens, 9 Heroon Polytechneiou,
Zographou Campus, Greece
e-mail: sxydis@microlab.ntua.gr

K. Pekmestzi
e-mail: pekmes@microlab.ntua.gr

D. Soudris
e-mail: dsoudris@microlab.ntua.gr

G. Economakos
e-mail: geconom@microlab.ntua.gr

N. Voros et al. (eds.), *VLSI 2010 Annual Symposium*,
Lecture Notes in Electrical Engineering 105, DOI: 10.1007/978-94-007-1488-5_7,

7.1 Introduction

One of the greatest challenges in modern IC design is how to manage the continuously growing circuit complexity, the so called *productivity gap* [1]. Figure 7.1 illustrates the productivity gap in terms of manufacturable complexity (number of transistors that we are able to manufacture) versus designer productivity (number of transistors that we are able to design). In order to shrink this productivity gap designers have moved towards higher abstractions levels, i.e. gate level, register transfer level (RTL), algorithmic level, system level etc. This work targets to the algorithmic level of abstraction. Algorithmic level design is also known as High Level Synthesis (HLS). HLS offers an automated and seamless path from high level behavioral specifications down to circuit level implementations. After extensive research, HLS is now in its mature phase [2], thus a continuously growing community of IC designers have adopted HLS techniques to tackle design complexity and meet tight time-to-market requirements.

The rise of the design abstraction exposed a large number of inter-dependent design parameters that have to be explored in order to discover the optimal solutions. Different trade-offs are associated with each solution and designers have faced the problem of reasoning on differing trade-offs among the set of design parameters. Thus, efficient exploration methodologies accompanied with automated tools are of great importance for a quick and concrete evaluation of the design space [3].

Design Space Exploration (DSE) is the procedure that evaluates the solution space in order to return a set of Pareto-optimal [4] design points according to some design criterions (execution delay, circuit area, dissipated power). Pareto optimality [4] specifies that one design solution dominates the other when it is at least as good in all of the criteria and also strictly better in at least one of them. Thus, designers are interested mostly on Pareto optimal configurations. Given that the size of the design space is usually huge, DSE methodologies are focused on the development of strategies for efficient traversal of the available design space.

In this contribution, we target the fundamental problem of exploring Performance-Area trade-offs during algorithmic level design [5]. A Performance-Area exploration curve can be generated by iteratively scheduling the behavioral description with different area constraints. Although the aforementioned exploration strategy produces the exploration curve for the specific set of parameters, it is far from delivering the most efficient design solutions. That is because code-level transformations (i.e. loop unrolling) and architectural level optimizations (i.e. operation chaining) are treated as user-guided pre-defined parameters during exploration, limiting the evaluation of their combined impact on the datapath. In a common case, such exploration strategies deliver suboptimal design points since the interaction between the structure of behavioral code and the architectural optimizations are silently assumed to be independent. We address this inefficiency by proposing an exploration framework which accounts for both the loop unrolling

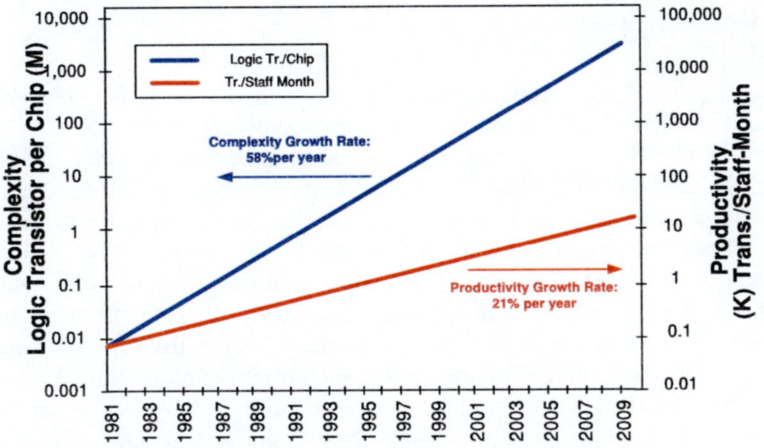

Fig. 7.1 The design productivity gap

code transformation and operation chaining architectural optimization as parameters of the design space.

Specifically, we propose the management of an extended design space during HLS exploration, through the incorporation of loop unrolling and operation chaining decisions as parameters of a unified solution space, in order to expose higher quality designs. Towards this direction, we developed a meta-heuristic strategy based on the iterative partitioning of the design space, integrating in a balanced manner exhaustive together with heuristic exploration. Heuristic exploration is performed through a gradient-based pruning technique which quickly explores the extended solution space. The proposed strategy is fully automated (supported through a CAD tool-flow) and it trades accuracy/quality of the derived curve for overall exploration's runtime and vice versa.

Extensive experimentation has been conducted based on real-life computationally intensive benchmarks in order to evaluate the efficiency of our approach. In each case, a shift of the exploration's curve towards more optimized solutions (Pareto-points) is reported in comparison to the existing design exploration methodologies. Additionally, the proposed exploration strategy delivers maximum average speedup of 34.1× in exploration's runtime and maximum average accuracy of 93.7% on the quality of the final curve compared with the exhaustive exploration of the design space.

The rest of the chapter is organized as follows. Section 7.2 discusses the current literature, while Sect. 7.3 provides the basic observations that motivated this work. In Sects. 7.4 and 7.5 the proposed exploration methodology is presented and analyzed. Section 7.6 reports experimental data evaluating the proposed methodology and finally Sect. 7.7 concludes the paper.

7.2 Related Work

Existing approaches in DSE during HLS concentrate on solving iteratively the resource allocation and operation scheduling problems [6–8]. Exploration is performed exhaustively traversing the entire defined solution space. Although Pareto-optimal design points are delivered, the exploration and evaluation of each design point imposes unaffordable execution runtime even for the case that heuristic scheduling/allocation is performed [7, 8]. Furthermore, neither the structure of the input code (i.e. loop unrolling) nor the architectural level optimizations (i.e operation chaining) are considered as exploration parameters, resulting in a large set of unexplored solutions which usually dominates the derived exploration curve. The duality between timing- and resource-constrained scheduling problems has been taken into account during exploration in [9, 10]. The exploitation of the duality between the scheduling problems reports higher quality results than the previous mentioned approaches. However, the impact of loop unrolling is ignored [9, 10] and operation chaining is only partially supported [10] during exploration.

Loop unrolling during HLS has great influence on the datapath implementation [11, 12]. In [13], an HLS exploration methodology is presented considering code transformations. However, the exploration loop is restricted only to the code level, without encountering neither the impact of code transformations on scheduling efficiency under various resource allocations nor the impact of operation chaining. Operation chaining in HLS [14] schedules data-dependent operations in a single control step by removing the intermediate storage logic (registers). It heavily depends on circuit's operating frequency and critical path of datapath's components. Traditional operation chaining targets data-flow behavioral descriptions [14, 15] for performance improvement. However, the relationship among operation chaining depth (OCD), loop unrolling factor (LUF) and resource allocation has only partially evaluated in existing exploration flows [9].

The aforementioned research works target the problem of DSE either excluding from the automated process the effects of loop unrolling and operation chaining during scheduling [6–10] or studying code transformations at a pre-synthesis level without considering their effects on operation scheduling under various resource allocation scenarios [12, 13]. We propose a differing exploration approach encountering operation scheduling under various resource allocation scenarios and decisions concerning (1) the LUFs and (2) the OCDs. Recent studies have shown that by this way the design space is explored in a more global manner exhibiting new Pareto configurations [16]. The proposed methodology is performed at a meta-level which makes it orthogonal with a large set of existing work on scheduling algorithms i.e. [6–10], since every core scheduling algorithm can be used inside the exploration framework.

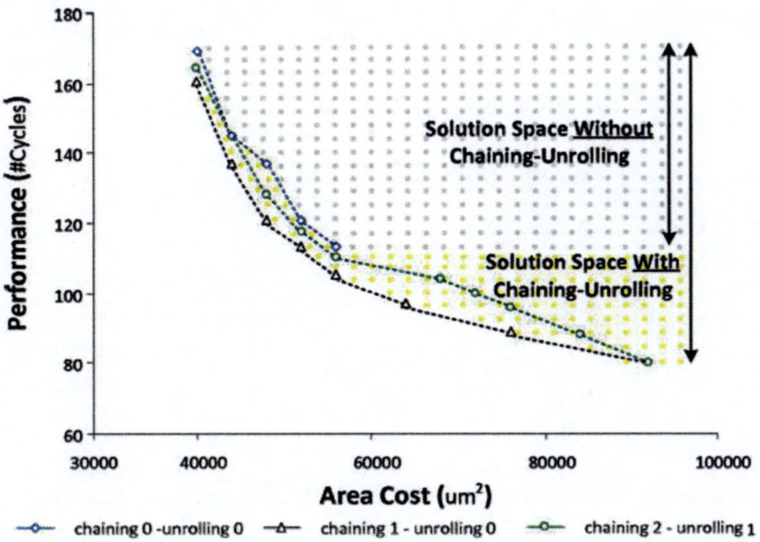

Fig. 7.2 Design space exploration considering different degrees of loop unrolling and operation chaining

7.3 Motivational Observations

As mentioned, we focus on the fundamental design problem of discovering the best trade-offs between the circuit's performance and its area cost. Regarding these trade-offs, each design solution can be included in 2D design space, where the x-axis accounts for the area-cost while the y-axis accounts for the performance. Through a high level area model (based on the number of allocated hardware resources [9, 10]), each solution point in the design space can be represented as a vector of $(Area, \#Cycles) \Rightarrow ((\#ALUs, \#Muls), \#Cycles)$.

The quality of the exploration curve is highly affected by the parameters of the design space that are explored. Considering a larger set of design parameters, the design space is explored more globally and solutions of higher quality are revealed. In the remainder of this section, we show through an illustrative example (Fig. 7.2) that the consideration of loop unrolling and operation chaining as design parameters during exploration, delivers solution points of higher quality. Even though loop unrolling [11] and operation chaining [14] are well known techniques for HLS tools, their incorporation in DSE methodologies is only partially supported and in many cases guided by the designer. However, pre-configuration of the LUFs and OCDs can lead in loosing Pareto solutions.

Fig. 7.2 illustrates exploration curves (derived through exhaustive DSE) for the case of 1D DCT kernel (part of 2D DCT kernel found in JPEG application [17]). Each exploration curve considers a different set for the LUF and the OCD. In case that neither loop unrolling nor operation chaining is considered, the exploration

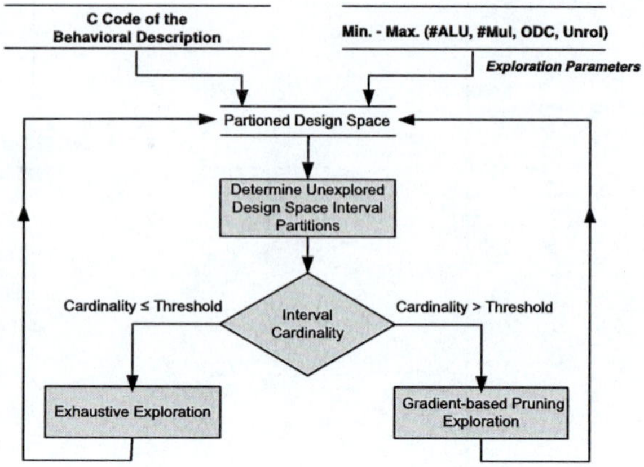

Fig. 7.3 Overall exploration flow

curve derived by evaluating only a small portion of the design space. Taking into account the impact of loop unrolling and operation chaining, a new unexplored design space is revealed which delivers new Pareto optimal design solutions (thus better performance-area trade-offs).

From Fig. 7.2, it is observed that arbitrarily increasing the LUF and/or the OCD parameters does not guarantee convergence to the exact Pareto frontier. For example, in the illustrated scenario the majority of points forming the final exploration curve are found considering $OCD = 1, LUF = 0$ rather than $OCD = 2, LUF = 1$. Furthermore, it is observed that operation chaining and loop unrolling has to be explored in a combined manner rather than pre-defined prior exploration, since the exact Pareto curve is formed by points generated with different unrolling and chaining configurations.

Motivated by the aforementioned observations, we target to the fast and efficient exploration of an augmented design space, considering the design parameters of loop unrolling and operation chaining in a combined manner.

7.4 Iterative Design Space Partitioning Exploration

In this section, we describe the proposed exploration methodology during HLS following a top-down approach. At first, we present the overall exploration scheme which is based on the synergetic incorporation of both exhaustive and heuristic techniques. Heuristic exploration is performed through a newly introduced gradient-based design space pruning (Sect. 7.5).

The overall exploration framework is depicted in Fig. 7.3. It is based on the iterative partitioning of the overall design space into smaller and disjoint subspaces (considering the area cost). Both exhaustive and gradient-based pruned heuristic exploration (Sect. 7.5) are used in a combined manner during the

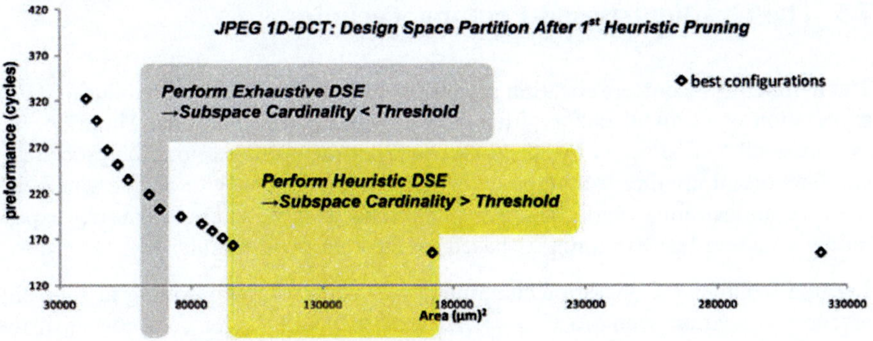

Fig. 7.4 Illustrative example of design space partitioning

exploration procedure. The adopted principle is that, each design (sub)space explored through the gradient-based heuristic is decomposed to a set of smaller subspaces than the initial one. "Large" design (sub)spaces are explored further heuristically while "small" design (sub)spaces are further explored in an exhaustive manner. Each design (sub)space is characterized as "large" or "small" according to its cardinality and a designer specified threshold. Cardinality is the number of different configuration/solutions existed in each (sub)space.

The user provides to the exploration framework (1) the behavioral description, (2) the set of exploration parameters and (3) their ranges. At the first iteration, no partition of the design space has been performed. Thus, without loss of generality, we handle the design space as a single partition. In case that the cardinality of the design space is smaller than the threshold specified by the designer, exhaustive evaluation is performed and the exploration completes. In the common case in which the cardinality is larger than the specified threshold, gradient-based heuristic exploration is performed to the whole design space. Gradient-based exploration returns a first set of pseudo-Pareto points. We call these points pseudo-Pareto since at the specific iteration they form true Pareto solutions of the explored (sub)space which can be replaced in a next iteration if a design solution that dominates them is found.

The initial design space is now partitioned in several subspaces. Design subspaces are defined dynamically during exploration, considering the neighboring pseudo-Pareto points as boundaries. The cardinality of each design subspace is further evaluated in order to be explored exhaustively or heuristically with gradient-based pruning. Every time a gradient-based pruned exploration is performed, new design subspaces are generated. The exploration procedure for each design space terminates when one of the following conditions are evaluated to true: (1) the subspace has been explored exhaustively or (2) two subsequent gradient-based explorations returns the same subspace boundaries. Figure 7.4 shows an illustrative example of the appliance of the iterative design space partitioning.

7.5 The Gradient-Based Pruning Technique

The number of resource allocation scenarios that have to be evaluated during HLS exploration is a critical factor of the overall exploration's runtime. The more the examined allocation scenarios, the larger the runtime of the exploration procedure. Gradient-based pruning technique defines a fast heuristic way to decide whether a resource allocation scenario has to be examined or to be excluded during exploration. Gradient-based pruning is based on the following lemma:

Lemma 1 Given the operation chaining degree and the loop unrolling factors, any resource allocation scenario $C = \{\#ALU, \#Mul\}$ with larger area cost than the cost found in the allocation scenario that matches exactly the maximum operation level parallelism (MOLP) of the application, $C_{MOLP} = \{\#ALU_{MOLP}, \#Mul_{MOLP}\}$, delivers either the same or worst latency results.

Proof Given the OCD and LUFs the minimum latency of a behavioral description is delivered through As-Soon-As-Possible (ASAP) scheduling [5] which exploits the whole operation-level parallelism. In case that a resource allocation configuration, C, with $\{\#ALU > \#ALU_{MOLP}\}$ and $\{\#Mul > \#Mul_{MOLP}\}$ is considered, then the remaining resources over the allocation of C_{MOLP} are idle since there is no available operation parallelism to be exploited. □

Lemma 1 provides a key insight for the exploration of resource allocation scenarios. It says that for a given OCD and LUF, the performance-area exploration curve presents zero-gradient for resource allocation scenarios which allocate a greater number of resources than the resources that match the maximum operation level parallelism. In other words, when a zero-gradient at the maximum operation level parallelism is occurred for a given set of resources, the increment of resources has no beneficial impact on the latency of the datapath, thus no Pareto point can be found and there is no need to evaluate these allocation scenarios.

Gradient-based pruning heuristically extends the above observations and proposes the following: "Each time during the design space exploration, a zero-gradient:

$$\frac{\Delta Latency}{\Delta Area}\Big|_{C_i}^{C_{i+1}} = 0, i \leftarrow \#ALUs \tag{7.1}$$

is occurred in the performance-area exploration curve, the solution with the lowest area cost forms: (i) a possible Pareto-solution and (ii) a "good" design point to alter the examined configuration mode (i.e. by increasing either the number of allocated Muls or the OCD or the LUF)". By this way, large portions of resource allocation scenarios are excluded from the exploration procedure improving the overall exploration runtime without degrading the finding of Pareto solutions.

However, zero-gradient segments can occurred also at resource allocations with lower operation level parallelism than the MOLP, when the specific allocations exhibit no performance improvement. These zero-gradient segments are local and

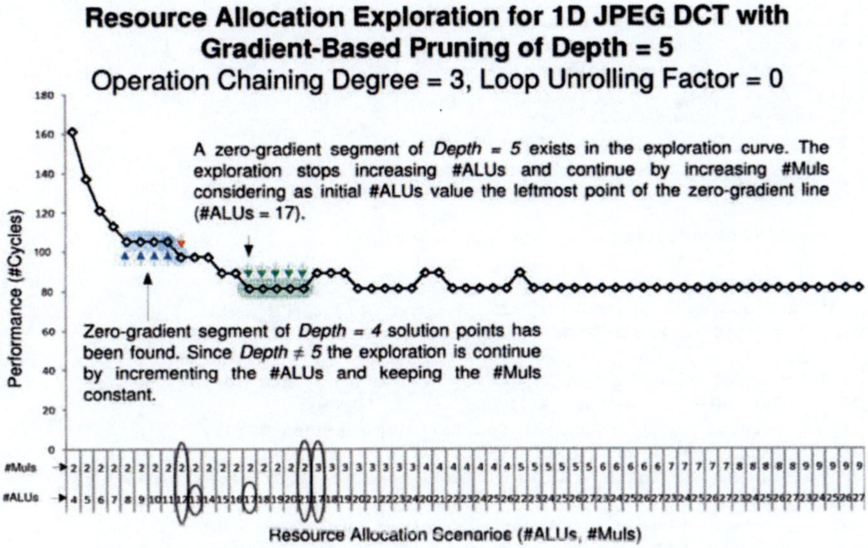

Fig. 7.5 Exploration of resource allocation scenarios based on gradient-based pruning

have to be de-characterized as "good" design points for altering configuration mode. Thus, examining zero-gradient only between two successive design solutions may lead to the loss of Pareto solutions. In order to avoid such situations, the zero-gradient segments are evaluated for *Depth* > 2 of successive design solutions. Thus, for *Depth* $> k$ the zero-gradient segment condition is formed as:

$$\frac{\Delta Latency}{\Delta Area}\Big|_{C_i}^{C_{i+1}} = 0)\bigcap \dots \bigcap(\frac{\Delta Latency}{\Delta Area}\Big|_{C_i}^{C_{i+k}} = 0, i \leftarrow \#ALUs \qquad (7.2)$$

High values of the *Depth* parameter guarantee high ratios of finding exact Pareto solutions in the expense of increasing exploration runtime. On the other hand, small *Depth* values improve the overall exploration runtime in the expense of degrading the accuracy of exploration curve.

Fig. 7.5 illustrates the application of gradient-based pruning during the exploration of various resource allocation scenarios for the 1D DCT kernel of the JPEG application. We consider a configuration setting of OCD $= 3$ with no loop unrolling, $(Min.\#ALUs, Min.\#Muls) = (4,2)$, $(Max.\#ALUs, Max.\#Muls) = (32,9)$ and $Depth = 5$. The explorative procedure evaluates resource allocation scenarios by increasing the $\#ALUs$ until a zero-gradient segment of $Depth = 5$ is occurred. Whenever such segments are found, the configuration mode alters by incrementing the $\#Muls$ until the $Max.\#Muls$. Gradient-based pruning derives that resource allocation scenarios for $\#ALUs > 17$ have to be excluded from the exploration. Thus, in a set of $(Max.\#ALUs - Min.\#ALUs) \times (Max.\#Muls - Min.\#Muls) = 196$ available resource allocation scenarios only 60 design points

```
1: input:     kernel.c: Behavioral Kernel Description in C;
2: input:     Space Boundaries in {#ALUs, #MULs, #OCD, #LUF_i, ...};
3: output:    Pareto Exploration_Curve;
4:
5: temp_min = Min_ALUs;
6: for(LUF_i= min_LUF_i ; LUF_i < max_LUF_i ; LUF_i++) {
7: ...
8 : for(LUF_n= min_LUF_n ; LUF_n < max_LUF_n ; LUF_n++) {
9:
10: CDFG ←C2CDFG(kernel.c, LUF_i, ..., LUF_n) ;
11:
12:  for(ODC = min_OCD ; OCD < max_OCD ; OCD++) {
13:   for(MUL_no = min_MULs ; MUL_no < max_MULs ; MUL_no++) {
14:    for(ALU_no = min_ALUs ; ALU_no < max_ALUs ; ALU_no++) {
15:
16:      sched_CDFG ←schedule (CDFG, ODC, MUL_no, ALU_no);
17:      xi.{ALU_no, MUL_no, ODC, LUF_i, ..., LUF_n};
18:      Contol_Steps ←Extract_Control_Steps (sched_CDFG);
19:      Area_Cost ←Area_Estimation (ALU_no, MUL_no);
20:      Grad ←Evaluate_Gradient ((Area_Cost, Control_Steps), Depth_k);
21:      switch Grad
22:        case  (Grad < 0) : Insert (Examined_Curve, xi); i++;
23:        case   (Grad ≥ 0) : temp_min = x[i-depth+1].ALU_no ;  i++; break ;
24:      }
25:     }
26:    }
27:  }
28: ...
29:}
30: Exploration_Curve ←Pareto_Extraction (Examined_Curve);
```

Fig. 7.6 Pseudocode for the gradient-based pruning heuristic

have been evaluated delivering a speedup of 3× in exploration's runtime (under the coarse assumption that the evaluation time of each design point is the same).

Figure 7.6 depicts the pseudocode of the gradient-based pruning technique. The various per parameter exploration loops are depicted in lines 6–14. The outer loops encounter for the loop unrolling factors per loop indexes (lines 6–8). After the LUFs are known the CDFG of the behavioral description is formed (line 10). CDFG depends only from the LUFs' values. The next exploration loop concerns the operation chaining degrees, OCDs (line 12), while the two inner loops (lines 13–14) concentrate on the exploration of resource allocation scenarios by incrementing the *#Muls* and *#ALUs*, respectively.

The core of gradient-based pruning is placed at the innermost loop since it targets the resource allocation level. At first, the CDFG is scheduled given the resource constraints and the OCD value (line 16) and a solution point x_i is constructed to handle the examined configuration vector (line 17). After scheduling the CDFG, the information about the number of control steps (line 18) and the area cost (line 19) of the examined solution, are extracted. We adopted the linear area estimation model used also in [10]:

$$Area\_Cost = (\#ALUs \times Area_{ALU}) + (\#Mul \times Area_{Mul}) \tag{7.3}$$

The gradient of the exploration curve, *Grad*, is evaluated according to a user specified *Depth* (line 20). In case that there is a negative gradient the examined solution forms a possible Pareto point and it is inserted in the *Examined_Curve* (line 22). In case that *Grad* is either positive or zero, the current ALU based exploration loop is broken and the last found "possible Pareto point" concerning the number of ALUs is retrieved $(x[i - depth + 1].ALU\_no)$ to form the initial point for the exploration loop of the next configuration (line 23). The points inserted in the *Examined_Curve* are characterized as "possible Pareto points", since there is a large possibility the real Pareto point to be generated in an exploration of a later configuration. Thus, at the end of the exploration procedure the points stored in the *Examined_Curve* are filtered and the final Pareto points are extracted (line 30).

7.6 Experimental Results

In order to evaluate the effectiveness of our approach, we have developed a fully automated HLS exploration framework implementing the proposed methodology by properly extending SPARK-HLS tool [18]. We used a representative set of computationally intensive DSP benchmark applications to evaluate the efficiency of the proposed DSE approach. The benchmark suite consists of 8 real-life DSP kernels in C including: (1) a 16th-order FIR filter (FIR16), (2) a 1D Discrete Cosine Transformation (1D DCT), (3) a YUB to RGBA filter (YUB2RGBA) [19], (4) a Fast Fourier Transformation (FFT) [20], (5) a Discrete Haar Wavelet Transform (DWT) [19], (6) a MESA Matrix Multiplication kernel (MatMul) [17], (7) the 2D DCT kernel found into the JPEG application (Jpeg DCT) [17], (8) the 2D IDCT kernel from the MPEG (Mpeg IDCT).

The HLS exploration tool was installed and run on a Linux Xeon server at 2.33 GHz with 4 GB RAM. For each kernel, a range of $(Min.\#ALUs, Min.\#Muls) = (4, 2), (Max.\#ALUs, Max.\#Muls) = (32, 16)$ resource allocation scenarios were explored. The area cost was of each component extracted through post-synthesis characterization [21] for a 0.13 um standard cell library [22]: $Area_{ALU} = 4,000\,\text{um}^2, Area_{MUL} = 12,000\,\text{um}^2$. The operation chaining degree ranged between $(Min.OCD = 0), (Max.OCD = 3)$. The minimum and maximum per loop index unrolling factors (LUFs) depend on the behavioral description of each kernel. In each case, the whole range of each LUF has been explored. Empirically, we considered $Max.Cardinality = 100$ for the designer specified cardinality threshold which guides the invocation of exhaustive or gradient-based pruned exploration.

Due to space limitations, Fig. 7.7 depicts two representative exploration curves derived from the proposed exploration methodology. The depth parameter for the gradient-based pruning was set to $Depth = 10$, thus 10 solutions are

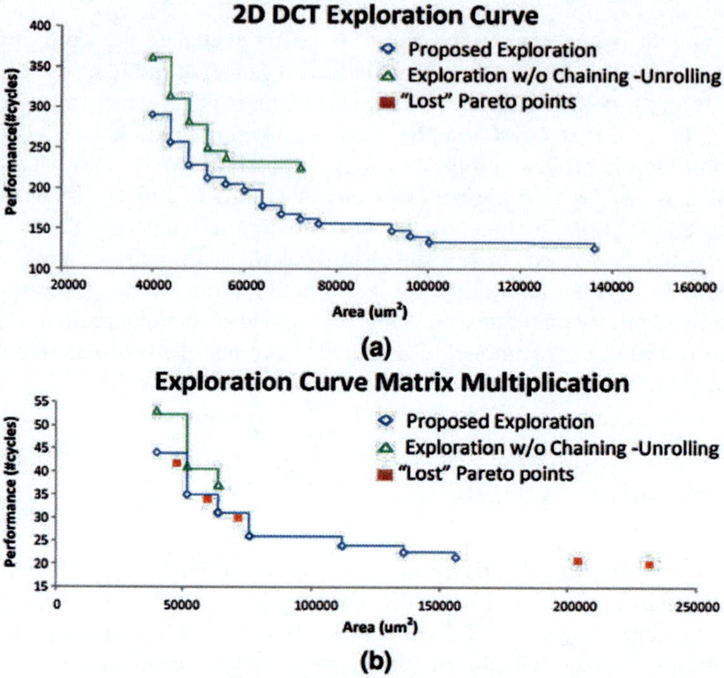

Fig. 7.7 Exploration curves produced by the proposed methodology vs. exploration curves of the non-augmented design space (without chaining-unrolling). Solution shifting towards higher quality solutions. The Pareto points that did not examined by the proposed methodology, were derived through exhaustive exploration and they are also depicted as "lost" Pareto points

examined to decide whether a zero-gradient segment is global or local. In the exploration diagrams of Fig. 7.7, the exploration curve of the non-augmented design space (exploration unawareness of loop-unrolling and operation chaining parameters considering only the various resource allocation scenarios) has been overlapped. The Pareto curve shifting towards higher quality performance-area trade-offs is depicted, confirming our motivational observations. Figure 7.7 depicts also the non-discovered ("lost") Pareto points from the proposed methodology (solution points which did not lay onto the exploration curve). The proposed methodology discovered the 100% of Pareto solutions in the 2D DCT benchmark with an average speedup of 10× in exploration's runtime. The loss of some Pareto points in MatMul kernel is a side-effect of the heuristic nature of gradient-based pruning.

Figure 7.8 depicts the speedup gains of the proposed methodology versus an exhaustive DSE approach and the impact of the gradient's *Depth* on the

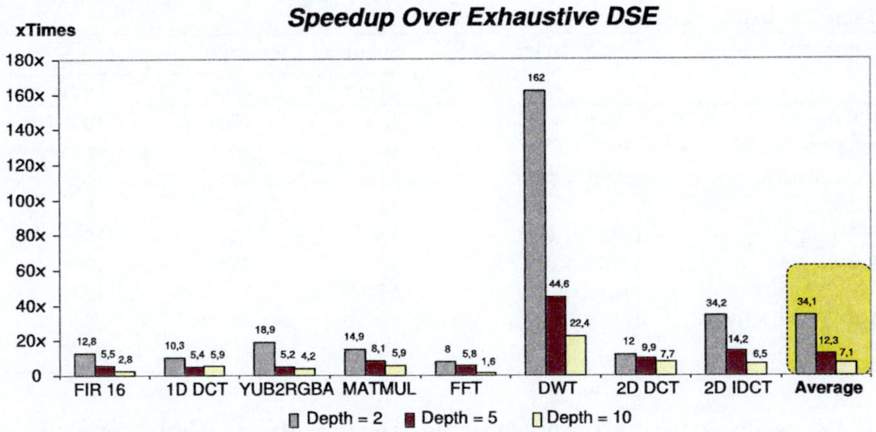

Fig. 7.8 Exploration speedup over the full exhaustive DSE approach

exploration's runtime. Speedup is defined as the ratio of the solutions explored from the full exhaustive DSE over the solutions explored with the proposed methodology. We considered three values of the gradient *Depth* parameter, namely $Depth_1 = 2, Depth_2 = 5, Depth_3 = 10$. Average speedups of $34.1\times, 12.3\times, 7.1\times$ in exploration's runtime are reported for each *Depth* value, respectively, Although, it is expected that smaller values of gradient *Depth* deliver higher speedup (which is a logical assumption for the average case), this is true only under the false assumption that the maximum number of exhaustive explorations performed by the proposed approach is the same for all the DSP kernels. Actually, there is a trade-off concerning exploration's runtime between the *Depth* and the *Max. Cardinality* parameters. This type of trade off is exposed in the 1D DCT case, in which the exploration with $Depth = 5$ requires 11 exhaustive explorations and so it reports lower speedup than the exploration with $Depth = 10$ which requires 6 exhaustive explorations.

We measure the accuracy of the generated exploration curves using the solution coverage metric, which refers to the ratio of the non-dominated solutions found in each approximate Pareto curve generated by our approach in respect to the exact Pareto curve generated by the full exhaustive exploration of the design space. Table 7.1 reports the size of examined design space and the values of the solution coverage for each examined benchmark. The average accuracy is 64.6, 82.9, 92.9% for the gradient $Depth \in \{2, 5, 10\}$. As expected, the accuracy increases for the larger values of the *Depth* parameter in the expense of higher exploration runtime (Fig. 7.8).

In summary, the proposed DSE methodology in the worst accuracy case ($Depth = 2$) delivers a 64.6% solution coverage with exploration speedup of

Table 7.1 Exploration's accuracy results

Benchmark Kernels	Size of design space	Solution Coverage(%)		
		Depth = 2	Depth = 5	Depth = 10
FIR16	27841	100	100	100
1D DCT	13920	80	90	90
YUB2RGBA	13920	33.3	92	100
MatMul	7840	42	60	60
FFT	1740	100	100	100
DWT	334080	50	75	100
Jpeg 2D DCT	111360	66.6	83	100
Mpeg 2D IDCT	111360	80	80	100
Average	–	**68.9**	**85.0**	**93.7**

34.1×, while at the worst speedup case ($Depth = 10$) it delivers an solution coverage of 93.7% with 7.1× speedup.

7.7 Conclusion

We presented an HLS exploration approach based on iterative partitioning. We identified the need to perform exploration in a global manner and we introduced a new heuristic algorithm for quick and efficient traversing of the design space. Experimental results have shown quality improvements over the conservative DSE approaches along with significant reductions in exploration runtime in comparison to the exhaustive approach without degrading the quality delivered design solutions.

References

1. International Sematech (2005) International Technology Roadmap for Semiconductors, http://www.sematech.org
2. Coussy P, Morawiec A (2008) High-level synthesis: from algorithm to digital circuit. Springer, Berlin
3. Gries M (2004) Methods for evaluating and covering the design space during early design development. Integr VLSI J 38(2):131–183
4. Pareto V (2008) Manuale di Economia Politica. Picola Biblioteca Scientifica, Milan, 1906, Translated into English by Ann Schweir (1971). Manual of Political Economy, MacMillan London
5. De Micheli G (1994) Synthesis and optimization of digital circuits. McGraw-Hill Higher Education
6. Blythe SA, Walker RA (1999) Efficiently searching the optimal design space. In: GLS '99: Proceedings of the Ninth Great Lakes Symposium on VLSI, p 192
7. Balakrishnan M, Marwedel P (1989) Integrated scheduling and binding: a synthesis approach for design space exploration. In: DAC '89: Proceedings of the 26th ACM/IEEE Design automation conference, pp 68–74

8. Dutta R, Roy J, Vemuri R (1992) Distributed design-space exploration for high-level synthesis systems. In: DAC '92: Proceedings of the 29th ACM/IEEE design automation conference. pp 644–650
9. Chaudhuri S, Blythe SA, Walker RA (1997) A solution methodology for exact design space exploration in a three-dimensional design space. IEEE Trans Very Large Scale Integr Syst 5(1):69–81
10. Wang G, Gong W, DeRenzi B, Kastner R (2007) Exploring time/resource trade-offs by solving dual scheduling problems with the ant colony optimization. ACM Trans. Design Autom. Electr Syst 12(4)
11. Kurra S, Singh NK, Panda PR (2007) The impact of loop unrolling on controller delay in high level synthesis. In: DATE '07: Proceedings of the conference on Design, automation and test in Europe. pp 391–396
12. Dragomir O, Panainte E, Bertels K, Wong S (2008) Optimal unroll factor for reconfigurable architectures. In: Proceedings of ARC. pp 4–14
13. Gerlach J, Rosenstiel W (2000) A methodology and tool for automated transformational high-level design space exploration. In: ICCD. pp 545–548
14. Marwedel P, Landwehr B, Domer R (1997) Built-in chaining: introducing complex components into architectural synthesis. In: Proceedings of the ASP-DAC. pp 599–605
15. Corazao M, Khalaf M, Guerra L, Potkonjak M, Rabaey J (1996) Perfomance optimization using template mapping for datapath-intensive high-level synthesis.. IEEE Trans. Computer-Aided Design Integrated Circuits Syst 15(2):877–888
16. Xydis S, Skouroumounis C, Pekmestzi K, Soudris D, Economakos G (2010) Designing efficient DSP datapaths through compiler-in-the-loop exploration methodology. In: Proceedings of ISCAS
17. The ExPRESS group (2009) http://express.ece.ucsb.edu
18. Gupta S, Dutt N, Gupta R, Nicolau A (2002) Coordinated parallelizing compiler optimizations and high-level synthesis. ACM Trans. Des. Autom. Electron. Syst 9:2004
19. C to Verilog, Circuit Design Automation, http://c-to-verilog.com/
20. point FFT source code, http://www.mit.edu/emin/source_code/fft/fft.c
21. Synopsys Inc. (2009) http://www.synopsys.com/products/
22. Artisan Components, TSMC 0.13 Library Databook

Chapter 8
A Scalable Bandwidth-Aware Architecture for Connected Component Labeling

Vikram Sampath Kumar, Kevin Irick, Ahmed Al Maashri and Vijaykrishnan Narayanan

Abstract This chapter discusses the design and implementation of a streaming-based Connected Component Labeling architecture. The architecture implements a scalable processor, which can be tuned to match the available I/O bandwidth on the computing platform that hosts the hardware. In addition, the chapter presents the hardware performance measurements when implemented on an FPGA platform.

8.1 Introduction

Connected Component Labeling (CCL) [2] is the process of identifying disjoint pixel regions in an image and assigning a unique label to each of these regions. A disjoint pixel region is an image patch in which each foreground pixel has zero

The work presented in this chapter is an extended version of [1], with more focus on the internal design of the Sliced Connected Component Labeling architecture, including a discussion on how the Association FIFO and Coalescing Unit are updated and a description of the CL RAM and Global RAM, which can be found in Sect. 8.5.

V. S. Kumar · K. Irick · A. Al Maashri · V. Narayanan (✉)
Microsystems Design Laboratory (MDL), Department of Computer Science and Engineering, The Pennsylvania State University, University Park, PA 16802, USA
e-mail: vijay@cse.psu.edu

V. S. Kumar
e-mail: vus119@psu.edu

K. Irick
e-mail: irick@cse.psu.edu

A. Al Maashri
e-mail: maashri@cse.psu.edu

N. Voros et al. (eds.), *VLSI 2010 Annual Symposium*,
Lecture Notes in Electrical Engineering 105, DOI: 10.1007/978-94-007-1488-5_8,

or more adjacent foreground pixels. CCL has a wide range of applications in image analysis, including blob detection and tracking.

In addition, CCL can be used to obtain some additional information from the connected regions; including:

- Coordinates of connected region's bounding box.
- Coordinates of connected region's geometric center (i.e. *centroid*).
- Area of connected region.

A number of studies have proposed a variety of CCL hardware architectures based on single pass algorithms. However, these architectures impose some input constraints that are unsuitable for real-time systems. This chapter, on the other hand, presents the design and implementation of a real-time, streaming-based CCL architecture that detects bounding boxes of each connected region in an image.

The CCL architecture, proposed in this chapter, is implemented using a Field-Programmable Gate Array (FPGA) computing platform. FPGA's offers a high-performance implementation of the Connected Component Labeling algorithm. Additionally, the ability to implement complex, single-cycle control structures makes FPGAs the ideal choice for control-intensive CCL implementations.

The next section summarizes previous work done with regard to connected component labeling.

8.2 Background

A number of studies have investigated Connected Component Labeling. For instance, Bailey [3] discusses the advantage of raster scan based image segmentation. Moreover, Ranganathan et al. [4] describes a parallel hardware implementation of Connected Component Labeling. The design represents a two pass algorithm capable of processing a single 128 × 128 image in 900 microseconds; well above real-time constraints. On the other hand, several other high speed parallel—but resource intensive—algorithms are discussed by Alnuweiti et al. [5]. Conversely, a resource-efficient, iterative algorithm implemented on FPGA is presented by Crookes et al. [6] but requires an indeterminate number of passes as pointed out by Bailey et al. [7], who developed a single pass labeling algorithm amenable to a streaming architecture. The FPGA implementation of this algorithm is discussed in [8]. However, the architecture has the following drawbacks:

- The pipeline of the architecture is limited to process only a single pixel per cycle. This restricts the means by which the input can be streamed into the hardware.
- The architecture relies on the presence of video blanking periods to perform costly merge operations at the end of each row. While video blanking periods are common to most CCD and CMOS camera interfaces, they do not exist when interfacing directly to a memory subsystem for which data can be accessed back-to-back with negligible delay.

This chapter presents an enhanced version of the single pass algorithm motivated by the drawbacks discussed above. The new architecture offers a support for systems that require more diverse pixel input rates, interfaces and protocols. However, before delving into the details of the proposed architecture, it is necessary to briefly describe the single pass algorithm to the reader.

8.3 Single Pass Algorithm and Implementation

Rosenfeld et al. [2] describes a simple algorithm for labeling connected components. The algorithm determines the label of the current pixel by looking at its four neighboring pixels; namely, left (L), upper-left (UL), upper (U) and upper-right (UR) neighbors. Table 8.1 summarizes the labeling algorithm.

Based on the algorithm described above, Bailey et al. [7] and [8] proposed an algorithm and an implementation for CCL using a single pass technique. In the remaining of this section, we will focus on describing the architecture of the single pass algorithm. This will help the reader in appreciating our proposed architecture that we will describe later in this chapter.

The architecture of the Single Pass Algorithm is illustrated in Fig. 8.1. In this architecture, the neighboring pixel labels are stored in registers D, A, B, and C corresponding to the 'L', 'UL', 'U' and 'UR' neighbors, respectively, as shown in Fig. 8.1b. The input to the CCL pipeline is a binary image representing the background (binary value '0') or foreground (binary value '1') classification of an image. Note that each bit in the image represents a pixel. In other words, in a byte accessible memory, a single byte represents the state of eight pixels. This means that if a byte is fetched per cycle, it would take eight subsequent cycles for the CCL pipeline to process each pixel. More realistically, in a system with a 64-bit system bus supporting 16-cycle burst transactions, 1024 (i.e. 64 × 16) status bits can be delivered in a single transaction. We refer to this unit of transfer as a Fetch Line. Consequently, the memory subsystem would remain idle for 1008 (i.e. 1024–16) cycles before being requested to supply an additional Fetch Line. If no other memory requests are pending much of the memory subsystem's bandwidth would be unused.

As CCL processes a row of an image, the labels from the previous row are cached in a buffer that exhibits a single-cycle latency during a read or write access. This behavior requires extra care when considering the merging scenario as depicted in case (d) in Table 8.1, where all existing references to label '3' in the row buffer should reference label '2' once the merge completes. Therefore, instead of searching the entire row buffer and replacing all occurrences of label '3' with label '2', the architecture uses a lookup table to store and retrieve aliases of a given label. This lookup table, known as Merge Table, is indexed by a label referred to as '*Lreference*', and returns the actual label referred to as '*Lactual*', for which '*Lreference*' has merged with since the row buffer was initialized.

Table 8.1 A summary of how the connected component labeling algorithm determines the label of the current pixel. In the column labeled "Example", white boxes represent current pixel, while gray boxes represent neighboring pixels

Case	Example
(a) **IF** the current pixel is a background pixel, **THEN** it is assigned label '0'	2 2 2 2 ? — No Labeling → 2 2 2 2 0
(b) **IF** all neighboring pixels are background **AND** the current pixel is a foreground pixel **THEN** a new label is assigned	0 0 0 0 ? — New Labeling → 0 0 0 0 2
(c) **IF** only a single label is used among the labeled neighbors **AND** the current pixel is a foreground pixel, **THEN** the current pixel inherits the label	4 0 0 0 ? — Inherited Labeling → 4 0 0 0 4
(d) **IF** different labels are assigned to neighbors **AND** the current pixel is a foreground pixel **THEN** the current pixel will merge the labeled regions and the current pixel and all labeled pixels will be labeled with the smallest label in the neighborhood	2 0 3 2 ? — Merging → 2 0 2 2 2

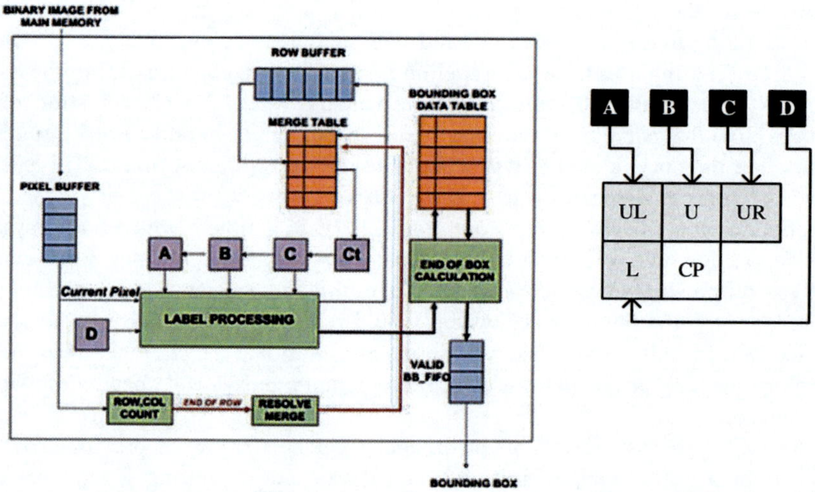

(a) **(b)**

Fig. 8.1 Streaming Single Pass Architecture; **a** Single Pass Pipeline, **b** a map showing registers mapped to neighbors of the current pixel

A close observation to the pipeline reveals that the lexicographic dependency of pixels in the labeling process makes it impractical to label more than two sequential pixels at a time. However, this inherited dependency can be overcome by operating on pixels that are lexicographically independent. This can be

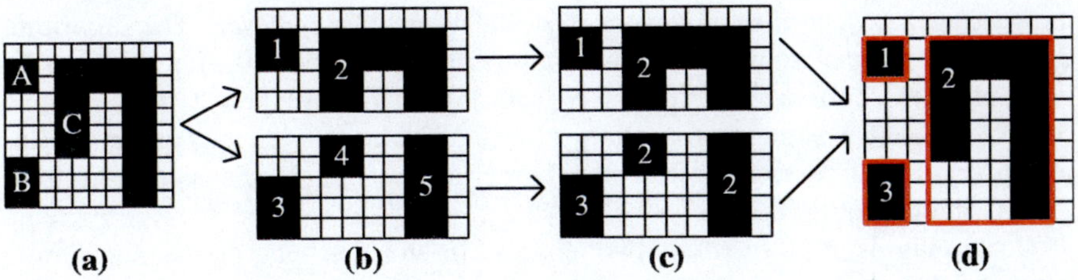

Fig. 8.2 SCCL algorithm example with number of slices is 2, **a** Original image contains three connected region, **b** SCCL slices the image into two slice, each of which is operated on independently, **c** Slices are merged by analyzing the last row of the top slice and the first row of the bottom slice, labels are updated accordingly, **d** Bounding boxes (1, 2 and 3) are detected for the three connected regions. It is to be noted that a bounding box is represented by the coordinates: {min_x, max_x, min_y, max_y}

accomplished by partitioning the image and performing CCL independently and simultaneously on each partition. This approach allows us to extract parallelism that is proportional to the number of independent image partitions. The following sections present an alternative algorithm and implementation for CCL that is based on operating on independent image partitions concurrently. We refer to this algorithm as "Sliced Connected Component Labeling".

8.4 Sliced Connected Component Labeling Algorithm

This section outlines the proposed Sliced Connected Component Labeling (SCCL) algorithm. This algorithm slices an input image into 's' number of slices, each of which is operated on independently and concurrently.

8.4.1 Slice and Merge Algorithm

The SCCL algorithm starts by slicing the input image into 's' slices. Each slice will be processed and assigned labels independent from other slices. Once all slices are processed, the algorithm coalesces the slices and merges any connected regions that extend across slices. Figure 8.2 illustrates an example of how SCCL algorithm works.

As shown in Fig. 8.2a, the original image contains three connected regions; labeled 'A', 'B' and 'C'. The SCCL algorithm slices the image into two slices as shown in Fig. 8.2b. We refer to these slices as top and bottom henceforth. Depending on the location of a connected region, it is determined if further processing is required. For example, connected region '1' does not lie on either first or last row of the slice, hence, no further processing will be required when merging the slices. On the other hand, labels 2, 3, 4 and 5 are stored for later processing at the merge stage to determine if any of these labels would coalesce. The algorithm establishes an association between the labels from the last row of the top slice and the first row of the bottom slice. It is clear from Fig. 8.2b that labels 2, 4 and 5 must coalesce since they stretch across slices. Note that the labels indicated in Fig. 8.2 are for illustration purposes only. The actual order of assigning labels is different from the one shown in Fig. 8.2

Coalescing is performed in two phases. In the first phase, the association between labels lying at the last row of top slice and the first row of bottom slice are recorded. In the second phase, the recorded associations are resolved. This is accomplished as follows:

- If the connected region does not stretch across slices, then its bounding box is detected and is committed to a table that is referred to as 'Global Bounding Box' table.
- If the connected region stretches across slices, then the bounding box of the coalesced region must be updated in the 'Global Bounding Box' table.

Although the process described above refers to the example shown in Fig. 8.2, yet the same process is applied to SCCL algorithm with more than two slices. In this case, the process is repeated for all slice boundaries until all bounding boxes are committed.

The following sub section outlines the mechanisms by which a bound box is actually detected.

8.4.2 Bounding Box Detection

The process of detecting the bounding box of a connected region is performed as the stream of pixels is being processed. This is accomplished through two tables: The Bounding Box Coordinate Table (BBCT) and the Bounding Box Coordinate Cache (BBCC). Both of these tables interact with the four registers discussed earlier and depicted in Fig. 8.1b.

The BBCT table stores the bounding box coordinates of a label in the following format {min_x, max_x, min_y, max_y}. On the other hand, The BBCC table stores the maximum column coordinate of a label in the current row. The BBCC is written the address of pixel in register 'D' when the current pixel is a background pixel and the previous was a foreground pixel (i.e. 'D' != 0, 'CP' = 0). Conversely, the BBCC is written the address of pixel in register 'A' when the following pattern is detected {'CP' = 0, 'A' != 0, 'B' = 0 and 'D' = 0}. This pattern indicates that the pixel at location of pixel 'A' is the lower rightmost pixel in the region containing pixel 'A'. This pattern enables a state machine that reads the BBCC location corresponding to label 'A' and the BBCT location corresponding to label 'A'. The current column coordinate is compared with the output from the BBCC and the current row coordinate is compared with the maximum row value of the label 'A' from the BBCT output. If the current column and row

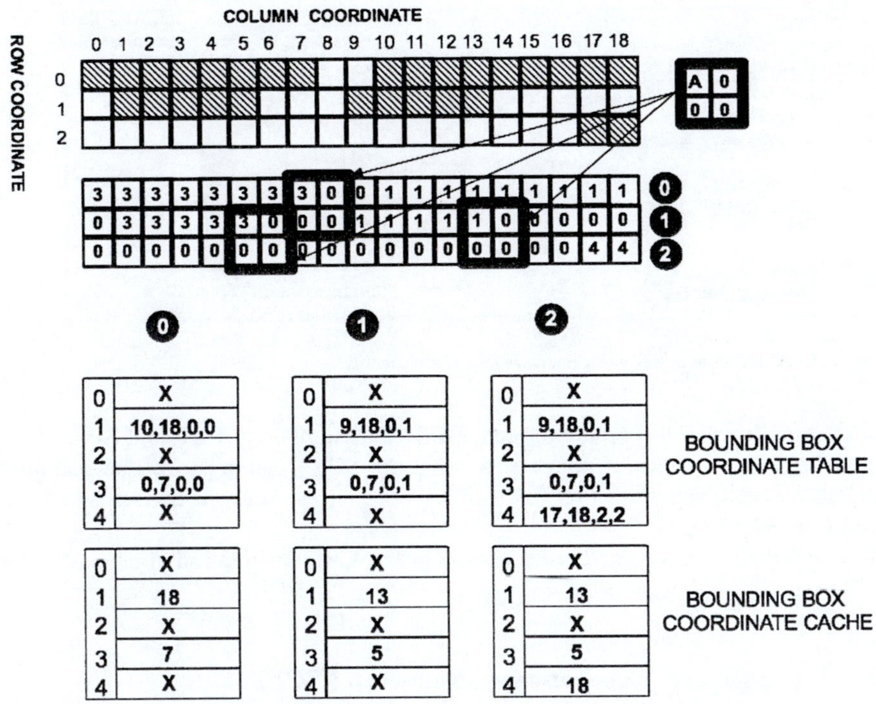

Fig. 8.3 An example that demonstrates how the bounding box detection works

are both greater than the corresponding pair of outputs from the two tables, then it is the end of the bounding box with label 'A'.

Figure 8.3 illustrates an example of bounding box detection process. The example shows that when processing row 0, the BBCT records the {min_x, max_x, min_y, max_y} coordinates of labels 3 and 1 as {0, 7, 0, 0} and {10, 18, 0, 0}, respectively. The BBCC records the column coordinates 7 and 18 for labels 3 and 1, respectively. Similarly, the pattern {3, 0, 0, 0} is detected at column 8 in row 1, and the coordinates corresponding to label 3 in both tables are retrieved. The BBCC table returns 5 (i.e. the last updated column coordinate) and the BBCT returns a maximum row coordinate of 1, which is equal to the current row.

8.5 SCCL Architecture

This section describes the hardware architecture of the SCCL. The architecture, shown in Fig. 8.4 , is composed of one or more of Connected Component Processors (CCP) and Slice Processors (SP). Also, the architecture contains a

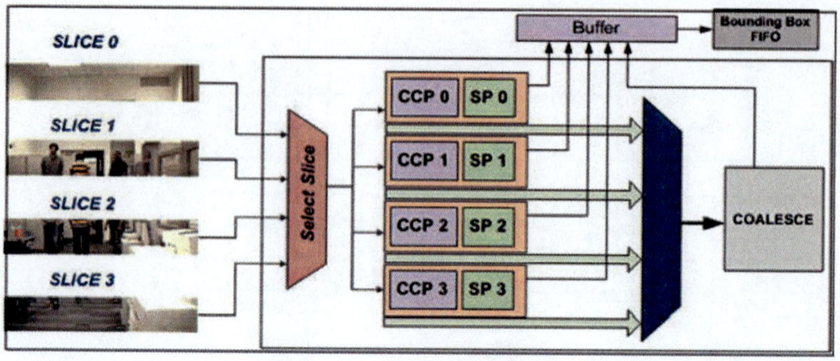

Fig. 8.4 Sliced connected component labeling architecture

coalescing logic and a bounding box FIFO. The number of CCP and SP units is determined by number of image slices that the SCCL can support. For example, the architecture shown in Fig. 8.4 supports four slices; hence it houses four CCP and four SP units.

The following sub sections discuss the internal implementation of each of these units.

8.5.1 Connected Components Processor (CCP)

Each CCP handles a slice that is composed of a range of rows. Consider an input image of size 240 × 320 that is split into 8 slices. In this case, rows 0 to 29 are handled by CCP_0, rows 30 to 59 are handled by CCP_1, and so on up to CCP_7 which handles rows 210 to 239. Each CCP fetches memory to fetch rows of its designated slice. Following our example, CCP_0 starts by fetching row 0. Once CCP_0 is done fetching and while it is processing row 0, CCP_1 starts fetching row 30. Similarly, other CCPs will fetch and process the first row of their designated slice. This mechanism allows the architecture to process a row from each slice in parallel. When CCP_7 is done fetching row 210 (i.e. the first row of its designated slice), memory access is arbitrated to CCP_0, which fetches and processes row 1 (i.e. the second row of its designated slice). This process continues until all the CCPs have fully processed all rows in their designated slice.

8.5.2 Slice Processor (SP)

When a CCP detects the end of a bounding box, it will send the box coordinates to the corresponding SP. The SP checks if the box boundary lies on the top or bottom of the slice by analyzing the minimum and maximum row coordinates, respectively. If the bounding box does not stretch across either slice's boundary, then the box coordinates are enqueued in Bounding Box FIFO. This indicates that bounding box requires no further processing. On the other hand, if the bounding box

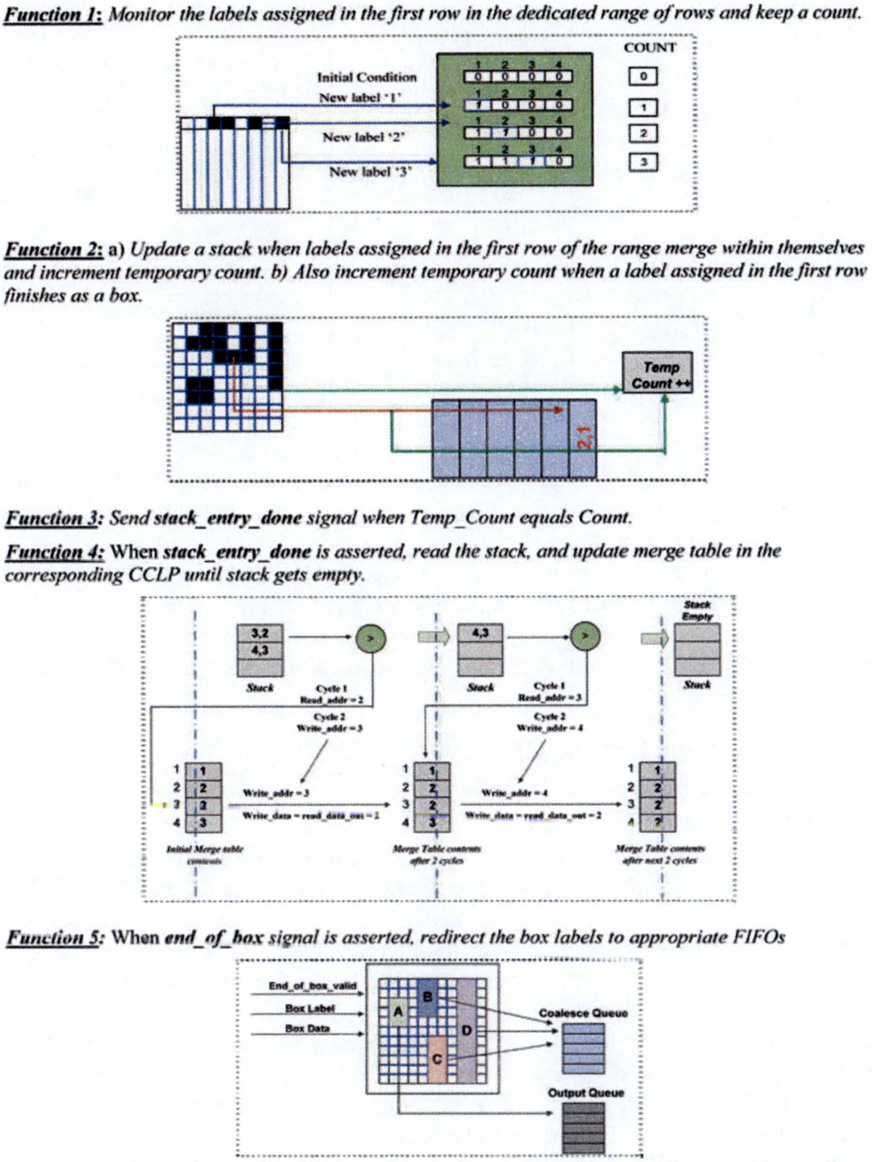

Fig. 8.5 Operations performed by the slice processor unit

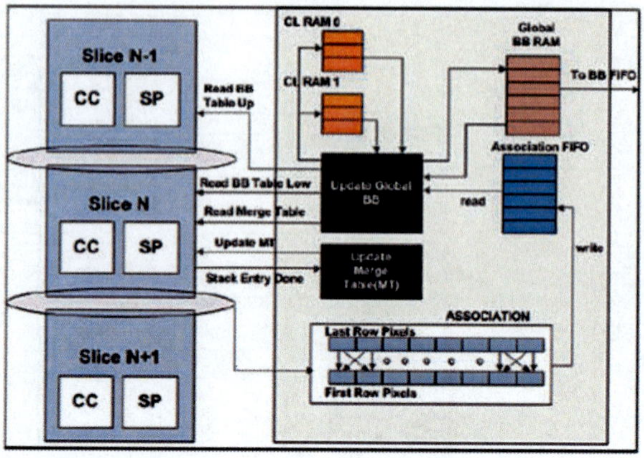

Fig. 8.6 The architecture of the coalescing unit

stretches across either the top or bottom slice boundary, then the SP will place the box coordinates and label in the Coalesce Queue. This indicates that further processing is required in the coalescing stage.

The SP maintains a record of labels assigned in the top row of its designated slice. This is necessary because the labels assigned in the top row may merge amongst themselves, which may result in relabeling connected regions as bounding box coordinates are detected. This scenario is illustrated in Fig. 8.5.

8.5.3 Coalescing Unit (CU)

The CU is responsible for coalescing regions across slices. Figure 8.6 demonstrates the main components of the CU. The unit consists of the Association FIFO, Common Label (CL) RAM0 (discussed later), CL RAM1 and Global Bounding Box RAM (denoted as Global BB RAM in Fig. 8.6). Consider the scenario in which the last row of slice N − 1 is completely labeled. This is when the CU begins the coalescing process. As shown in Fig. 8.6, when the CU reads the last row of slice N, it coalesces the boxes in slices N − 1 and N. At the same time, the unit records an association between the boundary boxes in slice N and the labels in slice N + 1. At the end of the coalescing process, the Global Bounding Box RAM is read along with a valid bit to determine all bounding boxes that are not connected any further and are ready to be sent to the next stage of processing. Note that the Association FIFO does not hold entries for those boundary labels that are not connected to any label in the adjoining slice (top or bottom). As a result, the Global Bounding Box RAM does not have entries for these boundary labels.

On the other hand, the Coalesce Queue has labels of all boxes in slice N − 1 that lie either on the top boundary, bottom boundary or both. The labels in this queue are read one by one and a connectivity check is done as discussed later to find out if the labels are connected at the top or bottom. If they are unconnected at both ends, the labels are ready to be committed.

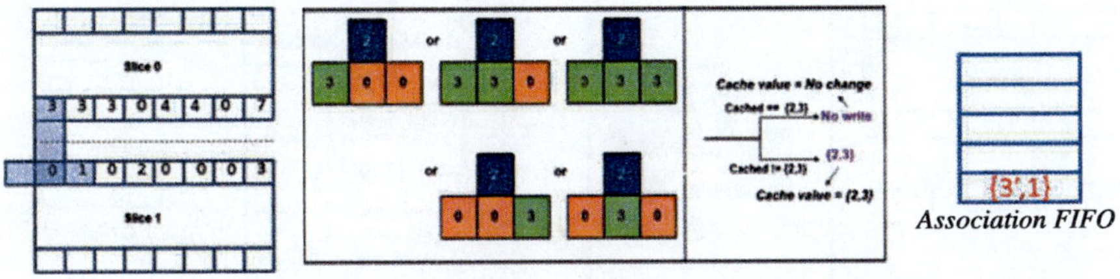

Fig. 8.7 Status of the Association FIFO with the first pattern

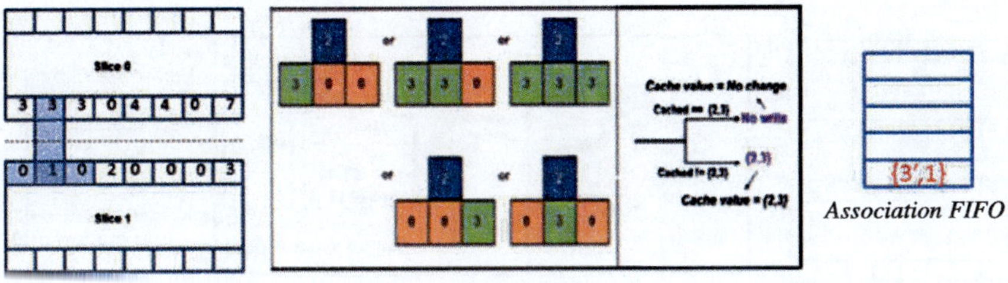

Fig. 8.8 Status of the Association FIFO with the second pattern

8.5.3.1 Writing to the Association FIFO

The CU needs to establish an association between slices N − 1 and N. The SP that handles slice N stores the first row labels in a buffer as discussed earlier. The labels of the last row of slice N − 1 are already stored in the row buffer of the CCP handling that particular slice and the Merge Table of this CCP has the updated values of these labels. The values from the first row of slice N are read from the first row buffer, whereas the values from the last row of slice N − 1 are read from the Merge Table—indexed by the value from the last row buffer as discussed earlier—in order to obtain the updated values. The two values that are read are checked and written as {top_label, bottom_label} to an Association FIFO. The architecture mitigates the necessity to perform redundant writes by caching the previous write value. This cached value is compared with the current value before writing, and only if the current value is different, a write takes place.

Figure 8.7 shows an example of a sequence of writes that occur at a hypothetical boundary between slice 0 and slice 1. The 'inverse T shaped' rectangles in Fig. 8.7 (leftmost) highlight the association pattern formed by the two slices. The value {3′, 1} is written into the association FIFO, where 3′ is the resolved value of label 3 obtained from the merge table of CCP_0. This value is also cached to avoid redundant writes in the following cycles.

Figure 8.8 highlights the next pattern, where an extra write to the Association FIFO is avoided because the current write value {3′, 1} is same as the cached value.

The next pattern, Fig. 8.9, exhibits a merge case where two labels from slice 1 merge with one label in slice 0. The two values that are to be written to the association FIFO are {3′, 1} and {3′, 2}. These are written in two consecutive

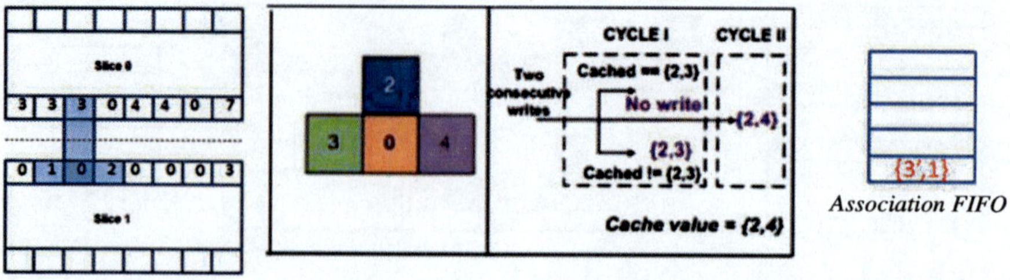

Fig. 8.9 Updating the Association FIFO; the first cycle of the merge process

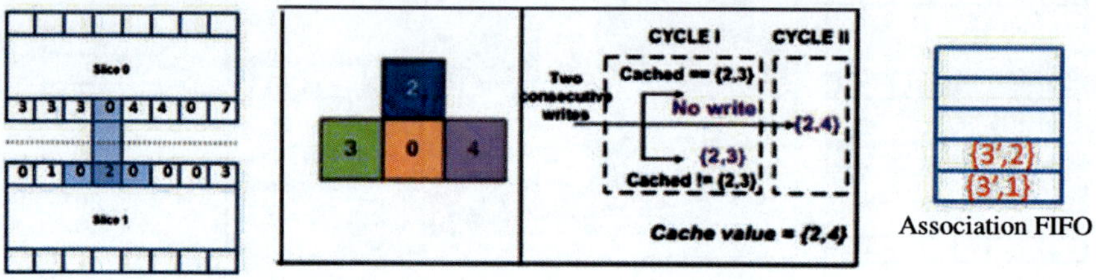

Fig. 8.10 Updating the Association FIFO; the second cycle of the merge process

cycles. In the first cycle, the value {3′, 1} is compared with the cached value (i.e. {3′, 1}). As a result, no write takes place in this cycle. In the second cycle, Fig. 8.10, since the value {3′, 2} is not equal to the cached value, a write to the Association FIFO takes place. Note that the pattern in the cycle following a merge does not affect the value that has to be written in the Association FIFO. This is because if the upper slice had a non-zero label, then that would be a redundant write. On the other hand, if the upper slice had a zero label, then there would be no new association. Other write patterns can be interpreted similarly. However, it is worth noting that at the end of a row, the cached values are reset to {0, 0}.

Furthermore, while writing to the association FIFO at the boundary between slices N and N + 1, the architecture maintains a status monitor register referred to as the *bottom_monitor_current*. The size of this register is M/2, where M is the number of columns in the image. When a write occurs to a particular label in the last row of slice N, the corresponding bit is set to 1 indicating that the label is connected to some other label in the bottom slice. The reason for doing so is explained later.

8.5.3.2 Reading from the Association FIFO

The coalescing of slices N − 1 and N takes place when the last row of slice N is being read from memory. Therefore, reading from the Association FIFO between slices N − 1 and N and writing to the Association FIFO between slices N and N + 1 might occur simultaneously. The reading starts once the stack corresponding to slice N becomes empty. A counter is needed to store the number of previous writes based on which of the association values are read. This is because a single FIFO is used to carry out the association between slices in order. The value read from the association FIFO is of the form {A0, A1}, where A0 is an updated label from the last row of slice N − 1, whereas A1 is a stale label from the first row of slice N. The updated values of the first row labels of slice N are stored in the merger table as discussed earlier. Therefore, indexing the merger table with A1 would return the updated value of A1. Let us call the returned value A1′. To meet cycle requirements, A0 has to be registered. Let us also call the registered version of A0 as A0′. Status registers called "entry registers" are maintained to track connectivity. There exists one M/2 bit register for the upper slice (N − 1) termed *upper_entry_register*, and one M/2 bit register for the bottom slice (N) termed *bottom_entry_register*. Note that the entry registers are updated during the read phase of the Association FIFO and the *bottom_monitor_current* register is updated during the write phase of the Association FIFO.

8.5.3.3 Common Label (CL) RAMs

The CU hosts two CL RAMs; namely, upper and lower. These CL RAMs store a common reference label—starting from decimal 1—for the connected labels between any two slices. The entry registers for A0′ and A1′ are retrieved and the subsequent processing is determined by the following algorithm:

(1) If both are 0, then a new common label is written both in the upper CL RAM indexed by A0′ and lower CL RAM indexed by A1′. Also, the new common label value is incremented, while the entry registers corresponding to these labels are set to 1.
(2) If entry register corresponding to A0′ is 0 and entry register corresponding to A1′ is 1, then this means that the label A1′ in the bottom slice is already connected to some other label in the top slice, however, the label A0′ in the upper slice is not connected to any other label in the bottom slice. Therefore, the corresponding entry registers are validated and the common label value output indexed by A1′ in the lower CL RAM is written into the upper CL RAM indexed by A0′.
(3) If entry register corresponding to A0′ is 1 and entry register corresponding to A1′ is 0 then it means the label A0′ in the upper slice is already connected to some other label in the bottom slice but the label A1′ in the bottom slice is not connected to any other label in the top slice yet. Thus, the corresponding entry registers are validated and the common label value output indexed by A0′ in the upper CL RAM is written into the lower CL RAM indexed by A1′.
(4) If both entry registers are 1 then it means both the labels A0′ and A1′ are already connected to some other label(s) in the bottom slice and top slice respectively. A comparison of the output common labels from both CL RAMS is made to ensure they are the same. If they are not the same, then the lower common label has to replace the higher common label. If not, no update is needed.

Note that after processing every boundary, the two CL RAMs are switched. The entry registers of the current upper CL RAM are zeroed (current *upper_entry_register* = 0) and becomes the lower CL RAM while coalescing slices N and N + 1. The current lower CL RAM becomes the next upper CL RAM but its entry registers (current *upper_entry_register* = previous *bottom_entry_register*) are kept as is, to maintain the connectivity history.

8.5.3.4 Updating Global Bounding Box RAM

The Global Bounding Box RAM stores the Bounding box values of the common labels. While reading the CL RAMs, the Bounding Box data tables from the connected regions are also read. A0′ is used to read the bounding box of the upper slice and A1′ is used to read the bounding box of the lower slice. The 2 bit entry is registered, and in the next cycle, the Global Bounding Box RAM is read with the common label as index. Based on the registered 2 bit value, the update of the Global Bounding Box RAM takes place as explained in the following algorithm:

(1) If both values are 0, then there is no entry for the new common label. So the two bounding box values from the connected region are compared and the updated bounding box is found. With the new common label as index, the value is written to the bounding box. When a new common label is assigned, a box counter is incremented to keep track of the number of boxes that have coalesced.
(2) If 01, then the values to be compared are the bounding box from top slice connected region and the bounding box from global Bounding Box RAM. The final bounding box is found and with the common label as index the Global Bounding Box RAM is updated.
(3) If 10, the values to be compared are the bounding box from bottom slice connected region and bounding box from Global Bounding Box RAM. The final bounding box is found and with the common label as index the Global Bounding Box ram is updated.
(4) If both values are 11 and the common labels are different, then both the values are read from Global Bounding Box RAM and the smaller of the two labels is updated with the final bounding box. A valid bit has to be set for the higher label, which indicates it is not a valid label anymore. Also the box counter has to be decremented.

Figures 8.11 and 8.12 illustrate the merging process using CL RAMs, Entry Registers and the Global Bounding Box RAM for two different cases. The status of the association FIFO is assumed to be that of the one shown in Fig. 8.10

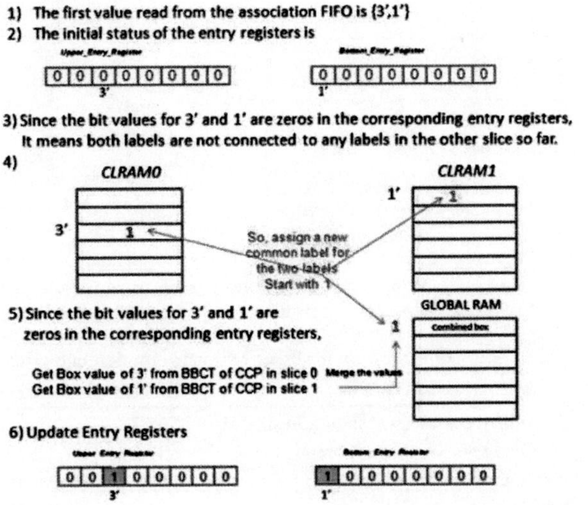

Fig. 8.11 Updates to CL RAMs, Entry Registers and Global Bounding Box RAM with first value read from the Association FIFO

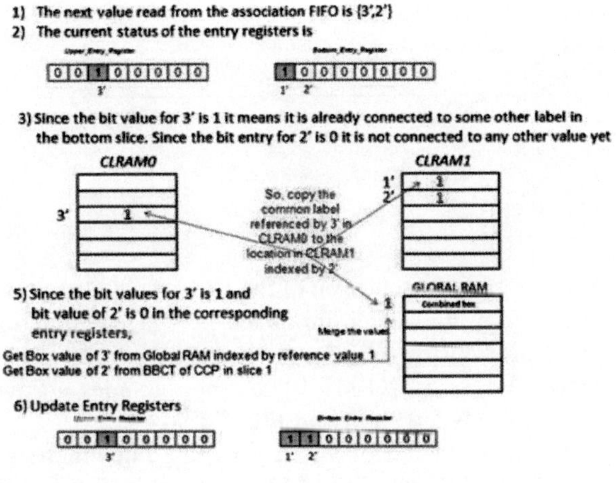

Fig. 8.12 Updates to CL RAMs, Entry Registers and Global Bounding Box RAM with second value read from the Association FIFO

8.5.3.5 Bounding Box Update

At the end of the coalescing stage, the Global Bounding Box RAM is updated with a validity bit for each label either set or reset. Before reading the contents of the Global Bounding Box RAM, the Coalesce Queue is read until empty. If the ID read from the queue is unconnected both at the top and the bottom—obtained from the corresponding entry register and *bottom_monitor_current* register respectively—then it is committed immediately by fetching the value from the corresponding bounding box table.

The following section discusses the performance measurements and results obtained from running the hardware implementation of the SCCL architecture.

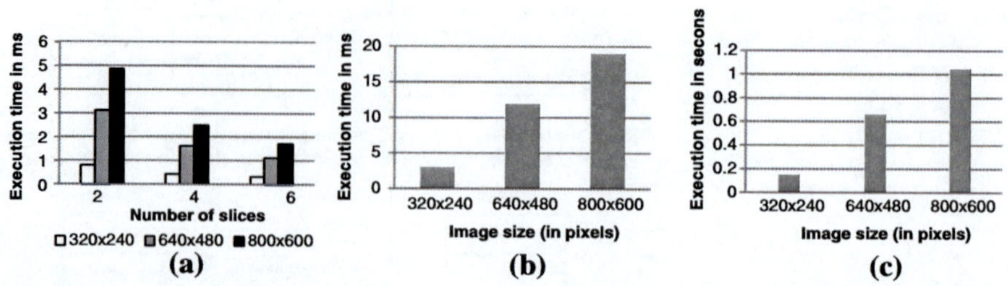

Fig. 8.13 Execution time as a function of image size, **a** SCCL, **b** Dual Core Workstation **c** Software implementation running on embedded PowerPC

Table 8.2 Comparison with other hardware implementations

Algorithm	Author	Clock (MHz)	Image size	Frame rate (frames/s)
Systolic array	Ranganathan et al. [4]	66.67	128 × 128	~1000
8-adjacency	Jablonski et al. [9]	60	512 × 512	~25
Run length	Appiah et al. [10]	156	640 × 480	~253
Iterative			640 × 480	~81
SCCL	Presented in this chapter			
Slices 2, 4, 6		100	320 × 240	~1200, ~2300, ~3000
Slices 2, 4, 6		100	640 × 480	~318, ~617, ~880
Slices 2, 4, 6		100	800 × 600	~200, ~400, ~580

8.6 Results

A Verilog implementation of the Connected Component Labeling architecture was simulated using ModelSim mixed-language simulation environment and validated on a Xilinx Virtex-5 FPGA development platform. The ML510 development board hosts a Virtex-5 FX130T FPGA that includes a PowerPC processor. The performance results are shown in Fig. 8.13.

The performance of SCCL hardware implementation was measured for varying image sizes and number of Slice Processors. As a reference, a multi-pass version Connected Component Labeling was implemented in C#, running on a 3.18 GHz Dual Core Xeon workstation. A similar implementation in C ++ was implemented and ran on an embedded PowerPC in the FPGA fabric. The results have shown that SCCL architecture—running at 100 MHz—outperforms the PowerPC reference implementation running at 300 MHz by a factor of approximately 215 and the workstation by a factor of 4. Note that the results presented here account for the bandwidth and latency characteristics of a realistic embedded system accessing a DDR2 memory device.

In addition, the results show that SCCL architecture scales favorably with the input image size as additional slice processors can be added when desired.

Furthermore, SCCL performance was compared against other hardware implementations summarized in Table 8.2. The table clearly shows how SCCL

Table 8.3 Resource utilization of the SCCL hardware according to number of slices

Number of slices	BRAM usage	LUT slices
2	21 (~7%)	15777 (~20%)
4	36 (~13%)	22317 (~28%)
6	51 (~17%)	28618 (~35%)

outperforms other hardware implementation, especially when number of slice processors is increased.

As for the resource consumed by the SCCL hardware; it was found that the architecture consumes around 20% of LUT slices and 7% of Block RAM of that available on the XC5VFX130T device for a design with 2 slices (2 SP + 2 CCP + CU). Synthesis reports show that with every additional slice (SP + CCP), a 4% increase in LUTs and 3% increase in BRAM resource utilization.

Table 8.3 lists the resource utilization of the SCCL hardware when synthesized on a Virtex 5 FX130T device for different number of slices.

Acknowledgments This work was supported in part by NSF awards #0702617 & #0916887, and a scholarship funding from the Government of the Sultanate of Oman.

References

1. Kumar VS, Irick K, Al Maashri A , Narayanan VK (2010) A Scalable bandwidth aware architecture for connected component labeling, Proceeding ISVLSI
2. Rosenfeld A, Pfaltz J (1966) Sequential operations in digital picture processing. J ACM 13(4):471–494
3. Bailey DG (1991) Raster based region growing. In: Proceedings of the 6th New Zealand image processing workshop, Lower Hutt, New Zealand, August , pp 21–26
4. Ashley R, Ranganathan N (1997) C3L: A chip for connected component labeling. In: the Proceedings of the 10th international conference on VLSI design: VLSI in Multimedia Applications, p 446. 4–7 Jan
5. Alnuweiti HM, Prasanna VK (1992) Parallel architectures and algorithms for image component labeling. IEEE Trans Pattern Anal Machine Intell 14(10):1014–1034
6. Crookes D, Benkrid K (1999) An FPGA implementation of image component labeling. In: Reconfigurable technology: FPGAs for computing and applications, SPIE 3844:17–23. Aug
7. Bailey DG, Johnston CT (2007) Single pass connected components analysis. In: Image and vision computing, New Zealand, Hamilton, New Zealand, pp 282–287. 6, 7 Dec
8. Johnston CT, Bailey DG (2008) FPGA implementation of a single pass connected components algorithm, In: IEEE international symposium on electronic design, test and applications (DELTA 2008), Hong Kong, pp 228–231. 23–25 Jan
9. Jablonski M, Gorgon M (2004) Handel-C implementation of classical component labeling algorithm. In: Euromicro Symposium on Digital System Design (DSD 2004), Rennes, France, 387–393 (31 August-3 September 2004)
10. Appiah K, Hunter A, Dickinson P, Owens J (2008) A run-length based connected component algorithm for FPGA implementation. In: International Conference on Field-Programmable Technology, Taipei, Taiwan. 7–10th Dec

Chapter 9
The SATURN Approach to SysML-Based HW/SW Codesign

Wolfgang Mueller, Da He, Fabian Mischkalla, Arthur Wegele, Adrian Larkham, Paul Whiston, Pablo Peñil, Eugenio Villar, Nikolaos Mitas, Dimitrios Kritharidis, Florent Azcarate and Manuel Carballeda

Abstract The main obstacle for the wide acceptance of UML and SysML in the design of electronic systems is due to a major gap in the design flow between UML-based modeling and SystemC-based verification. To overcome this gap, we present an approach developed in the SATURN project which introduces UML profiles for the co-modeling of SystemC and C with code generation support in the context of the SysML tool suite ARTiSAN Studio®. We finally discuss the evaluation of the approach by two case studies.

W. Mueller (✉) · D. He · F. Mischkalla
University of Paderborn/C-LAB, Fuerstenallee 11, 33102 Paderborn, Germany
e-mail: wolfgang@acm.org

A. Wegele
Atego, Major-Hirst-Street 11, 38442 Wolfsburg, Germany

A. Larkham · P. Whiston
Atego, 701 Eagle Tower, Montpellier Drive, GL50 1TA Gloucestershire, UK

P. Peñil · E. Villar
Avenida de los Castros s/n. Edif. ETSIIT Department of TEISA,
Universidad de Cantabria, Santander, 39005 Cantabria, Spain

N. Mitas · D. Kritharidis
Intracom Telecom, 19.7 km Markopoulo Avenida Peania,
190 02 Athens, Greece

F. Azcarate · M. Carballeda
Thales Security Solutions and Services, 20-22 rue Grange Dame Rose,
78141 Velizy Cedex, France

N. Voros et al. (eds.), *VLSI 2010 Annual Symposium*,
Lecture Notes in Electrical Engineering 105, DOI: 10.1007/978-94-007-1488-5_9,

9.1 Introduction

For a wider industrial applicability of UML, it is essential to close the gap between UML-based modeling and the execution of the models for their verification. The SATURN project developed an efficient approach to close this gap by providing code generation from UML/SysML to synthesizable SystemC and C/C++. For this we defined a SATURN methodology covering the entire design flow from co-modeling via co-simulation to co-synthesis. This chapter introduces the SATURN UML 2.0 profiles for synthesizable SystemC and C. The profiles are defined to customize the SysML tool suite Artisan Studio® for SystemC/C co-modeling. Based on these profiles, we also customized the code generation capabilities of Studio to generate synthesizable SystemC for simulation including makefiles for C/C++program compilation and scripts for design flow automation. A co-simulation is realized by means of the QEMU software emulator for the execution of the native software on the target operating system. After co-simulation, SystemC can be synthesized to VHDL as input for the ISE/EDK framework to configure the FPGA. The OS image which includes the SW executable is finally loaded to the target CPU for a complete system configuration. The remainder of this chapter is structured as follows. The next section discusses related work. Section 9.3 gives an overview of Artisan Studio® before Sect. 9.4 introduces the SATURN approach. In Sect. 9.5 we present the evaluation by two industrial case studies. Finally, Sect. 9.6 closes with a conclusion.

9.2 Related Work

Several approaches for UML profiles for hardware modeling were presented in the last years. As such, there were published several articles in the context of FPGA synthesis [1]. The UML MARTE profile defines the modeling of a specific set of HW components in [2]. Most recently, we can also find approaches for more specific hardware profile like IP-XACT based UML profiles [3–5] and we can identify approaches dedicated to SystemC UML profiles like [6]. The OMG UML profile for SoC also falls into the latter category as it is very much oriented towards SystemC [7]. However, it lacks details and does not come with a standard meta-model implementation. Both approaches cover complete SystemC and do not address properties for synthesizable SystemC [8].

Different methodologies cover communication and time at different abstraction levels. Methodologies, such as SystemC-AMS [9] and the synthesizable RTL subset [8], use specific signal channels for communication. The OSSS+R methodology [10] raises the level of abstraction of communication by means of shared objects. All these approaches have also raised the level of abstraction of time handling from discrete event (DE) to timed-clocked. However, they still handle time in too many details for virtual platform and system-level specification. The growing interest in the development of virtual platforms determined the development of the TLM-2.0 OSCI standard [11]. In terms of time, TLM-2.0 is more abstract than time-clocked approaches, but still needs to assign a time dimension to transactions to enable platform models with more than one master (i.e. multiprocessor platforms) and simultaneous transactions. Although untimed modelling is beyond the scope of TLM-2.0, other works point out that transaction level virtual platforms can also be built as untimed functional models, what is called pure Programmer's View or PV level [12]. Moreover, a productive electronic system level (ESL) design methodology requires system-level specifications at the most abstract untimed level. Several works have taken up this challenge in SystemC, such as SysteMoC [13], SystemC-H [14], and HetSC [15]. The latter is applied by the SATURN approach.

9.3 ARTiSAN Studio®

Artisan Studio® is an all-in-one integrated development tool suite which provides systems and software modeling and component based development targeted for technical systems. It is an ideal tool for complex mission-critical systems and software engineering. Artisan Studio provides comprehensive support for the leading industry standards, including OMG SysML, OMG UML and Architectural Frameworks. Studio delivers an integrated collaborative development environment—allowing systems and software engineering teams together. Artisan Studio is highly scalable and suitable for use on small and large technical projects with a proven multi-user repository providing a stable, robust working environment to ensure both the high availability of model information, while securing all valuable data. Engineering teams using Studio's powerful suite of tools can model systems and software, document legacy systems and generate new code with complete control. Geographical distribution of teams can create a multitude of issues for product development. With Artisan Studio sharing a full model is simplified and allows for control from a common repository, increasing benefits in time and process simplification that is difficult with file based systems. However, one model can still be split into logical packages, by means of configuration management 'sandboxes' to meet the needs of the project and team dynamics.

Artisan Studio® supports SysML and UML methodologies. It is also is powerful and easy to use customization to meet the needs of particular industry, company and individual preferences. Ergonomic profiling can be used to create an industry domain specific profile for the tool, allowing custom views and tools to meet specific needs. Using Artisan Studio® and extending and adding to the UML and SysML profiles project specific information can be captured using an industry standard. Explorer panes, icons and diagrams are included to display specific elements and the user interface is modified to include new context menu items and behavior and the APIs are used to extract model data for analysis.

Artisan Studio® comes with the Automatic Code Synchronizer (ACS) which is an innovative and unique round-trip engineering tool that generates and synchronizes C, C++, C#, Ada, SPARK Ada and Java code. Driven by UML designs within Artisan Studio, the ACS speeds development by automatically generating software system code. It then ensures the UML design and code remain synchronized - immediately ready to support ongoing development, maintenance, enhancement and integration tasks. This design and development integration allows designers, developers and testers to work together more readily. As well as the typical UML class and relationship information, the Automatic Code Synchronizer also uses dynamic information from UML—such as state diagrams—to generate code logic. The resultant code is also instrumented to animate the appropriate diagram within.

Artisan Studio®, when the application is executed either on the host or the target. Code instrumentation lets users interact with the live application using model-level debug capabilities. This simulation approach builds in quality at a very early stage, allowing low cost error correction during design, rather than more costly fixes later in the development process. The ACS technology is user configurable and using the Transformation Development Kit it is possible for users to build their own code generators for domain specific languages such as SystemC.

9.4 SysML Based HW/SW Codesign

As given in Fig. 9.1, the SATURN design flow starts with Artisan Studio® which is customized by SATURN UML profiles for SystemC/C co-modeling. ACS has been configured for code generation for SystemC/C co-simulation. ACS generates SystemC models for simulation as well as interface software for full system mode co-simulation with the QEMU SW emulator. The C code compilation is taken by the generated makefiles. The additionally generated scripts implement the design flow automation like the OS image generation. We additionally support the co-simulation with other simulators like Simulink by means of the EXITE ACE co-simulation environment, e.g., for testbench simulation. After co-simulation we integrated SystemC compilers, currently Agility and SystemCrafter, to generate VHDL which is synthesized for FPGA configuration. The same OS image including the application SW which was executed under QEMU can be finally loaded to the microcontroller of the FPGA for a configuration of the complete system.

9.5 SysML Based HW/SW Co-modeling

For SystemC-based modeling, SATURN defined three UML profiles. i.e., for (i) Synthesizable SystemC, (ii) Synthesis-Specific definition, and (iii) C. The first one is based on SysML and summarized in Table 9.1. It assigns stereotypes with SystemC specific constraints to SysML objects like blocks, parts and ports. Graphical symbols are inherited from SystemC OSCI drawing conventions.

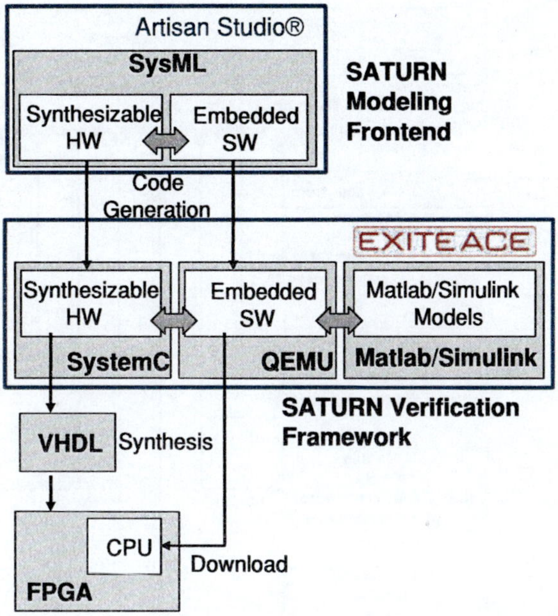

Fig. 9.1 SATURN design flow

Table 9.1 SATURN UML profile for systemC

SystemC	Stereotype	Metaclass	Notation
Module	<<SC Module>>	Class	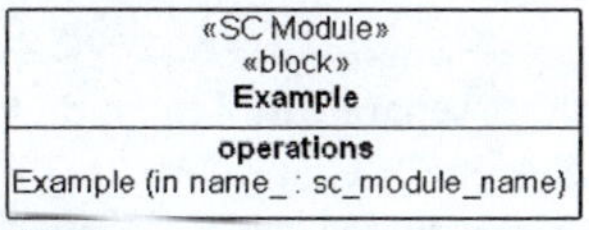
Interface	<<SC Interface>>	Interface	
Port	<<SC Port>>	Port	
Primitive Port	<<SC In ≫/≪SC Out>>	Port	
Prim. Channel	<<SC Signal ≫/≪SC Fifo>>	Property, Connector	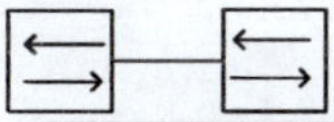
Process	<<SC Method ≫/≪SC Thread>>	Action Node	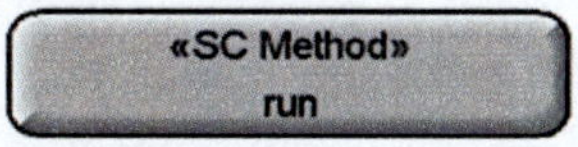
Main	<<SC Main>>	Operation	None
Clock	<<SC Clock>>	Class	

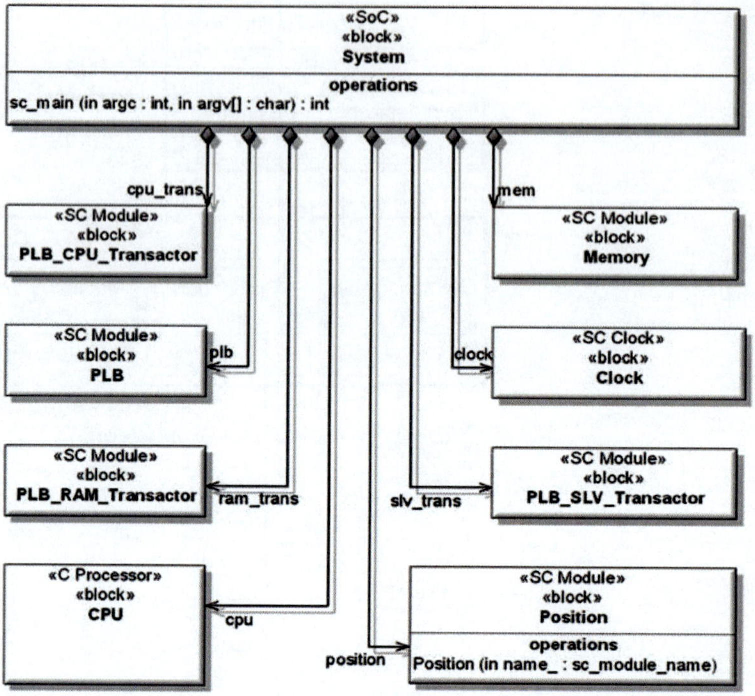

Fig. 9.2 Block definition diagram (BDD)

A separate tool-specific UML profile is implemented for synthesis. For the Agility SystemC Compiler [21], we defined an Agility profile with <<ag_main>>, <<ag_blackbox>>, <<ag_add_ram_port>>, <<ag_constrain_ram>>, <<ag_constrain_port>>, and <<ag_ global_reset>> as stereotypes.

Additional C stereotypes had to be defined for SW integration with SytemC. Here, we simply apply <<Processor>> on SysML blocks to define the CPU target platform by its architecture and operating system. A reference to software executables is introduced by <<Executable>>.

In application we start co-modelling with a SysML Block Definition Diagram (BDD) as shown in Fig. 9.2. The BDD is based on UML structured classes and describes the composition of a whole system. Each block in a BDD represents either a SystemC module (HW) or a Processor component (SW). As the SATURN profile is targeting at synthesizable SystemC code generation, we support the composition of blocks with different multiplicity.

In a next step the system architecture is defined by a set of Internal Block Diagrams (IBDs) which compare to classical SystemC block diagrams as given in Fig. 9.3. Though the content of Figs. 9.2 and 9.3 are both hardly readable they give an idea of principle BDD and IBD structures of a PLB bus example with SystemC and C parts.

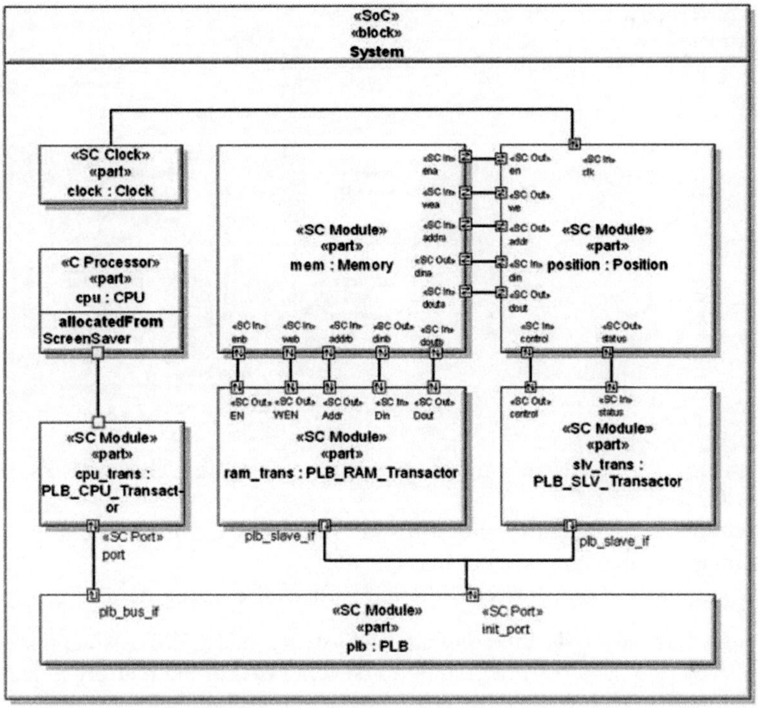

Fig. 9.3 Internal block diagram (IBD)

9.6 SATURN Code Generation

One goal of SATURN is automatic code generation from SysML-based models. For this, we implemented our code generator by means of the ARTiSAN's Transformation Development Kit (TDK). ACS loads UML/SysML Model into the Dynamic Data Repository (DDR), which stores it in an internal representation. A Dynamic Link Library (DLL) generates the code files. Each time the User Model is modified, ACS detects the changes, updates the DDR, and regenerates the code. For reverse engineering, ACS can also be triggered by modifications in the generated code.

9.7 Co-simulation

SATURN comes with a uniform verification framework covering a heterogeneous set of simulators supporting different types of execution platforms:

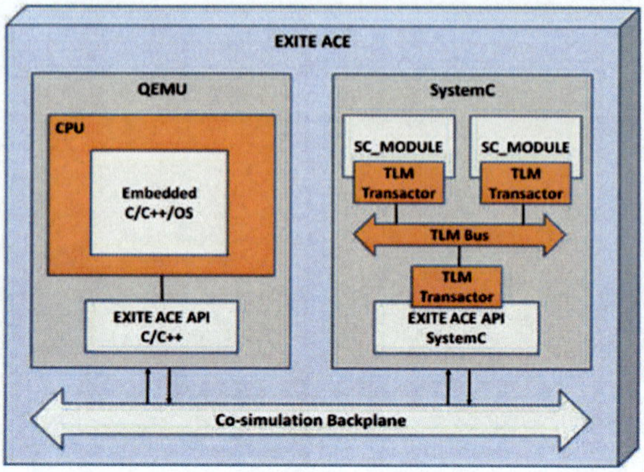

Fig. 9.4 SATURN simulation framework

- Simulators for hardware and test environments like SystemC, Simulink, Dymola,
- Instruction Set simulators and software emulators for full system emulation including the operating system like QEMU,
- Hardware-in-the-Loop for real-time integration of existing hardware.

Though currently only the coupling of SystemC and QEMU is supported, the final goal is the generic integration of several execution platforms including Matlab/Simulink with EXITE ACE as shown in Fig. 9.4

9.8 HetSC for Architecture Exploration

Current embedded systems are composed of an increasing number of components in order to implement a huge amount of different functionalities covering a wide variety of behavioural semantics at different levels of abstraction. In this context, HetSC was defined as a system specification methodology for concurrent *HET*erogeneous embedded systems in *S*ystem*C* [16, 17]. HetSC makes a clear separation between the computation and the communication aspects of the system, supporting the creation of formal executable specifications of the system since HetSC is based on the ForSyDe formal metamodel [18]. Additionally, HetSC provides mechanisms required by specific model of computations (MoCs) what enables the creation of heterogeneous system specifications; each different component that composed the system can be described under different MoCs. The support of heterogeneity facilitates software generation and hardware synthesis since each system components are syntactically and/or semantically closer to the platform entities. Therefore, the application of HetSC in addition to validating the system's behavior provides a link to system implementation [16].

The HetSC methodology is covered by the SATURN methodology in early design stages where the designer identifies the different concurrent entities that composed the system. Once the functional architecture has been decided, the designer can explore different behavioral semantics in order to characterize the system. As such, HetSC supports different approaches to build untimed concurrent models, which significantly reduce simulation time and specification complexity. Additionally, HetSC supports fast and automatic generation of embedded software implementations over different embedded RTOS-based platforms through SWGen [16].

For SystemC model refinement in Artisan Studio®, an additional UML profile for HetSC has been defined. The stereotypes of this profile represent a set of HetSC channels that implements the semantics of different Untimed MoCs. In untimed systems, there is only a partial order and a causality relation among events. In addition, the computation and the communication take an arbitrary and unknown amount of time. The application of the MoCs on the creation of the system specification guarantees specific properties that determine the system behavior, e.g., determinism, protection against deadlocks.

In order to represent processes networks (PNs), HetSC provides two kinds of channels. <<uc_inf_fifo>> represents a channel with an unlimited buffering capacity whereas the channel <<uc_fifo>> has limited this capacity. The first one implements the semantics of the Kahn Process Network MoC and the second one the Bounded Kahn Process Network MoC. In order to create systems with the semantics of the Synchronous Data Flow (SDF) MoC, HetSC provides the channel <<uc_arc_seq>>. Finally, it covers three different rendezvous channels: <<uc_rv_sync>>, <<uc_rv_uni>> and <<uc_rv >> for the Communicating Sequential Process (CSP) MoC. They specify three different cases of communication. The first one implements the synchronization only between asynchronous concurrent elements. The second one implements synchronization with unidirectional data communication and the last one synchronization with bidirectional data transfer.

9.9 Evaluation

The SATURN approach was evaluated in two case studies by two industrial partners:

- An IEEE 802.16e base station for broadband wireless communications undertaken by INTRACOM Telecom
- Smart Camera for Automatic License Plate Recognition (ALPR) undertaken by Thales Security Solutions and Services

9.10 IEEE 802.16e Base Station

For the evaluation of SATRURN we have selected the last part of the base station Tx PHY chain of an IEEE 802.16e system [19]. This consists of the IFFT, the Cyclic Prefix and the Preamble Insertion blocks. The first was mapped to HW and the last two ones to SW.

As a target platform, the Virtex 5 ML510 [20] development board was used, which consists of a dual PowerPC and Xilinx FPGA Virtex-5 XC5VFX130T. Modeling of the system was initially done in Artisan Studio®. The approach that we followed was a top-down modeling of the system:

We started with creating the Block Defintion Diagram (BDD) of the system. The top-level block contained the blocks PPC (software part) and *FPGA* (hardware part). The FPGA block included:

1. The FFT block, which carried out the main bulk of the computations for the transformation as well as the direct communication from/to the Block RAM allocated for this purpose.
2. The Block RAM to transfer data from/to the hardware as well as for the twiddle coefficients used in FFT and various other configuration parameters such as the size of the FFT and the mode (inverse FFT/direct FFT).
3. The PLB (Processor Local Bus) with the respective components (plb_cpu_transactor, plb_ctrl_status_trans-actor) which handles the communication from software to the BRAM through the BRAM.

In a second step, we designed the Internal Block Diagram (IBD) of the top-level block which included HW and SW parts and their communication. Thereafter, the FPGA's IBD was created. The subsystem was composed of:

a) The FFT Peripheral.
b) The dual port Block RAM.
c) The PLB bus and the three transactors for the Block RAM, the FFT and the PowerPC
d) The global clock Clk of the system.

Blocks a), b) and the CPU interface were connected on the PLB through PLB Transactors. The fact that the FFT and the PLB were connected to the two ports of the BRAM allowed us to have a unified memory map with the BRAM address space visible to the CPU as well as access to the BRAM directly from our FFT peripheral. The FFT's two control registers *ctrl* and *status* were also mapped on the memory map of the CPU so they are accessible from software.

In the next step, we created the FFT activity diagram using the SATURN profile. We have used an SC_METHOD sensitive to the *clock* and to the *ctrl* register to model the FFT peripheral.

After modeling the system and the hardware processes, we generated the SystemC code for the FFT peripheral through ACS/TDK in Artisan Studio®. Subsequently, VHDL for the FFT was generated from the respective SystemC code through the Agility Compiler [21].

The code generation was used to produce SystemC simulation traces which enabled us to have an early testing environment. We wrote self-contained test benches in SystemC where we applied ACS/TDK to generate the remaining environment. Thereafter, we ran extensive simulations to make sure that our FFT peripheral worked correctly in all modes. We compared test vectors taken from the real system with the output of the SystemC simulation. In a second step of simulation we used the VHDL which was generated from Agility with manually written test benches in VHDL to ensure the correctness of the SystemC to VHDL translation. After importing the FFT we were able to synthesize and place & route the design through XST and ISE. After tuning of the timing constraints to cater for multi-cycle paths created by Agility, the design was downloaded to the Xilinx Virtex-5 ML510.

On the software side we implemented the state machine in C to control the hardware. Currently, the software application code is written outside Artisan Studio and co-simulation was carried out through EXITE ACE. The SATURN profile generated the drivers to access the registers and the BRAM from the software. The drivers relied on Linux operating system and used the*/dev/mem* node (which represented the whole memory of the system) and *mmap* in order to transfer data from/to the memory mapped BRAM and FFT registers.

Therefore, we had to build a Linux kernel image for the PPC 440 in order to bring-up the system and execute our user space application with the SATURN generated drivers. In order to create a kernel image we needed:

1. A DTS file which contains the memory map for the system that we had created through EDK BSB. In order to generate the DTS file we used the *device tree generator* addon of EDK.
2. A file system containing the user space application built for the PPC440.
3. The standard Linux kernel. After the generation of the file system and the DTS file we were ready to build the Linux kernel.

We finally downloaded the Linux image to the PowerPC and the bit file to the FPGA. After booting Linux we could run the user space application and we could verify that data were provided from SW to the FPGA and correctly passed through the FFT. They were then transferred back to SW and processed through the Cycle Prefix and the Preamble Insertion running on the PPC.

9.11 Smart Camera

In a second case study, we have applied the SATURN tool suite and methodology to design FPGA blocks for an Automatic License Plate Recognition (ALPR) system which is an image-processing technology for license plate identification. This technology is used in various security and traffic applications, such as in the access control system for car parks as given in the following (Fig. 9.5).

The system uses illumination such as Infra-red and a camera to take the image of the front or rear of the vehicle, then image-processing software analyses the images and extracts the plate information. This data is used for enforcement, data collection, and can be used to open a gate if the car is authorized or keep a time record on the entry or exit for automatic payment calculations.

The license plate detection is mainly a sequence of filters designed to detect regions with a high density of small elements which is characteristic of text. For

Fig. 9.5 License plate recognition for airport parking

our studies in the context of SysML, we focused on the implementation of morphological operators which are heavily used in the plate detection stage of the ALPR algorithms covering the image acquisition from a video bus, automatic image contrast correction, automatic exposure time. They are designed as a pixel pipeline which makes computation on the fly and loops back to the camera control.

Like the previous case study, this case study also followed the lines of the SATURN design starting with the SysML tool suite for modeling and cosimulation after code generation. The co-simulation is used for the communication between the main processor and FPGA. In contrast to the previous application, the smart camera is an embedded system composed of a complete board equipped with a PowerPC or ARM and a separate FPGA for the image processing which has some dedicated external memory controlled by the main processor. The Virtex 5 FPGA was configured by Xilinx ISE and the Linux image was separately loaded to the PowerPC.

9.12 Conclusions

In this chapter, we presented the SATURN approach to SysML-based HW/SW co-modeling, -simulation, and -synthesis by means of Artisan Studio®. The two case studies demonstrated the applicability for industrial designs. They also indicated a few subjects for improvements which are planned to be resolved until the end of the SATURN project. We applied Artisan Studio® with the SATURN profiles in replacement of a traditional C++ IDE to write SystemC code. The module composition is defined as a Block Definition Diagram. However, in the first step the definition of members and methods was a little bit slower compared to a traditional IDE as it requires going through a lot of windows pop-ups and tabs. Then, Internal Block Diagrams was used to define the connections of the module's ports. They allow drawing the wiring visually, which is an improvement over the normal coding of connections since it is faster and leads to less errors in connecting ports. Once the system's structure was defined, the behavior and logic had to be coded as usual in a textual editor. Here for efficient capturing, an adequate support of user friendliness, refactoring abilities and code parsing is mandatory for an UML/SysML editor in order to compete with current C++IDEs. The previous benefit of the graphical wiring is mitigated by the time lost for the definition of the behavior without this support as, from our experience, the complexity of behavior and signal timings is higher than structure definition. When iterating through the verification process, most of the changes affect the behavioral logic whereas the module's structures are more stable. The graphical modeling suffers as usual from its lack of compactness and precision compared to textual languages. Although it gave a comprehensive overview of the network-like relations, it partly lacks semantics. One has to explore the graphical editor tabs associated to any item in order to exactly know how it is defined. In contrast, a textual language is quite more synthetic and gives all details. Another important point is that, when VHDL is not directly generated, the SystemC based approach requires an additional SystemC to VHDL compiler. Thus, a modeling environment with integrated SystemC and VHDL code generation would be appreciated to support the existing SystemC-VHDL tool chain.

The major benefit of the SATURN environment is the ability to simulate both the SystemC modules and the target operating system of the embedded platform. This is due to the extension of QEMU to exchange data with the SystemC modules. It shortens the verification time as it allows debugging hardware software interfaces faster than usual: it can be tested by the emulator. Additionally, it allows to easily moving between the different hardware platform and operating system from within the modeling environment to explore different HW/SW alternatives.

In the UML/SysML based approach, a real collateral benefit is having ready to use diagrams of the modules for the documentation. It usually requires drawing them with a generic drawing tool or an UML editor which is not adequately integrated into the design flow. With SATURN tools this time is moved from the specification or post-design phase to the design phase. However, the time saving compared to classical drawing tools depends on the user friendliness and support of the individual UML editor, on the number of times a design is changed, and on how many diagrams are really needed for the documentation. Regardless to this, the main gain here is having always up-to-date and precisely defined diagrams in the documentation.

Acknowledgments The work described herein is supported by the ICT Project SATURN (FP7-216807).

References

1. Kangas T et al (2006) UML-based multiprocessor SoC design framework. ACM Trans Embedded Comput Syst (TECS) 5(2)
2. Object Management Group (2009) A UML profile for MARTE, www.omgmarte.org
3. André C, Mallet F, Mehmood A, de Simone R (2008) Modeling SPIRIT IP-XACT with UML MARTE. In: Proceedings of the DATE workshop on modeling and analysis of real-time and embedded systems with the MARTE UML profile

4. Arpinen T, Salminen E, Hännikäinen M, Hämäläinen TD (2008) Model-driven approach for automatic SPIRIT IP integration. In: Proceedings of 5th international UML-SoC DAC workshop. Anaheim, USA
5. Xie T et al (2009) A UML frontend for IP-XACT-based IP management. In: Proceedings of DATE 2009
6. Riccobene E et al (2009) SystemC/C-based model-driven design for embedded systems. ACM Trans Embedded Comput Syst (TECS) 8(4)
7. Object Management Group (2006) UML profile for system on a chip (SoC). OMG formal/06-08-01
8. OSCI (2009) SystemC synthesizeable subset
9. OSCI (2009) SystemC-AMS. www.systemc-ams.org
10. Gruettner K, Oppenheimer F, Nebel W, Colas-Bigey F (2008) SystemC-based modelling seamless refinement, and synthesis of a JPEG 2000 decoder. In: Proceedings of DATE'08
11. OSCI (2008) OSCI TLM-2.0 Manual
12. Ecker W, Esen V, Hull M (2006) Execution semantics and formalisms for multi-abstraction TLM assertions. In: Proceedings of MEMOCODES'06. Napa, CA
13. Keinert J, Streubhr M, Schlichter T, Falk J, Gladigau J, Haubelt C, Teich J, Meredith M (2009) Systemcodesigner: an automatic ESL synthesis approach by design space exploration and behavioral synthesis for streaming applications. In ACM TODAES 14(1):1–23
14. Patel HD, Shukla SK (2005) Towards a heterogeneous simulation kernel for system-level models a systemc kernel for synchronous data flow models. IEEE Trans CAD Integr Circ Syst 24:N8
15. HetSC homepage: www.teisa.unican.es/HetSC/
16. Herrera F, Villar E (2008) A framework for heterogeneous specification and design of electronic embedded systems in systemC. ACM Trans Des Autom Electron Syst Spec Issue Demonstrable Softw Syst Hardw Platforms 12(3):N22
17. Herrera F, Villar E (2006) A framework for embedded system specification under different models of computation in systemC, Annual ACM IEEE design automation conference proceedings of the 43rd annual conference on design automation
18. Jantsch A (2004) Modeling embedded systems and SoCs. Morgan kaufmann, Elsevier
19. IEEE: IEEE Standard for local and metropolitan area networks, Part 16: Air interface for fixed and mobile broadband wireless access systems
20. Xilinx 2008 ML510 Embedded development platform, user guide, UG356 (v1.1) December 11, 2008
21. Agility SC Compiler. www.msc.rl.ac.uk/euro practice/software/mentor. html

Chapter 10
Mapping Embedded Applications on MPSoCs: The MNEMEE Approach

Christos Baloukas, Lazaros Papadopoulos, Dimitrios Soudris, Sander Stuijk, Olivera Jovanovic, Florian Schmoll, Peter Marwedel, Daniel Cordes, Robert Pyka, Arindam Mallik, Stylianos Mamagkakis, François Capman, Séverin Collet, Nikolaos Mitas and Dimitrios Kritharidis

Abstract As embedded systems are becoming the center of our digital life, system design becomes progressively harder. The integration of multiple features on devices with limited resources requires careful and exhaustive exploration of the design search space in order to efficiently map modern applications to an embedded multi-processor platform. The MNEMEE project addresses this challenge by offering a unique integrated tool flow that performs source-to-source transformations to automatically optimize the original source code and map it on the target platform. The optimizations aim at reducing the number of memory accesses and the required memory storage of both dynamically and statically allocated data. Furthermore, the MNEMEE tool flow parallelizes the application's source code and performs optimal assignment of all data on the memory hierarchy of the target platform. Designers can use the whole flow or a part of it and integrate

---

C. Baloukas (✉) · L. Papadopoulos · D. Soudris
Institute of Communications and Computer Systems (ICCS),
9, Iroon Polytechniou Street, 15773 Athens, Greece
e-mail: cmpalouk@ee.duth.gr

S. Stuijk
Eindhoven University of Technology, Eindhoven, The Netherlands

O. Jovanovic · F. Schmoll · P. Marwedel
Design Automation for Embedded Systems Group, TU-Dortmund, Germany

D. Cordes · R. Pyka
Informatik Centrum Dortmund e.V, Dortmund, Germany

A. Mallik · S. Mamagkakis
IMEC vzw, Leuven, Belgium

F. Capman · S. Collet
Thales Communications France, 160 Boulevard de Valmy, 92704 Colombes, France

N. Mitas · D. Kritharidis
Intracom Telecom, Athens, Greece

N. Voros et al. (eds.), *VLSI 2010 Annual Symposium*,
Lecture Notes in Electrical Engineering 105, DOI: 10.1007/978-94-007-1488-5_10,

it into their own design flow. This work gives an overview of the MNEMEE tool flow. It also presents two industrial case studies that demonstrate how the techniques and tools developed in the MNEMEE project can be integrated into industrial design flows.

10.1 Introduction

In today's embedded systems market there is a trend towards integrating more and more services on mobile devices. These systems combine applications from various domains like communications and multimedia (e.g., HD-video coders, wireless protocols, image processing and 3D games). All these examples make heavy use of data transfers and computation, which makes multi-processor platforms the perfect candidate for their implementation. However, Multiprocessor systems-on-chip (MPSoCs) include complex memory hierarchies and synchronization systems, which in turn makes it difficult to map an application on an MPSoC. For such complex design targets, the design choices will have a high impact on the performance of the system and the success of the device on the market.

There are several challenges that MPSoC designer faces. First, the application should be parallelized to take advantage of the multiple processor system. This is achieved by breaking the execution into several tasks that can be performed in parallel. Second, the statically and dynamically allocated data structures must be optimized to take advantage of the memory hierarchy of the target platform. Third, the final optimized parallel code should be mapped and compiled onto the MPSoC.

In current embedded system design flows, the optimization of the source code is done mainly manually. This process is becoming very tedious and error prone as the complexity of the system and the applications are constantly increasing. This makes it necessary to develop a systematic methodology and tooling that will deal with the aforementioned challenges automatically.

The MNEMEE project [1] addresses the challenge of mapping an application onto an MPSoC platform. Several state-of-the-art source-to-source optimization methodologies and tools have been developed that automatically extract possible parallelization from a target source code, while they also optimize statically and dynamically allocated data structures to reduce the time needed for memory accesses. Additionally, the code is mapped onto the various processors and memories of the MPSoC platform. Each methodology and tool has a very clearly defined interface that allows the whole set of tools developed under the MNEMEE project to be integrated into a common tool flow or used as individual optimization steps. It is this exact feature that renders the MNEMEE tools easy to use in industrial applications, where companies already work with certain tool chains. By integrating the MNEMEE tools into their tool chain, industrial designers can automate their flows. As will be demonstrated in the Sect. 10.3, parts of the tool

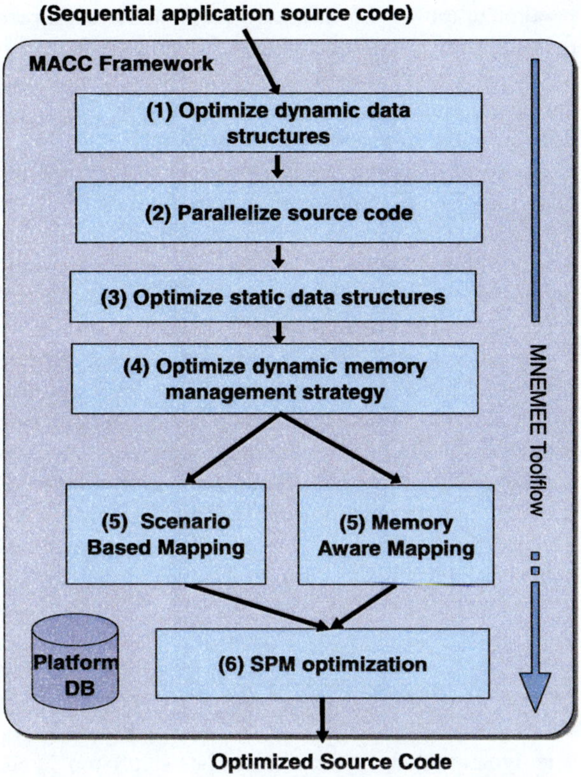

Fig. 10.1 MNEMEE tool flow

flow can be used to replace existing manual or less efficient steps in an industrial design flow. Two examples are provided, one from the communications domain and one from the multimedia domain.

10.2 MNEMEE Tool Flow

10.2.1 Overview

Figure 10.1 shows the MNEMEE tool flow. The input of the tool flow is sequential source code written in C. The output of the flow is source code that is parallelized and optimized for the target MPSoC and its memory hierarchy.

The first step targets the optimization of all dynamically allocated data structures by changing their implementation. These changes may affect the parallelization and therefore they are performed first. Step 2 identifies any potential parallelization in the source code and implements it by breaking the code into several tasks. Step 3 optimizes the statically allocated data structures. Since the parallelization of the original source code has already taken place, the tool can map the statically allocated data structures efficiently onto the memory hierarchy, as their size and behavior is known. The next step, maps the dynamically allocated data structures onto the memory hierarchy. Step 5 maps the parallelized application onto the processors and memories. The MNEMEE tool flow offers two alternative mapping techniques. The first technique called scenario-aware mapping, tries to exploit the dynamic behavior of an application in order to save resources. The second technique, called memory-aware mapping, explore the trade-off between performance and energy consumption of the processors and memories. Finally, step 6 further optimizes the scratchpad allocation of each processor in the target platform.

To combine the large number of required processing steps into a single tool flow, the MACC framework for source level optimization development has been used [21]. This framework exploits the abstract syntax tree code representation provided by the ICD-C compiler development framework. MACC provides a common platform model along with a well-defined interface for integration of analysis and optimization tools. This framework was partially developed in the MNEMEE project [1]. Furthermore, a graphical user interface is provided to enhance the usability of the tool flow.

10.2.2 Dynamic Data Type Optimizations

The dynamic data type optimizations step (step 1) changes the implementation of all dynamic data structures like dynamic arrays, linked lists and trees, based on the Dynamic Data Type Refinement (DDTR) methodology [2, 3]. All dynamic data types (DDTs) are profiled to identify their access patterns and allocation behavior. The optimized implementation is customized to fit this particular access pattern. The objective is to boost performance and restrain the memory consumption. This customization of the data structures is achieved by restructuring these data structures using components from a library of DDTs. As an example, we can consider a list that is accessed sequentially. The use of a memory pointer to store the last accessed element can greatly boost the performance without having to resort to an array solution where we would have to plan for the worst case. The linked list can be restructured adding this memory pointer component.

10.2.3 Parallelization

The parallelization tool [4] is the second one in the MNEMEE tool flow. Its purpose is a fully automatic exploitation of parallelism for the given sequential application's source code, which is provided by the MACC framework. Therefore, the application is transformed in a so called hierarchical task graph, based on the one presented by Girkar in [5]. Once this intermediate representation is extracted from the source code, an ILP based parallelization approach is used to search for parallelism on each hierarchical level of the task graph. Only by introducing hierarchy into the task graph representation it is possible to make usage of very complex ILP formulations. The whole graph is analyzed for possible parallelism by a bottom-up search strategy, which extracts a balanced, parallelized version of the application's source code. Though, the algorithm deliberates about whether it should extract new tasks on the current level of hierarchy or whether it should just use the tasks, extracted earlier deeper in the hierarchy. In addition, the parallelization tool also considers special requirements for embedded systems and is e.g., able to limit the number of extracted concurrently executed tasks to the number of available processor units.

Once the whole graph is processed by the parallelization algorithm, the taken decisions are annotated on the application's source code and a parallelization specification is generated, which complies with the input specifications of the MPMH tool [6]. This tool is then used in step 3 of the tool flow to implement the extracted parallelism.

10.2.4 Optimize Static Data Structures

The MPSoC Parallelization and Memory Hierarchy (MPMH) tool [6] generates parallel source code based on the directives specified by the previous step in the tool flow. This step has annotated the source code with parsections. These parsections specify the sequential code segments that must be parallelized. A directive has been added to each parsection that specifies the number of threads that will execute this parsection in parallel. The parallelization tool (step 2) further indicates whether the work that is done in the parsection must be divided over these threads in terms of functionality (i.e., to split the parsection based on functional parallelism), or in terms of loop iterations (i.e., to split the parsection based on data parallelism), or a combination of both depending on what is the most appropriate for a given parsection. Given the source code and the parallelization directives, the MPMH tool generates a parallel version of the code and insert FIFOs and synchronization mechanisms where needed. Hence, the designer does not need to worry about dependencies between threads. This is taken care of automatically by the tool. This ensures that the tool will always generate correct-by-construction parallel code. MPMH provides the optional mechanism of shared variables. Shared variables need explicit synchronization and communication between threads. The tool will check for possible inconsistency in the synchronization for the shared variables. However it cannot guarantee correct-by-construction parallel code for these shared variables. This option is therefore not used in the MNEMEE tool flow.

The MPMH tool is also capable of performing static data optimizations by optimizing the accesses to static data arrays. The tool uses compile-time application knowledge and profiling data to find out which data copies can be made and whether these copies are beneficial. It also determines how the data and the copies have to be mapped onto the various memory layers in a memory hierarchy such that the energy consumption is minimized and/or the performance is maximized. The result of the optimization is a transformed parallelized application with data copies and block transfers explicitly expressed in the source code, automatic handling of synchronization of data, and a mapping of the data and copies to the various memory layers.

10.2.5 Optimize Dynamic Memory Management

The dynamic memory management methodology is responsible for deciding where a data type or an individual dynamic variable should be placed on a certain memory layer in the memory hierarchy of the targeted MPSoC. The decision is based on the allocation behavior of all dynamic data in the source code. A profiling report reveals the allocation/deallocation timeline of dynamic objects. This report is analyzed to identify the most frequently used objects. These objects are placed closer to the processors in the memory hierarchy, while other data structures are placed in a higher level of the memory hierarchy.

This step does not finalize the mapping of dynamic objects to the memories. Instead it provides the remainder of the flow with hints about the preferred placement. These hints are used to guide the mapping decisions that are made in step 6.

10.2.6 Task Mapping

Two alternative techniques are available in the tool flow to map the parallelized application onto the processors and memories in the MPSoC. The first technique, called scenario-aware mapping, tries to exploit the dynamic behavior of an application in order to save resources. The second technique, called memory-aware mapping, focuses on finding a mapping that minimizes the energy consumption of the system while considering memory requirements of tasks.

10.2.6.1 Scenario-Aware Mapping

Existing mapping techniques (e.g., [7, 8]) model these applications using relatively simple and static models, such as (homogeneous) synchronous dataflow graphs [9]. These models abstract from the dynamic behavior of an application which may lead to a large overestimation of its resource requirements. The dynamic behavior of an application can be taken into account in a mapping technique by using a scenario-based design approach [10]. In this approach, the dynamic behavior of an application is viewed upon as a collection of different behaviors (scenarios) occurring in some arbitrary order, but each scenario by itself is fairly static and predictable in performance and resource usage. Therefore, resource allocation can be performed for each scenario using existing mapping techniques. However, these mapping techniques can only provide timing guarantees per scenario. They cannot guarantee the timing behavior when switching between scenarios. For many streaming applications it is however important that timing guarantees are provided when switching between scenarios. The scenario-aware mapping technique can provide such guarantees.

The input of the scenario-aware mapping step is a set of synchronous dataflow graphs (i.e., one for each scenario). These graphs are automatically derived from the parallelized application source code. The scenario-aware mapping step allocates processing, memory, and communication resources for all these graphs. The output of this mapping step is a set of mappings that provide different trade-offs between the amounts of processing, memory, and communication resources that are used from the MPSoC. (More details on the scenario-aware mapping step can be found in [11]). A run-time mechanism (e.g., [12, 1]) can use this set of mappings to adapt the mapping of an application to the available resource in different use-cases. The development of such a run-time mechanism is also studied in the context of the MNEMEE project. At this moment, the run-time library does however not support run-time configuration. Therefore, the MNEMEE tool flow now selects one of the mappings generated by the scenario-aware mapping step (i.e., the mapping which minimizes the memory usage). This mapping is then used by the last step in the tool flow.

10.2.6.2 Memory-Aware Mapping

The memory-aware mapping tool provides a static assignment of tasks to processors. The focus of the tool lies in the integration of the memory subsystem in the mapping optimization decision. The memory hierarchy and the impact on energy, runtime and communication are often disregarded by other mapping tools. The mapping tool considers the memory requirements of tasks and maps them to available memories in the hierarchy. However, the mapping tool does not decide on the final mapping of memory objects to the memories. It only provides hints to the last step of the flow. This step finalizes the mapping of data objects to memories.

The memory-aware mapping tool is based on the DOL framework [13]. A multi-objective optimization is implemented which balances the load on the processors and communication channels, and minimizes the overall energy of the system. An evolutionary algorithm based on EXPO [14] and PISA [15] performs this optimization. In a first step, several mapping solutions (genes) are generated. In the next step, an energy and performance evaluation is accomplished. The best solutions are stored and used for the generation of new solutions. These last steps are repeated until a maximum number of generations are reached. At the end a set of Pareto optimal solutions is provided. Since the goal of the MNEMEE project is to minimize the energy consumption of the system, the solution with the minimal energy consumption is selected and provided to the next step in the MNEMEE tool flow.

10.2.7 Scratchpad Memory Allocation

The scratchpad memory allocation tool enables a system designer to implement an efficient memory allocation for statically allocated data within a very short time. It exploits scratchpad memories that are known for fast memory accesses consuming little energy. By allocating frequently accessed data objects to these memories the runtime and energy consumption of a system can be reduced significantly. Unfortunately, beside the favored properties scratchpad memories have a small size. Therefore, the tool applies a knapsack-based approach [16] to calculate the set of data objects that, once allocated to scratchpad memories, will lead to the most savings.

By using integer linear programming, the scratchpad memory allocation tool can solve this problem optimally in a very short time. Finally, the tool implements the resulting optimal non-overlaying allocation by applying source-to-source transformations. The allocation at source-code level and the fact that the tool collects all required information about the memory hierarchy from the MACC platform model make the tool very flexible and portable to new platforms. This facilitates an easy integration into tool flows. Another major advantage of the tool is, that it operates fully automated, thus a system designer can reduce runtime and energy consumption at the push of a button.

10.3 Industrial Application

The MNEMEE tool flow can be used to automate existing industrial embedded systems design tool flows, by replacing manual steps in these flows with steps from the MNEMEE flow. This section presents two examples from companies working on different domains, namely the communication and multimedia domain. Both examples demonstrate the integration of the MNEMEE tool flow into their tool flows.

10.3.1 Communication Domain

In the context of the MNEMEE project Intracom Telecom targets the PHY layer of the IEEE 802.16e system for broadband wireless communications, for fixed, nomadic and mobile users. IEEE 802.16e is a broadband wireless solution that enables convergence of mobile and fixed broadband networks through a common

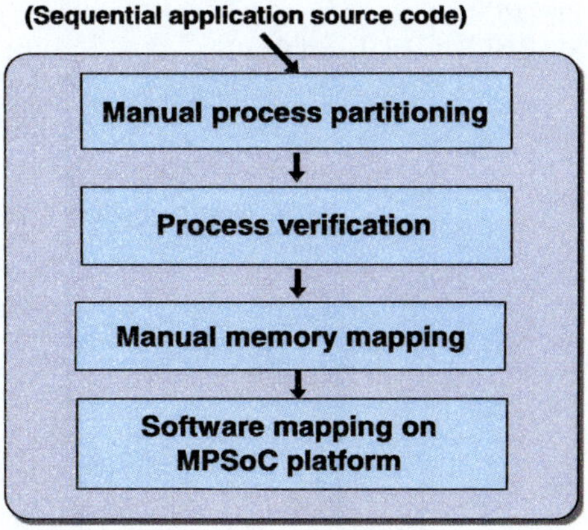

Fig. 10.2 Communications: pre-MNEMEE design flow

wide area broadband radio access technology and flexible network architecture. IEEE 802.16 and WiMAX are designed as a complementary technology to Wi-Fi and Bluetooth.

The target platform is the Freescale MSC8144 [17], a high-performance multicore DSP device. It includes four extended cores, each one comprising the DSP core with dedicated instruction cache, data cache, memory management unit (MMU), interrupt controller (EPIC) and timers. The complexity of the platform demands large design effort to map a streaming application like 802.16e. The preMNEMEE design approach from Intracom can be seen in Fig. 10.2. The application's source code has to be manually partitioned to take advantage of the four DSP cores. Then the processes should be verified so that the partitioning does not break any timing constraints. Finally, manual memory mapping of data objects needs to be performed.

The 802.16e PHY layer incorporates a wide spectrum of DSP algorithms. The following blocks from the 802.16e PHY layer were implemented:

1. Randomization
2. Interleaving
3. FEC encoding
4. Constellation Mapping
5. Burst Mapping–Frame Construction
6. Space–Time Coding
7. Fourier Transformation
8. Cycle prefix insertion
9. Preamble insertion

The above DSP algorithms are inherently sequential. As such they do not provide significant room for improvements within the design space of each processing block. The architecture of the system though is such that parallelization opportunities exist on a coarser level. The MNEMEE tools were able to automatically identify such opportunities and give parallelization directives on this higher level of abstraction. The MNEMEE tools were able to make use of all 4

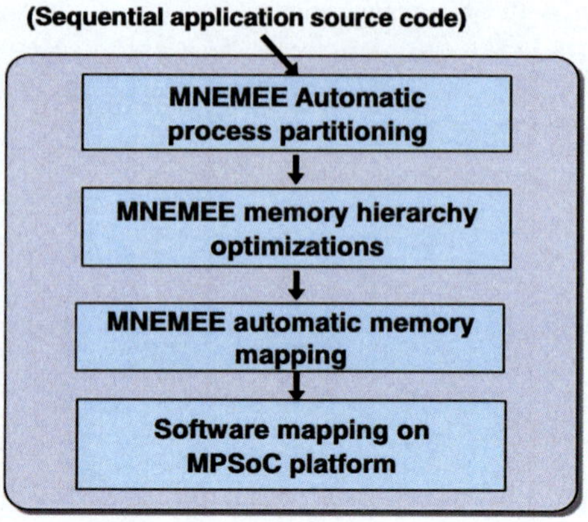

Fig. 10.3 Communications: MNEMEE based design flow

cores by spawning 3 threads (+1 master thread) for each of the 2 distinct parallel sections. The usage of the parallelization tools did not require any user interaction making the tools even more useful in situations where the code is provided from 3rd parties. Nevertheless if the application requires higher speed-up, the user can assist the tools and specify more sections of the code that could be parallelized. The MNEMEE tools were able to produce results with 1 more parallel section in addition to the 2 identified automatically from the tools.

The mapping of data objects is also performed automatically when using the MNEMEE tools. This further reduces the design effort. The MNEMEE based design flow is shown in Fig. 10.3. The time taken on the traditional flow that has been followed prior to MNEMEE can be broken down to:

1. Identifying objects that are good candidates to be moved to faster memories.
2. Moving objects from where they are originally instantiated to a separate header file.
3. Including header files from source files where the objects are used.
4. Writing pragmas on the header files to be mapped to special segments.
5. Writing linker scripts to allocate memory for the extra segments
6. Test the application to ensure functional correctness.
7. Retry with different candidates (step 1).

For a sequential application, the procedure outlined takes around 2 days of manual labour. With the usage of the MNEMEE tools equal or better quality results have been obtained within 4 h.

As the quality of the results for this step depends on the optimal selection of the objects, the above iteration has to be done several times in order to identify the trade-offs for each object and select the appropriate ones afterwards. Thus the majority of the time is spent in the iteration of the mentioned above sequential flow steps. Furthermore, debugging when manually managing the memories is error-prone and results in increased debugging effort. The MNEMEE tools avoid this problem. When using the MNEMEE tools, the process outlined above can be broken down into the following steps:

1. Running the tools over the code.
2. Writing pragmas for the header files.
3. Writing linker scripts to allocate memory for the extra segments.
4. Test the application to ensure functional correctness.

The MNEMEE tools automate the majority of the process required to map data objects of the applications into the memories. The output they provide can be directly fed back into the development tools used at Intracom. The debugging required is also minimized as there are no bugs inserted by moving the objects to different memories as this is now done automatically by the MNEMEE tools.

The above has been done for the sequential application and parallelized application where the objects can also be shared between threads. This requires an even more complicated manual object selection, which could successfully be automated using the MNEMEE tool flow. Therefore, usage of the MNEMEE tools was considered an efficient solution as it considers both performance and design time.

10.3.2 Multimedia Domain

The multimedia application is a state of the art low bit rate speech coder based on the enhanced Mixed Excitation Linear Predictive (MELPe) algorithm (NATO standard STANAG 4591 [18, 19]). The target platform is the OMAP-L137 [20], a dual core low-power application processor comprising an ARM92EJ-S and a C674x floating point DSP core. Both processors have their own instruction and data cache, but also share some internal and external memory.

Before integrating the MNEMEE tool flow in the design flow, the required steps were those depicted on Fig. 10.4. The source code must be split manually between the two cores, followed by a series of manual optimizations and code rewriting to adapt the code to the target processors. Then code and data are mapped to memories. To do the mapping, first the cache is activated, if then the application doesn't meet timing requirements, critical parts of code and data are identified by profiling the application. Critical data that are most often used are mapped into the internal memory. Critical functions that do not meet the necessary constraints are rewritten in assembly code. This process iterates until timing and memory requirements are met.

Integrating the multimedia application on the OMAP-L137 platform led us to face two main challenges: (1) how to efficiently split the code between the two processors and (2) how to make the most of the four available memory levels. Manually addressing these challenges will require a lot of design and integration efforts. The different optimization tasks depicted in Fig. 10.4 have an impact on each other thus leading to an iterative and error prone optimization process. Moreover the optimal solution could hardly be obtained manually. Using the MNEMEE tool flow, every step is performed and linked to the next in an automated way. This new design flow is given in Fig. 10.5.

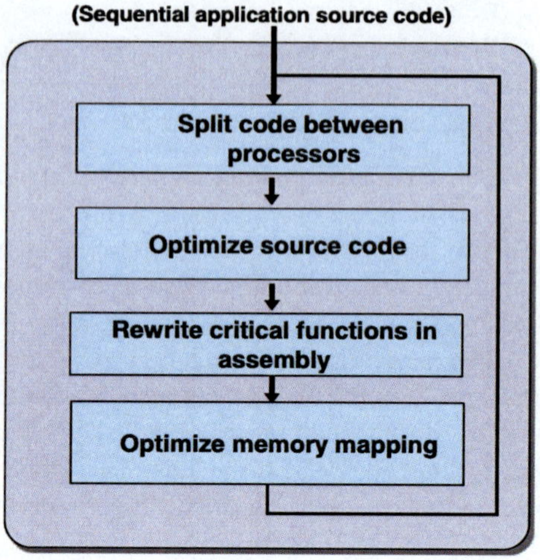

Fig. 10.4 Multimedia: pre-MNEMEE design flow

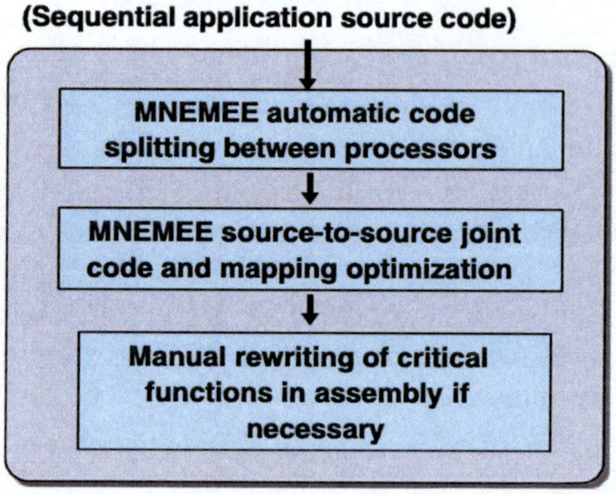

Fig. 10.5 Multimedia: MNEMEE based design flow

Using the MNEMEE parallelization tools, the application is split between the two processors. Although the MELPe algorithm is mostly sequential, the Parallelization tools successfully found some low-grain (data-level) parallelism. In addition, a high-grain (function-level) parallelism has been manually specified for the MPMH tool. The latter is based on the designers' knowledge of the application. (This shows how the MNEMEE tool flow can effectively be used to assist designers when developing a new system.) This first step results in a source code containing several threads to be run in parallel as well as the mechanisms to handle threads (creation, data exchange ...) and lasted about a day—one day to specify function-level parallelism and around 10 min to execute the parallelization tools.

On an OMAP like platform, the GPP usually manages the application and handle I/O; while the DSP does most of the processing. This leads to an unbalanced load between the processors. The mapping tools from the MNEMEE toolflow allow going beyond this typical partitioning.

First, the Scenario Based Mapping tool has been used. As it required precise information about the architecture (memories, buses width and speed …) and the application (scenarios being used, execution time and memory use for each task of the scenarios); about a week has been spent to gather these parameters and generate the inputs for the tool then 5 min to execute the tool. The output mappings seem promising as they do share the overall load between the processors for all scenarios.

The Memory Based Mapping tool has not been used at the moment of writing, but we expect it to take the most of the four available memory levels to have an optimal task mapping and a decrease in the energy consumption.

Using the Dynamic Memory related tools required some modification of the input source code, which has yet not been realized at the moment of writing. However, several parts of code are using some small temporary dynamic buffers which management may be optimized. We expect the Dynamic Memory tools to reduce the overall amount of Heap required which may then be mapped into internal memory thus reducing the memory footprint and the execution time.

As no mapped parallel source code has yet been obtained, the Scratchpad Memory tool has been used only on the sequential source code. Only 2 min are required to run the tools as not previous step is required. The mapped application is 33% faster and Stack memory has been reduced by 82%. However the overall memory footprint for data has grown by 21%. This growth in footprint is due to a lack of dynamic management of static data, which is done by MPMH.

The main advantages of using the MNEMEE tool flow are that each step is optimized in two ways: (1) the tools can generate and evaluate several alternative solutions; and (2) the tools can work simultaneously on multiple optimization targets—execution time, energy consumption or memory footprint—to find out the best solution. Furthermore, as most of the tools provide source-to-source optimization, the output of each tool can be viewed and analysed. Of course, some parts of the source code may still require manual optimization, but it should be far less than with a full manual optimisation process.

It should also be noted that if the application needs to be modified or completed with additional functionalities, the use of an automated tool flow such as the MNEMEE tool flow will greatly facilitate the re-factoring process.

10.4 Conclusions

The complexity of novel embedded systems is increasing rapidly. These systems combine many different streaming applications in a single system. To meet the processing and memory requirements of these applications, multiprocessors systems-on-chip with a memory hierarchy must be used. The complexity of these architectures and applications make the design of these systems very challenging. A decreasing time-to-market and the need to differentiate products add to this design challenge. Manual optimization and mapping of the application source code is becoming prohibitively slow. Therefore, a structured methodology and automated tools are needed to map the application source code onto the target hardware. The MNEMEE project provides a set of techniques that fills the need for automation in the respective industry's design flows. In this work, the MNEMEE techniques have been presented, along with two real-world industrial examples, demonstrating the applicability and the gains that are possible by their exploitation.

Acknowledgments This work was supported in part by the EC through FP7 IST project 216224, MNEMEE.

References

1. MNEMEE Project (IST-216224)–http://www.mnemee.org/
2. Baloukas C, Risco-Martin JL, Atienza D, Poucet C, Papadopoulos L, Mamagkakis S, Soudris D, Ignacio Hidalgo J, Catthoor F, Lanchares J (2009) Optimization methodology of dynamic data structures based on genetic algorithms for multimedia embedded systems. J Syst Softw 82(4):590–602
3. Young M (1989) The technical writer's handbook. University Science, Mill Valley, CA
4. Cordes D, Marwedel P, Mallik A (2010) Automatic parallelization of embedded software using hierarchical task graphs and integer linear programming, CODES+ISSS'2010, October 2010
5. Girkar M, Polychronopoulos CD (1994) The hierarchical task graph as a universal intermediate representation. Int J Parallel Program 22(5):519–551
6. Iosifidis Y, Mallik A, Mamagkakis S, De Greef E, Bartzas A, Soudris D, Catthoor F (2010) A framework for automatic parallelization, static and dynamic memory optimization in MPSoC platforms. In the design automation conference, DAC 10, Proceedings (June 2010), ACM
7. Moreira O, Mol J-D, Belooij M, van Meerbergen J (2005) Multiprocessor resource allocation for hard-real-time streaming with a dynamic job-mix. In: 11th real time and embedded technology and applications symposium, RTAS 05, Proceedings (March 2005), IEEE, pp 332–341
8. Stuijk S, Basten T, Geilen M, Corporaal H (2007) Multiprocessor resource allocation for throughput-constrained synchronous dataflow graphs. In: 44th design automation conference, DAC 07, Proceedings (June 2007), ACM, p 777–782
9. Lee E, Messerschmitt D (1987) Static scheduling of synchronous data flow programs for digital signal processing. IEEE Trans Comput 36(1):24–35
10. Gheorghita S, Palkovic M, Hamers J, van de Cappelle A, Mamagkakis S, Basten T, Eeckhout L, Corporaal H, Catthoor F, van de Putte F, Bosschere KD (2009) Systemscenario-based design of dynamic embedded systems. ACM Trans Des Autom Electron Syst 14(1):1–45
11. Stuijk S, Geilen MCW, Basten T (2010) A predictable multiprocessor design flow for streaming applications with dynamic behaviour. In: digital system design, 13th euromicro conference, DSD 10 Proceedings (September 2010), IEEE, pp 548–555
12. Shojaei H, Ghamarian A, Basten T, Geilen M, Stuijk S, Hoes R (2009) A parameterized compositional multi-dimensional multiple-choice knapsack heuristic for CMP run-time management. In: 46th design automation conference, DAC 09, Proceedings (June 2009), ACM, pp 917–922
13. Thiele L, Bacivarov I, Haid W, Huang K (2007) Mapping applications to tiled multiprocessor embedded systems. In: application of concurrency to system design, ACSD 07, Proceedings (July 2007), IEEE, 2007
14. Thiele L, Chakraborty S, Gries M, Künzli S (2002) A framework for evaluating design tradeoffs in packet processing architectures. In: 39th annual design automation conference, DAC 02, Proceedings (June 2002), ACM, pp 880–885

15. Bleuler S, Laumanns M, Thiele L, Zitzler E (2003) PISA—A platform and programming language independent interface for search algorithms. In: Fonseca CM, Fleming PJ, Zitzler E, Deb K, Thiele L (eds) Evolutionary multi-criterion optimization (EMO 2003), vol 2632/2003 of LNCS. Springer, Heidelberg, pp 494–508
16. Steinke S, Wehmeyer L, Lee B, Marwedel P (2002) Assigning program and data objects to scratchpad for energy reduction. In: design, automation and test in europe, DATE 02, Proceedings, IEEE, 2002 p 409
17. MSC8144 Reference Manual, http://www.freescale.com/files/dsp/doc/ref_manual/MSC8144RM.pdf
18. The 600 bits/s, 1200 bits/s and 2400 bits/s NATO interoperable narrow band voice coder, NATO standard STANAG No. 4591 edition Y (amendment W) Ratification Draft 1, Jan 2006
19. Guilmin G, Capman F, Ravera B, Chartier F (2006) New NATO STANAG narrow band voice coder at 600 bits/s. Proceedings of IEEE international conference on acoustics, speech, and signal processing, Toulouse, May 2006
20. OMAP-L137 Low-power applications processor, sprs563c, November 2009
21. Pyka R, Klein F, Marwedel P, Mamagkakis S (2010) Versatile system-level memory-aware platform description approach for embedded MPSoCs, LCTES 2010, April 2010

Chapter 11
The MOSART Mapping Optimization for Multi-Core ARchiTectures

Bernard Candaele, Sylvain Aguirre, Michel Sarlotte, Iraklis Anagnostopoulos, Sotirios Xydis, Alexandros Bartzas, Dimitris Bekiaris, Dimitrios Soudris, Zhonghai Lu, Xiaowen Chen, Jean-Michel Chabloz, Ahmed Hemani, Axel Jantsch, Geert Vanmeerbeeck, Jari Kreku, Kari Tiensyrja, Fragkiskos Ieromnimon, Dimitrios Kritharidis, Andreas Wiefrink, Bart Vanthournout and Philippe Martin

Abstract MOSART project addresses two main challenges of prevailing architectures: (1) The global interconnect and memory bottleneck due to a single, globally shared memory with high access times and power consumption; (2) The difficulties in programming heterogeneous, multi-core platforms MOSART aims to overcome these through a multi-core architecture with distributed memory organization, a Network-on-Chip (NoC) communication backbone and

---

B. Candaele · S. Aguirre · M. Sarlotte
THALES Communications, Colombes Cedex, France

I. Anagnostopoulos (✉) · S. Xydis · A. Bartzas · D. Bekiaris · D. Soudris
Institute of Communications and Computer Systems, Athens, Greece
e-mail: iraklis@microlab.ntua.gr

Z. Lu · X. Chen · J. M. Chabloz · A. Hemani · A. Jantsch
Royal Institute of Technology-KTH, Stockholm, Sweden

G. Vanmeerbeeck
IMEC, Interuniversity Micro-electronics Center, Leuven, Belgium

J. Kreku · K. Tiensyrja
VTT Communication Platforms, Oulu, Finland

F. Ieromnimon · D. Kritharidis
INTRACOM S.A. Telecom Solutions, Peania, Greece

A. Wiefrink · B. Vanthournout
SYNOPSYS, Leuven, Belgium

P. Martin
ARTERIS, Guyancourt Cedex, France

I. Anagnostopoulos
Microprocessors and Digital Systems Lab, School of Electrical and Computer Engineering, National Technical University of Athens (NTUA),
9 Heroon Polytechneiou, 15780 Athens, Greece

N. Voros et al. (eds.), *VLSI 2010 Annual Symposium*,
Lecture Notes in Electrical Engineering 105, DOI: 10.1007/978-94-007-1488-5_11,

configurable processing cores that are scaled, optimized and customized together to achieve diverse energy, performance, cost and size requirements of different classes of applications. MOSART achieves this by: (1) Providing platform support for management of abstract data structures including middleware services and a run-time data manager for NoC based communication infrastructure; (2) Developing tool support for parallelizing and mapping applications on the multi-core target platform and customizing the processing cores for the application.

11.1 Introduction and Motivation

The widening gap between power and performance requirements of applications and what is afforded by technology scaling and architectural techniques clearly points to multi-processor architectures as the solution. As an example, even the present day wireless standard 802.11a requires more than 5 GIPs (IST—Project E2R) of conventional DSP processing for its physical layer. The challenge going forward is to be able to sustain several applications that are at least an order of magnitude more demanding than the 802.11a/n/m.

Memory dominates the cost, power and performance of heterogeneous multi-processor architectures. The need for large amount of storage and a high bandwidth access to it comes from two ends. The primary need comes from the applications becoming more complex and data intensive (high resolution, higher bandwidth communication etc). The secondary need comes from the requirement to hide the latency of accessing slower off chip memory. To comprehensively optimize both the aspects, the challenge is to treat the memory question at system level where decisions are made about how to map complex and abstract data structures to efficient distributed memory hierarchy and provide runtime support for memory management and scheduling.

To address the memory and interconnect challenges, MOSART has developed a distributed memory architecture that is tightly integrated with a Network-on-Chip (NoC) interconnect backbone. Physically and architecturally NoC is an enabling technology that addresses the memory and interconnect challenge. Such NoC based distributed architecture enables arbitrary communication pattern among applications and also significantly lowers the interconnect latency, memory latency and energy requirement for accessing data. Developing appropriate design methods and tools, we have explored within affordable time budgets, various NoC interconnection topologies and multi-layer memory structures resulting into high performance and low energy NoC architecture.

To effectively utilize the distributed architecture and make the development cycle more modular, MOSART has developed middleware services for memory management for runtime data allocation and access scheduling. This middleware provides an abstract data type library offering optimized data types to the applications running on the platform. Additionally, a run-time data allocator is in charge of the data allocation over the distributed memory of the NoC platform.

Present in the middleware are APIs that interface to the data transfer services (e.g., block transfers over the communication infrastructure).

Key characteristics of the developed architecture and the methodology are flexibility, scalability and modularity. The flexibility comes from a library of system level building blocks, both functional and infra-structural. The scalability comes from the ability to logically combine resources for increased performance, storage and/or bandwidth. The modularity comes from the way the building blocks are architected and harnessed at the chip level and how the design methodology models and abstracts them.

MOSART attempts to solve some of the most vexing problems facing the SoC architectural and design community like (a) design productivity; (b) computational power; (c) low power; (d) domination of memory in terms of power and performance; (e) global interconnect latency; (f) bus scalability and (g) managing arbitrary concurrency. In MOSART, we use NoC with distributed memory architecture to provide us with scalability so that we can tune and customize the computational power, the interconnect bandwidth and the storage to the needs of applications at hand. MOSART is also developing a methodology to support the deployment for real life applications.

To summarize, the technical objectives are: (a) to develop a multi-core architecture with distributed memory organization, a NoC communication backbone and configurable processing cores that are scaled, optimized and customized together to achieve diverse energy, performance, cost and size requirements of different classes of applications; (b) to provide platform support for management of abstract data structures including middleware services and a run-time data manager for NoC based communication infrastructure; (c) to develop tool support for parallelizing and mapping applications on the multicore target platform and customizing the processing cores for the application, and (d) to validate and evaluate the architecture and tool support using applications from future high data rate wireless access.

11.2 Project Description

The MOSART project has developed a flexible modular multi-core on-chip platform architecture and associated exploration design methods and tools. The overall system level methodology by MOSART is depicted in Fig. 11.1. The methodology steps are (a) Applications and Performance Requirements (Sect. 11.2.1); (b) Parallelization and System-Level exploration (Sect. 11.2.2); (c) NoC Customization (Sect. 11.2.3) and (d) ASIP exploration (Sect. 11.2.4).

11.2.1 Applications and Performance Requirements

The credibility of the MOSART approach is demonstrated by means of illustrative applications that demonstrate a high degree of usability for the existing design base. Two such applications have been chosen for the purposes of

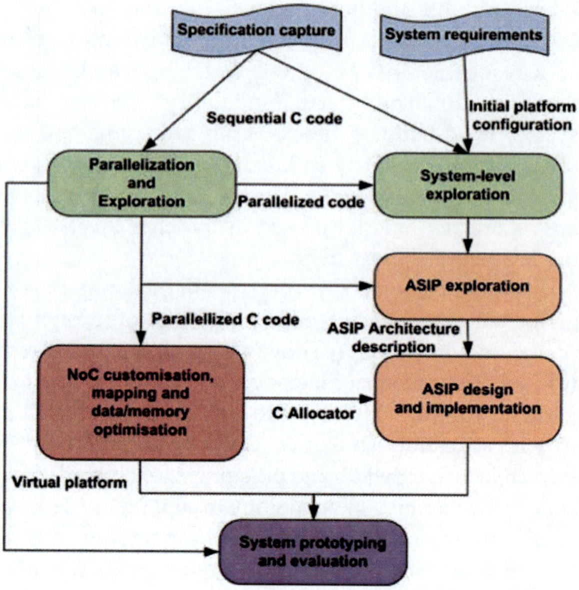

Fig. 11.1 MOSART framework overview

validation/evaluation. The first one is the implementation of a part of the cognitive radio application on the MOSART platform. Cognitive radio is a new concept, employed in order to optimize the frequency band usage. It will be integrated into the next generation of post-SDR wireless terminal. This test case has already been used to demonstrate the interest of the ASIP approach in a first step, that is followed by the implementation of porting and execution of the parallelized code on a combination of multi-core and multi-ASIP architecture.

The second is an implementation on the MOSART platform of selected parts of the PHY layer of an experimental prototype of an IEEE 802.16e based broadband wireless system. The 802.16e standard has been defined to support broadband mobile connectivity in urban environments. The standard places heavier processing requirements than the earlier fixed WIMAX standard of 802.16d, coupled with the ever-present need for low-power mobile terminals. The chosen application subset has been coded in C, and gone through the steps of parallelization. This step provides the necessary profiling information that will guide the ASIP exploration phase.

11.2.2 Parallelization and System-Level Exploration

Extraction of parallelism from the sequential model of applications is conventionally used by algorithm developers. The MPSoC Parallelization Assist (MPA) tool [1] analyzes the application and generates parallel source code based on directives specified by the designer. The general idea of parallelization with MPA is that the designer identifies parts of the (sequential) code as candidates for parallelization. Then via a separate directives file, he decides how to introduce parallelism into to code. This could be as a functional pipeline, distributed loop-iterations or a mixture of both. Given the input code and the parallelization directives, the tool will generate functionally correct parallelized code. This is achieved by doing lifetime analysis of all variables and automatic insertion of communication queues and/or synchronization points for variables that are used over multiple parallel threads. The MPA tool also comes with a simulator that allows not only functional verification, but also allows for time-annotation of the software. This way a first order approximation can be obtained for the performance gain of the parallelization vs the sequential execution.

The performance modeling and analysis approach is achieved with ABSOLUT [2] that is a model-based approach for system-level design. This approach takes service orientation into focus, and the execution platforms are modeled in terms of services provided (ASIPs, memories, interconnect, etc.). The layered hierarchical workload models represent the computation and communication loads the applications cause on the platform when executed. The layered hierarchical platform models represent the computation and communication capacities the platform offers to the applications. The workload models are mapped onto the platform models and the resulting system model is simulated at transaction-level to obtain performance data. The approach enables performance evaluation early, exhibits light modeling effort, allows fast exploration iteration, reuses application and platform models, and provides performance results that are accurate enough for system-level exploration.

11.2.3 NoC Customization

MOSART has developed new technologies for future MPSoC based upon Network on Chip and distributed memory and computing cores for multimedia and wireless communications. In our McNoC, memories are distributed but shared among network nodes. An example is shown in Fig. 11.2. The system is composed of 16 Processor-Memory(PM) nodes interconnected via a packet-switched mesh network. A node can also be a memory node without a processor, pure logic or an interface node to off-chip memory. As shown in Fig. 11.2, each PM node contains a processor, for example, a LEON3, hardware modules connected to the local bus, and a local memory. The key module, which we introduce as an engine for memory and data management, is the DME, able to simultaneously serve various requests from the local core and the remote ones via the network. A Data Management Engine (DME) [3] has been designed and implemented to handle all on-chip memory and data management tasks for a distributed shared memory architecture. A set of data management methodologies for future McNoC platforms is proposed too. The first methodology that is developed is the abstract data type optimization (ADT). Employing

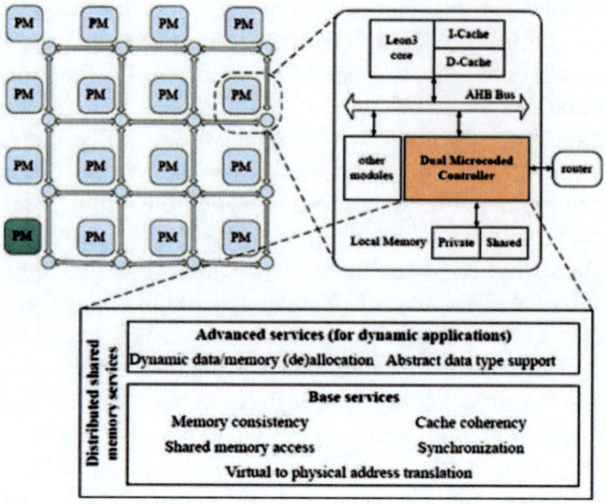

Fig. 11.2 A 16-node Mesh McNoC; Processor-Memory (PM) node and supported services

this technique, the designers will be able to change the way the dynamic data of applications are stored and accessed (MTh-DMM). Also, the mapping of abstract data types to a distributed memory architecture is managed by the runtime memory management.

A novel asynchronous communication scheme (GRLS = Globally Ratiochronous-Locally Synchronous) [4] has been developed. The GRLS paradigm is based on the observation that in SoCs all on-chip clocks are normally derived from the same master clock. The GRLS paradigm constrains all local frequencies to be rationally related, and uses clock dividers for the generation of the local frequencies. The asynchronous communication problem is inherently more complex compared to the ratiochronous counterpart, and we used the periodic properties of rationally-related systems to build efficient latency-insensitive communication interfaces, allowing maximum throughput, low latency and low overhead, coupled with low complexity and high flexibility. We have shown how GRLS communication does not require handshake and has overhead and performance figures which are close to those of mesochronous interfaces, while keeping a flexibility close to that of GALS. We used the GRLS paradigm as the basis for the MOSART power management scheme, which partitions the SoC into different clock regions, which can be optimized independently from each other by means of Dynamic Voltage and Frequency Scaling (DVFS). Voltage Scaling is realized using a quantized approach, in which multiple supply voltages are distributed throughout the chip and the regions can dynamically select which voltage to use for power supply. We have developed fully programmable Power Management Units to manage power services in the platform. The Power management Units allow to dynamically change the frequency and the supply voltage of any region and offers clock gating and shutoff services. Dynamic reconfiguration of the GRLS regions is also supported.

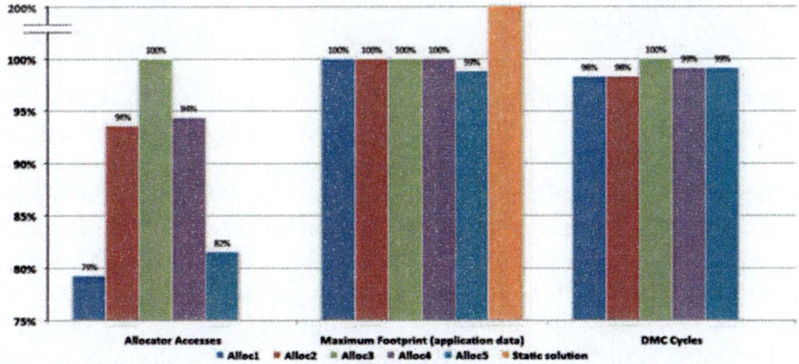

Fig. 11.3 Description and comparison of the different memory allocators

11.2.4 ASIP Exploration

The ASIP design space can be very complex, and the performance estimations become very late in the design process in traditional approaches. The amount of manual work is considerable and the design cycle takes so much time that the exploration of the ASIP architecture design space remains very weak. The proposed methodology and prototype tool is according to our knowledge the first attempt to raise the ASIP design abstraction level above a standalone ASIP. Adding the ASIP architecture exploration to front of an existing ASIP design flow will allow for finding a good architecture for the actual design of an ASIP. It facilitates evaluation of the ASIP performance early in the design process which results in a more systematic approach, increase automation and allow exploration of larger ASIP design space. The method and tool gives estimates of number of registers, number and types of functional units, number of pipeline stages and the instruction set of the ASIP core that would best satisfy the computational requirements of the types of algorithms it is targeted for. From the set of core models in the design space, the approach finds the most optimal for the given algorithm Fig. 11.3.

11.3 Experimental Results

In this section, examples of MOSART's are presented. According to the aforementioned methodology we show the results in the field of (a) Parallelization; (b) System level exploration; (c) Supporting distributed shared memory services and (d) ASIP exploration.

11.3.1 Parallelization

The aforementioned application of selected parts of the PHY layer of an experimental prototype of an IEEE 802.16e based broadband wireless system (Sect. 11.2.1) was used as input to the parallelization tool. The first step towards code mapping was the modification of the C sources, so that the coding style is conformant with MPA syntax and semantics requirements. Once the C source was cleaned-up, the sequential code was executed on a conventional PC platform to verify that the application functionality had been preserved. It was then annotated with standard C syntax labels, to facilitate parallelism extraction from the MPA tool and conversion of the original sequential code into a multi-threaded version. Instrumentation code had also been added by the MPA suite, to facilitate gathering of vital program statistics which are displayed in textual and visual form once the modified code is compiled and run onto the targeted MPSoC platform. The parallelization and optimization process was thus guided towards production of multi-threaded code, matching the capabilities of the multi-core platform.

During the learning phase of MPA use, trivial parallelization scenarios were run, where a single thread executes all functionality of the labeled code segments. Subsequently, more elaborate parallelization scenarios were tried, initially extracting the "easy" parallelism that is suggested by the application code outline: the iFFT/FFT blocks were assigned to individual threads, with the rest of code functionality assigned to a couple of additional threads. During the process of code parallelization, additional "unsafe" code features were identified and removed from the original sequential code, that is always the starting point of the exploration effort. Issues such as arrays of structures and inconsistent use of declared multi-dimensional arrays, which are forbidden by MPA although allowed by C semantics and able to go through production compilers such as gcc, were removed from the code. All such transformations of the sequential original sources were validated by runs on the PC platform.

11.3.2 System Level Exploration

The JPEG encoder was used to experiment and validate the GCC compiler-based workload generation tool in the context that takes MPA-parallelized source codes, creates respective workload models and maps them on the ABSOLUT platform model for transaction-level performance simulation in SystemC. The parallel versions of the JPEG encoder were created with the MPA tool. We generated four sets of workload models from the encoder. The first one was from the unmodified sequential application and the other three from parallelized versions of the application: Par-1 had two threads with the second thread executing Getblock and DCT algorithms. Par-2 consisted of three threads with the second and third one interleaving the execution of Getblock, DCT, and Quantization. Par-3 had also

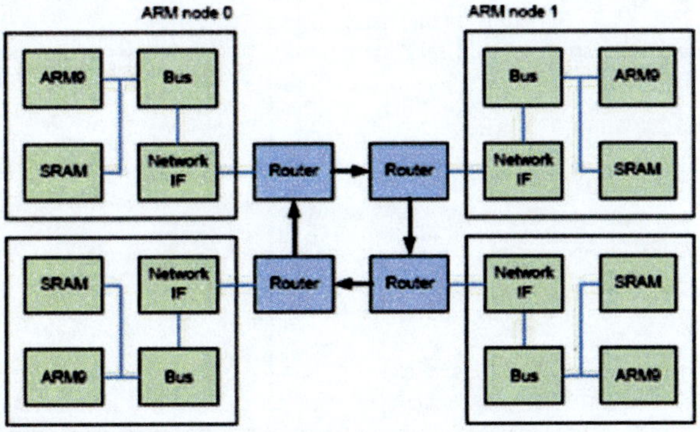

Fig. 11.4 Example platform consisting of four ARM nodes

three threads with the second and third thread executing just Getblock and DCT in an interleaved manner.

The execution platform model for the performance simulation of the JPEG application is depicted in Fig. 11.4. It consists of 4 ARM nodes connected by routers, which form a ring-like network. Each node has an ARM9 CPU, some local SRAM memory, a shared bus, and an interface to the other nodes. The accuracy of the ABSOLUT simulation approach has been evaluated with several case examples in [5, 6].

According to [6] both Par-1 and Par-3 have 100% utilization on the cpu of the ARM node 0. Par-1 has 44% cpu utilization in the second ARM node, whereas Par-3 has 21% utilization across nodes 1 and 2. Par-2 has 88% utilisation in the first node: it is idling at some point of simulation while waiting data from the other two threads. Since Par-2 has a shorter execution time and more work for nodes 1 and 2, the cpu utilisation in those nodes is considerably higher at 53%.

11.3.3 Supporting Distributed Shared Memory Services

11.3.3.1 Utilization of Base Services

We implemented two applications, matrix multiplication and 2D radix-2 DIT FFT, on the McNoC platform (See Fig. 11.2) with a range of sizes from 1 node to 64 (8–8) nodes. The matrix multiplication, which is computation intensive and does not involve synchronization, calculates the product of two matrices, A[64, 1] and B[1, 64], resulting in a C[64, 64] matrix. To vary the computation time, we consider both integer and floating point matrix multiplications. Figure 11.5 shows the system speedup for the two applications. As the system size increases from 1 to 64 cores, the speedup rises from 1 to 36.494 for the integer matrix multiplication, from 1 to 52.054 for the floating point matrix multiplication, and from 1 to 48.776 for the 2D FFT. The speedup for the floating point matrix multiplication is higher than that for the integer matrix multiplication. This is as expected, because, when

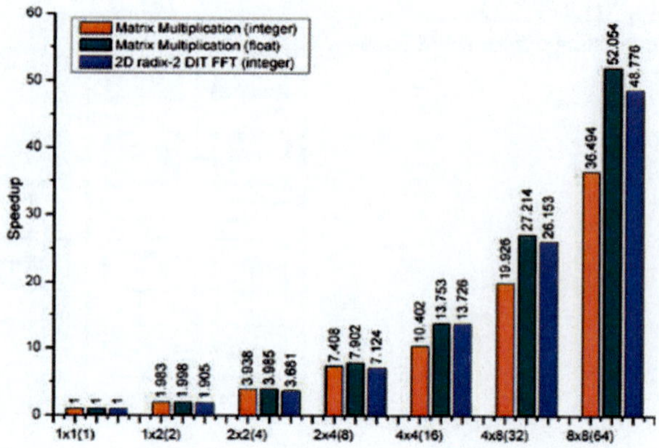

Fig. 11.5 Speedup of matrix multiplication and 2d DIT FFT

the computation takes more time, the portion of communication time becomes less significant, thus achieving higher speedup. That is to say, as the system size increases, communication becomes a more limiting factor for performance due to nonlinear increase in communication latency. In all cases, the DME overhead is insignificant.

11.3.3.2 Utilization of Advanced Services

The application we use as a test driver is a combination of several real-life kernels that are present in network applications [7, 8]. We triggered the system with a set of traces from a real wireless network. The software application is fully multi-threaded as it is increasingly common in computing systems: each kernel is executed in its own independent thread and communicates asynchronously with the other kernels through asynchronous FIFO queues. Through extensive application profiling we captured the allocation behavior of the application [8]. This information contains the block size distribution of the memory allocation requests. Based on the allocation behavior of the application the most appropriate allocator would be a pure-private one [9], offering the best performance in multi-processor environments. To evaluate our approach we use five different pure-private memory allocators, presented in Table 11.1.

The results are presented in Fig. 11.3, where a comparison is performed in terms of number of memory accesses, maximum requested memory footprint and DME cycles. Out of the five memory allocators Alloc1 is the one that offers the smaller amount of memory accesses (21% less than Alloc3, which is the most complex one) and DME cycles, as it has the simplest internal structure and thus needing few memory accesses to service the allocation requests. Since all memory allocators have their free-lists and mapping functions to match the application's requirements, they all have similar behavior regarding the requested maximum memory footprint. However, when the static memory solution is compared against the dynamic one it requires 100% more memory footprint (Fig. 11.3).

Table 11.1 Description of the five different allocators

Allocator	Description (free-lists)	Code size (Bytes)
Alloc 1	Free-list 0 (*blockSize* = 40)	4792
	Free-list 1 (*blockSize* = 1,460)	
	Free-list 2 (*blockSize* = 1,500)	
	Generic heap (holds blocks of other sizes)	
Alloc 2	Free-list 0 (*blockSize* ∈ [0, 40])	4792
	Free-list 1 (*blockSize* ∈ [1, 280, 1, 460])	
	Free-list 2 (*blockSize* ∈ [1,460, 1,500])	
	Generic heap (holds blocks of other sizes)	
Alloc 3	Free-list 0 (*blockSize* ∈ [0, 40])	7728
	Free-list 1 (*blockSize* = 1,460)	
	Free-list 2 (*blockSize* = 1,500)	
	Free-list 3 (*blockSize* ∈ [40, 92])	
	Free-list 4 (*blockSize* ∈ [92, 132])	
	Free-list 5 (*blockSize* ∈ [132, 256])	
	Free-list 6 (*blockSize* ∈ [256, 512])	
	Free-list 7 (*blockSize* ∈ [512, 1,024])	
	Free-list 8 (*blockSize* ∈ [1,024, 1, 500])	
	Generic heap (holds blocks of other sizes)	
Alloc 4	Similar to Alloc 1 with the addition of free-list 3 (*blockSize* = 92)	5824
Alloc 5	Similar to Alloc 2 with the addition of free-list 3 (*blockSize* = 92)	5824

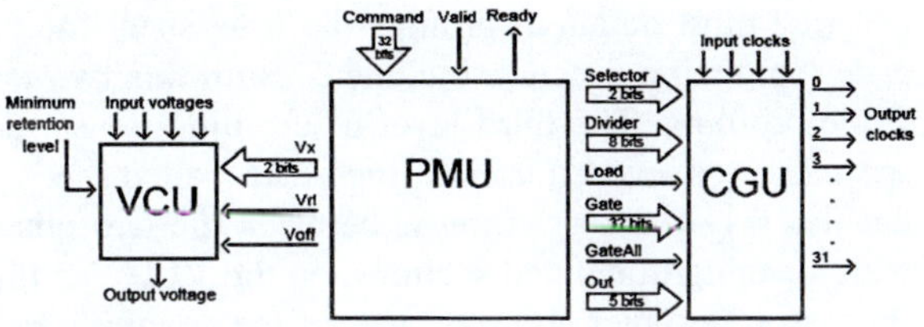

Fig. 11.6 Structure of the power management system

11.3.3.3 Power Management Services

The power services are accessed by the Power Management Intelligence software (PMINT) through what we call Power Management System (PMS) (Fig. 11.6). The PMS is made up of three separate blocks: the Power Management Unit (PMU), the Clock Generation Unit (CGU) and the Voltage Control Unit (VCU). The Power Management Intelligence PMINT communicates with the PMU, which is a complex set of state machines giving access to the power services. The power services are coordinated by the PMU and actuated by the CGU and the VCU, used respectively to generate the local clock(s) for the region and to regulate its supply voltage. While some of the power management services involve only CGU or VCU, the majority involve both units under the supervision of the PMU. The offered services are: (a) changing frequency; (b) changing voltage; (c) changing

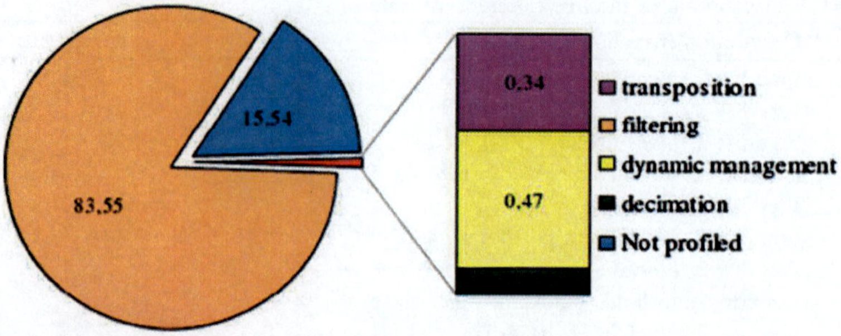

Fig. 11.7 Initial profiling results on the VLIW architecture template

DVFS point; (d) clock gating; (e) hybernation and (f) power off. The maximum frequency (post layout) of the PMU is 1ψ25 GHz.

11.3.4 ASIP Exploration

11.3.4.1 Initial Profiling

There are three profiling level. The highest one (the least detailed) only simulate the total cycles required to run the whole application. On the other hand, the lowest one (the most detailed) profiles the time spent in each functions called in the C code (emulation function included). Both this two levels are activated by default while profiling. The third level of profiling is user-defined, and profiles the time spent in a (or several) user-defined part (called section) of the C code; usually, the detail level of this profiling is between the two other level. The initial profiling, with meaningful defined sections, on the VLIW architecture template provided by Processor Designer showed most of the cognitive radio application runtime were required to perform the wideband filter, as shown in Fig. 11.7. The "not profiled" part is mainly composed of extra cycles introduced by the user-defined profiling section, for about 10% among 15%. The 5% last percents gathered mainly memory management (malloc, free, etc.), and the filter coefficient initialisation.

11.3.4.2 MAC Instruction

First, as the filtering required more than 85% of the total runtime, a MAC instruction has been added in the processor instruction set to speed this up. In order to not slow down too much the frequency, this MAC operation is implemented within two pipeline stages, i.e. within two clock cycles. During, the first cycle, operands are read and the multiplication is computed. Then, during the second cycle, the multiplication result is accumulated in the destination result, i.e. added

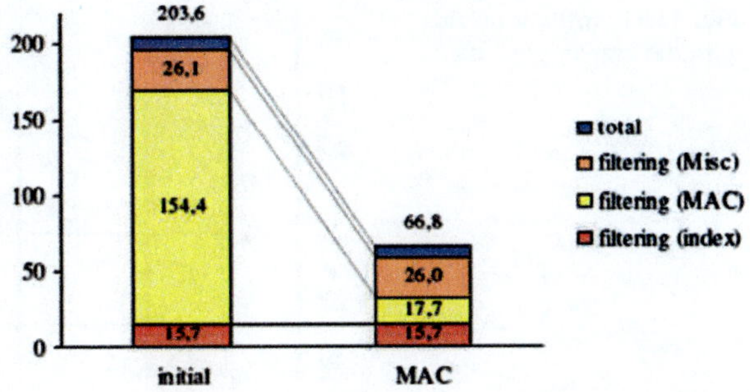

Fig. 11.8 Profiling result with the MAC instruction

with the destination register previous value. The implementation of this operation has been automatically mapped to a multiplication-accumulation from the C to the assembly code tested and validated. Then, the application has been profiled again on this optimized processor, and its results are shown on Fig. 11.8.

As a comparison, initial profiling results are shown on the left (figures are in millions of cycles). Moreover, the total runtime is cut in four different parts. The firstone ("MAC", represented in yellow) is the time spent computing the multiplication accumulation. The second one ("index", represented in red) is the time spent computing the memory address of the MAC operands (input sample and filter coefficient). The third one ("misc", represented in orange) is the remaining time spent in the filtering function; it is mainly composed of extra (useless) cycles introduced by the user-defined profiling sections, and of loop branching instruction and index computing. Eventually, the last part (represented in blue), is the time left (transposition, decimation, etc.). Among these four parts, the MAC instruction obviously optimized the "MAC" part; it speeds it up more than eight times.

11.3.4.3 Branch Prediction

The MAC instruction enhanced much the step 1 runtime. A second analyzing of these new performances showed many cycles were wasted in computing branch condition. As the application is based on nested loops, there are many conditional branch instructions, and each of them required a pipeline stall to compute whether the condition is true or not. The second ASIP optimization consists in implementing a loop-optimized branch prediction. It means that the processor would recognize a conditional branch that corresponds to a loop (defined by a branching address before the current program address). Then, it would automatically take the branch without computing the condition which is actually true most of the time. Then, the condition is computed to check the branching was right; if not, the processor goes back the instruction right after the branch instruction. Performances for this new ASIP have then been profiled, and the results are shown in Fig. 11.9.

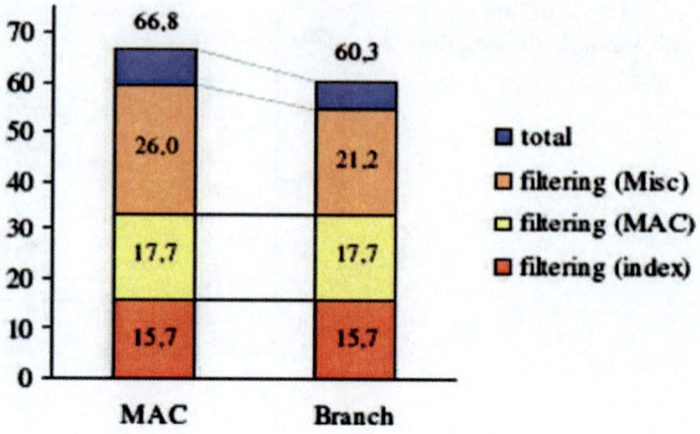

Fig. 11.9 Profiling result with the branch prediction

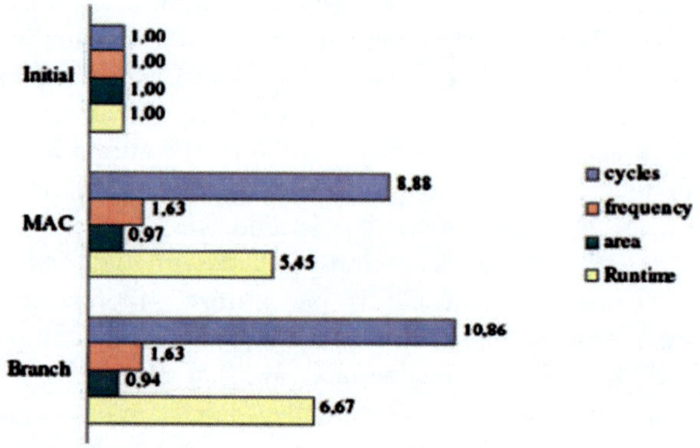

Fig. 11.10 Software/hardware trade-off

11.3.4.4 SW/HW Performances Trade-Off

Figure 11.10 provides a summary of the achieved results about the software/hardware trade-off. It represents the different gains from the initial VLIW architecture. The first gain (cycles) represents the software gain, i.e. how faster (in comparison with the initial version) the application runs considering the same chip frequency; for example, a gain of 5 means than the application required five times less cycles to run in comparison with the initial version. Then the next two figures (frequency and area) are related to the hardware impact. A figure greater than one means that the design is better than the initial one; for area, it means the chip is smaller, and for frequency, it means it run faster. Eventually, the last one (runtime) takes into account both software and hardware gains. Runtime represents the time (in seconds) required to run the application, and is determined from both the amount of cycles and the chip frequency. Actually, it shows than the great software gain afforded by the MAC instruction is partially counterbalanced by the frequency fall it introduces.

11.4 Conclusions

The objective of the MOSART project is to develop a flexible, modular multi-core on-chip platform architecture and associated exploration design methods and tools, to allow the scaling of the platform and optimization of its constituent elements for various embedded, multimedia and wireless communication applications. In MOSART, we have deployed a cluster of ASIPs to target a suite of applications and we enhance the efficacy of the MPSoC concept by using distributed memory architecture and use of NoC. By adopting such an architecture, we claim that we not only gain flexibility, scalability and modularity, we also improve the computational efficiency to the extent that in the ladder of computational efficiency, the proposed architecture would be only one notch below hardwired ASICs and yet largely retain the flexibility of programmable solutions.

Acknowledgments This work is supported by the E.C. funded FP7-215244 MOSART Project, www.mosartproject.org

References

1. Mignolet J-Y et al (2009) Mpa: Parallelizing an application onto a multicore platform made easy. IEEE Micro 29(3):31–39
2. Kreku J et al (2008) Combining uml2 application and systemc platform modelling for performance evaluation of real-time embedded systems. EURASIP J Embedded Syst 2008:1–18
3. Chen X et al (2010) Supporting distributed shared memory on multi-core network-on-chips using a dual microcoded controller. In: Proceedings of DATE. pp 39–44
4. Chabloz JM, Hemani A (2009) A flexible communication scheme for rationally-related clock frequencies. In: Proceedings of ICCD. pp 109–116
5. Kreku J et al (2004) Workload simulation method for evaluation of application feasibility in a mobile multiprocessor platform. In: Proceedings of DSD. IEEE Computer Society. pp 532–539
6. Kreku J et al (2010) Automatic workload generation for system-level exploration based on modified GCC compiler. In: Proceedings of DATE
7. Bartzas A et al (2008) Enabling run-time memory data transfer optimizations at the system level with automated extraction of embedded software metadata information. In: Proceedings of ASP-DAC. pp 434–439
8. Bartzas A et al (2010) Software metadata: Systematic characterization of the memory behaviour of dynamic applications. J Syst Software 83(6):1051–1075
9. Wilson PR et al (1995) Dynamic storage allocation: A survey and critical review. In: Proceedings Of IWMM. Springer-Verlag, Berlin, pp 1–116

Part III
Emerging Devices and Nanocomputing

Chapter 12
XMSIM: Extensible Memory Simulator for Early Memory Hierarchy Evaluation

Theodoros Lioris, Grigoris Dimitroulakos and Kostas Masselos

Abstract This paper presents a memory hierarchy evaluation framework for multimedia applications. It takes as input a high level C code application description and a memory hierarchy specification and provides statistics characterizing the memory operation. Essentially the tool is a specialized C++ data type library which is used to replace the application's data types with others that monitor memory access activity. XMSIM's operation is event driven which means that every access to a specific data structure is converted to a message towards the memory model which subsequently emulates memory hierarchy operation. The memory model is highly parametric allowing a large number of alternatives to be modeled. XMSIM's main advantage is its modularity allowing the designer to alter specific aspects of the memory operation beyond the predefined ones. The main features are the capability to: (1) simulate any subset of the application's data types, (2) user defined mapping of data to memories, (3) simultaneously simulate multiple memory hierarchy scenarios, (4) immediate feedback to code transformations effect on memory hierarchy behavior, (5) verification utilities for the validation of code transformations.

Keywords Memory simulation · Memory hierarchy evaluation tools · Computer aided design · Code transformations

T. Lioris · G. Dimitroulakos (✉) · K. Masselos
Computer Science and Technology Department, University of Peloponnese, Tripolis, Greece
e-mail: dhmhgre@uop.gr

N. Voros et al. (eds.), *VLSI 2010 Annual Symposium*,
Lecture Notes in Electrical Engineering 105, DOI: 10.1007/978-94-007-1488-5_12,

12.1 Introduction

Microprocessor speeds have risen dramatically during the last years in discrepancy to current memory operation speeds. This fact indicates a major bottleneck in the processor-memory intercommunication, deteriorating system performance and increasing time delay. Moreover, applications in the embedded system domain require drastic reduction of power consumption for long life battery operation. The use of cache memories has long provided a way to hide such latencies and reduce power consumption [1]. For the above reasons, it is highly acceptable today that memory hierarchy design is one of the major issues in modern embedded processors' implementation.

Nowadays system design flow is a two phase process. Initially the application is developed in a high level language (such as C/C++) by designers bearing an algorithmic background. More or less they provide a functional code which is by far not optimized. In the second phase the mapping of the application to a specific embedded processor is implemented and optimized. Today's mainstream practice requests collecting results of the application first and feeding the results to a separate profiling/optimization program afterwards. Our work facilitates the completion of the optimization process in the following ways: (1) application's code development and profiling are performed on the same development framework and (2) instant verification of the code transformation.

Usual practice is to develop algorithms in C language which is more close to hardware implementation. More specifically, a small subset of the C language's data types are used that is, arrays and scalars which are determined at compile time. Although there has been an extensive research on the automation of the optimization, it is still an area that human interaction is required. In most cases the designer should have a minimum knowledge about how the application works to perform optimizing decisions. Additional design effort can be saved if the algorithm's designer has an insight about how its decisions affect the performance of the memory system. Our work provides the designer with the ability to evaluate his decisions on the back-end of the hardware design flow.

In this paper, a memory hierarchy evaluation framework is proposed that unifies the algorithmic development and memory hierarchy optimization phase. It is an event-driven environment where the application's arrays and scalars are substituted by objects with the same behavior, providing profiling capabilities. Profiling is accomplished by means of event generation when memory accesses take place that is, on access to a data structure. These messages bear information about data structure accesses and are forwarded to a memory model that records and translates them to memory accesses. The output of the tool is a series of profiling data, characterizing the memory hierarchy performance.

The advantages of the tool are: (1) The memory model is highly parametric and models a large number of memory hierarchy scenarios. (2) The overall algorithmic development, debugging and optimization take place under a single environment which is the Microsoft Visual Studio IDE [2]. (3) Verifying the program's integrity after code transformation's is made easy by verification routines developed for this purpose. (4) Multiple memory hierarchy scenarios can be examined simultaneously. (5) The designer can fully control data mapping to memories and the subset of data structures that he wants to simulate. (6) The tool is extensible that is, the designer can built new classes inheriting the default ones providing user defined analysis as well as configure the memory's operation by means of callback functions and (7) It can be extended to cooperate with existing tools (CACTI [3, 4]) that provide power estimation.

The paper is organized as follows: Sect. 12.2 includes the related work while Sect. 12.3 explains the simulator architecture. Section 12.4 describes the development environment Sect. 12.5 presents the experimental results, and finally Sect. 12.6 concludes the paper.

12.2 Related Work

Most cache simulators nowadays target more on educational purposes than cache design. Although our work can be used on both disciplines, our main purpose is to facilitate code optimization towards cache efficient design. Currently there are two types of cache simulators: execution-driven and trace driven. Execution-driven simulators [5–7] display cache operation, as well as cache-processor communication, giving a more detailed view of the computer system at the expense of simulation time and clarity. Trace-driven simulators [8] simulate the result on cache state and contents of a specific memory access trace which is fed as input to the simulator. Moreover, execution-driven simulators bind on specific processor architecture while trace-driven are architecture independent. From this point of view, our work is categorized to the trace driven division.

One of the first attempts to build a cache simulator was the PC-Spim [9] which was extended with the appearance of SpimCache [10]. Unfortunately, the latter was only suitable for one cache level simulations and could not alter cache characteristics like the line size or different policies. An extension of SpimCache was the SpimVista tool [6] intended to represent cache hierarchy and the interaction between cache levels. Simplescalar [5] is another representative example of an execution-driven cache simulator, as well as the mlcache simulator [7]. On the other hand, trace-driven simulators are easier to understand. Dinero IV [8] is a trace-driven simulator written in C where various cache design parameters (like cache associativity, write-back or write-through, cache line size etc.) can be configured.

Our work differentiates in many aspects. In both types of existing simulators it is difficult to associate the C code statement executed and the particular action taking place in the cache because the input to all simulators is the assembly language. In our work the designer observes the results of every C statement in memory state and contents. Moreover, it is extensible such that new measurements

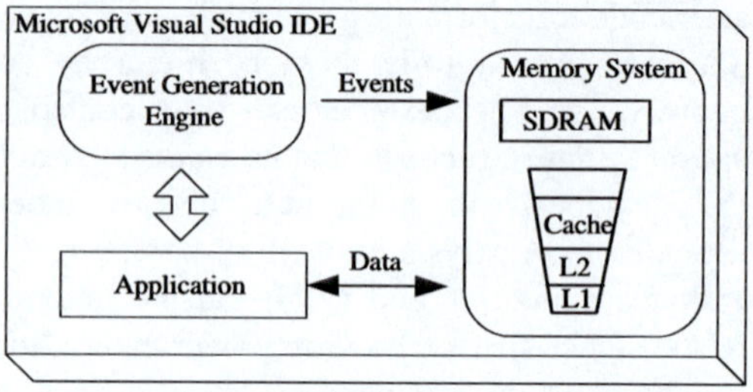

Fig. 12.1 XMSIM architecture

can be introduced by the designer. Finally, it is equipped with the capability to verify correctness of the application's transformed code.

12.3 Simulator Architecture

The simulator engine consists of two main parts: (1) the Event Generation Engine and (2) the memory model. The Event Generation Engine is the core of XMSIM. It is a library of two template classes emulating the C language's arrays and scalars. Each array or scalar can be fully emulated from a properly parameterized template class. The main purpose of these classes is to record data structure access activity and produce events to experimental objects such as a virtual memory hierarchy or CPU. Since, these classes can fully substitute the C native types the application executes in an identical way providing a trace of events able to trigger update in state and contents of the experimental objects. The overall state transition can be studied either step by step or in summary making the evaluation possible.

Figure 12.1 illustrates XMSIM architecture. The tool engagement to the application involves two steps: (1) substitution of the application's arrays and scalars with the tool's data types and (2) mapping of these data types to Memory Model. As the application is executed, events originating from data type accesses are directed to the Memory Model, which responds with data flowing to and from the application. The whole simulation/analysis process is based on a well established development IDE, the Microsoft Visual Studio which is exploited to extended the tool's capabilities. Each simulation context consists of the application and XMSIM libraries. Both are compiled and linked as one C/C++ project and the execution provides the experimental results.

The Event Generation Engine includes two classes one for arrays (called CArrayWrapper) and one for scalars (CScalarWrapper). It uses operator overloading [11] to emulate all native C language's operators by user defined ones functionally equivalent but with the ability to profile and driving experimental objects through the use of events. For this reason, each template class includes an array of pointers to objects corresponding to an abstract class [11] called CContext representing the interface of experimental objects to the Event Generation Engine. The CContext class consists of a set of member functions each corresponding to a

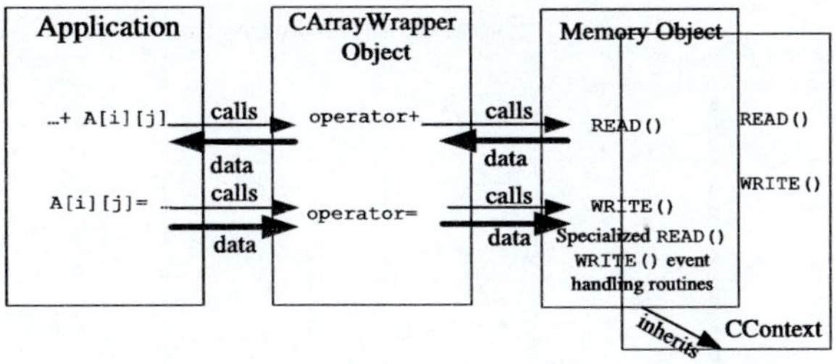

Fig. 12.2 Event generation and handling

type of event. To support event capture and service each class representing an experimental object should be derived from the class CContext.

The Memory Model is a set of classes each corresponding to different types of memory technology. All the classes are derived from the CContext class to support event handling and interface with the Event Generation Engine. Currently our framework supports parametric models for SDRAM, static RAM, cache and scratch-pad memory. Each, memory object includes pointers to other memory objects so as to model memory hierarchies. In addition, for the event handling each memory class provides specialized implementation of the member functions inherited from CContext according to its operation.

To summarize, events are generated upon access to any scalar or array data structure. That is in every operator function inside CArrayWrapper and CScalarWrapper. Events take the form of calls to CContext member functions inside each experimentation object. The objects handle these messages by altering their state and contents according to the information conveyed by each message. The carried information includes the type of data access (read or write), virtual address, etc. Figure 12.2 depicts the event generation and handling mechanism.

The overall tool architecture gives prosperous ground to extend its functionality. Every class in both the Event Generation Engine and the Memory Model can be inherited by a user defined class to widen the functionality and simulation utilities. Hence, inheritance [11] provides event handling support necessary to trace data type access events and gives user defined opportunities to extend memory access analysis and behavior. Figure 12.3 summarizes the extensibility options of XMSIM. The designer may combine one of the Event Generation Engine components (base or extended) with one of the Memory Model's. Totally there are 4 combinations by which the base and extended components can be combined. By extending the Event Generation Engine new opportunities of analysis emerge from the application's data types. On the other hand by extending the Memory Model classes, memories with different characteristics and functions can be introduced. Moreover, new analysis scenarios can be built based on the XMSIM event driven concept.

Moreover, having one conceived the organization of the project can alter or enhance at will the behavior of the memory models. The core of the functionality of the Memory Model Service lies within the four main routines of each memory class, which are the MAP, ADDRESS, WRITE and READ functions.

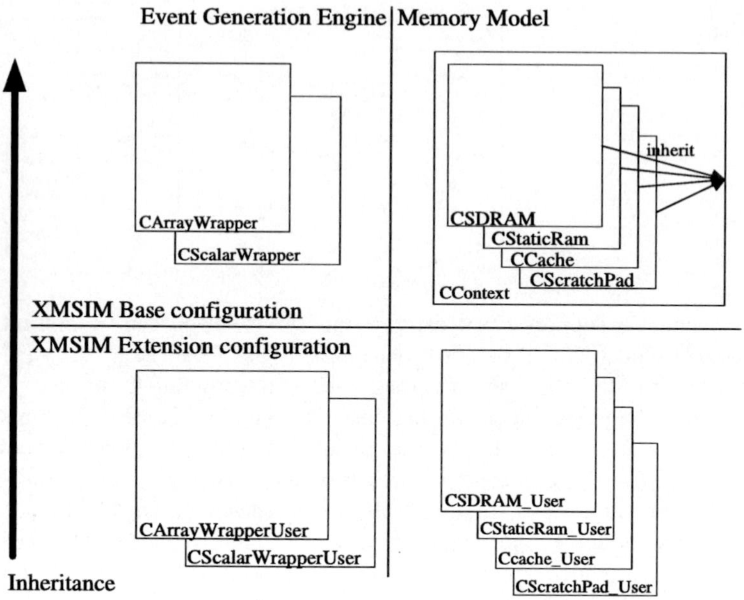

Fig. 12.3 XMSIM extensibility options

These functions can be re-written to fit one's needs (like time delays from data transfer or specific sorts of misses), ornamented with enhanced functionality (like split caches or write-around policy), or even extended by calling additional routines. Furthermore all the classes can be extended in order to implement brand new concepts of memory layout. Thus the researcher can model any memory unit and test any memory hierarchy, restricted merely by his/her programming aptitude and his ambition.

The following example illustrates how one aspect of XMSIM functionality can be configured by rewriting the callback function ADDRESS(). The default realization of XMSIM assumes row-wise mapping of array data to memory. The example shows how the row-wise mapping can be transformed to column-wise. In XMSIM the Physical Address (PA) of an array data value in memory is the sum of the array's Base Address (BA) and the Virtual Address (VA). The latter determines the ordering of array elements in memory. The ADDRESS() function calculates the virtual address and provides this value to the memory model to compute the physical address. The physical address is estimated from the following equation

$$PA = BA + VA, \tag{12.1}$$

The VA for row-wise mapping is given by the following equation

$$VA = I_n + \sum_{i=1}^{n-1} \left(\prod_{j=i+1}^{n} N_j \right) \cdot I_i \tag{12.2}$$

$$VA = I_1 + \sum_{i=2}^{n} \left(\prod_{j=1}^{i-1} N_j \right) \cdot I_i, \tag{12.3}$$

where *n* is the number of array dimensions and Nj is the size of the j-th dimension. On the other hand, column-wise mapping is realized by Eq. 12.3. Figure 12.4 shows one possible realization of the two versions of the ADDRESS() function corresponding to the row and column-wise mapping respectively. There isn't a one to one correspondence among the indexes in the loops in Fig. 12.4 and in Eqs. 12.2 and 12.3 but semantically are identical.

12.4 Development Environment

In the sequel the environment supporting the design, analysis and optimization phases will be described. The overall framework is based on a popular development environment the Microsoft Visual Studio. It is an extensible and mature environment used extensively in many contexts. Microsoft Visual Studio provides an extensive set of debugging capabilities step by step execution and trace of the application's variable values.

The experimental setup requires the memory architecture, the application and the variables' set to examine. Memory hierarchy scenarios are described in a flat input file in a very simple language format. The parser instantiates and links the memory objects to form memory hierarchies. The tool informs about the memories made, or about erroneous input. Many memory hierarchies can be imported in a single input file. Thus one can simultaneously challenge a code against alternative memory hierarchies.

The language used at the input file to depict the memory model is straightforward and easy to understand. Every command must take up a single line and be delimited by the semicolon (;) as in C/C++ grammar. All commands are of the form: "parameter: value" and the command repertoire hardly exceeds fifteen members. The script can include one-line comments after the hash sign (#) at the beginning of the line. Every memory unit is uniquely named and declared separately from the rest. The directives describe physical characteristics of every memory unit such as the number of blocks, words of every memory block, the size of every word in bits, number of banks in the case of SDRAM as well as operation policies like write-back/write-through and associativity. Energy consumption can be defined and the way different memory units are connected to form a multilevel hierarchy. Figure 12.5 shows an example of memory hierarchy description file.

The second stage involves two steps. The first concerns the replacement of the application's data types with the ones provided by XMSIM. At this point the

```
ROW-WISE MAPPING

template <typename T>
void CArrayWrapperT<T>::ADDRESS(unsigned int &VirtualAddress) const{

        unsigned int i,j;
        int icoef ;

        // MACROS
        // ARRAY_DIMENSIONS    : Returns the number of array dimensions
        // DIMENSION_SIZE(i)   : Size of the ith dimension
        // ARRAY_SUBSCRIPT(i)  : Value of the ith array subscript
        VirtualAddress =0;
        for ( i = ARRAY_DIMENSIONS-1; i>= 0 ; i-- ){
                icoef = 1;
                for ( j=ARRAY_DIMENSIONS-1; j>=i+1; j--){
                  icoef *= DIMENSION_SIZE(j);
                }
                VirtualAddress += icoef*ARRAY_SUBSCRIPT(i);
        }
        return;
}
```

```
COLUMN-WISE MAPPING

template <typename T>
void CArrayWrapperT<T>::ADDRESS(unsigned int &VirtualAddress) const{

        unsigned int i,j;
        int icoef ;

        // MACROS
        // ARRAY_DIMENSIONS    : Returns the number of array dimensions
        // DIMENSION_SIZE(i)   : Size of the ith dimension
        // ARRAY_SUBSCRIPT(i)  : Value of the ith array subscript
        VirtualAddress =0;
        for ( i = 0; i< ARRAY_DIMENSIONS-1 ; i++ ){
                icoef = 1;
                for ( j=0; j>=i-1; j++){
                  icoef *= DIMENSION_SIZE(j);
                }
                VirtualAddress += icoef*ARRAY_SUBSCRIPT(i);
        }
        return;
}
```

Fig. 12.4 ADDRESS() function transformation example

designer may decide to have two versions of the code the one engaged to XMSIM and the other being the original. This option is essential to verify code correctness when the designer applies code transformations. For this reason, XMSIM provides verification routines in CArrayWrapper and CScalarWrapper classes to

```
Name: Main memory;
Type: SDRAM;
Banks: 4;
Rows: 1024;
Columns: 1024;
Wordsize: 32;
end: Main memory;

Name: L2;
Type: Cache;
Upper memory level: Main memory;
Miss policy: write-through;
Associativity: 2;  /* 2-way associative */
Blocks: 256;
Words per block: 64;
Wordsize: 32;
Energy consumption per access: 1.3 nJ;
end: big;

Name: L1;
Type: Cache;
Upper memory level: L2;
Miss policy: write-back;
Associativity: 1; /* direct mapped */
Blocks: 16;
Words per block: 2;
Base level memory: Yes; /* closest to CPU */
Wordsize: 32;
Energy consumption per access: 0.5 nJ;
end: L1;
```

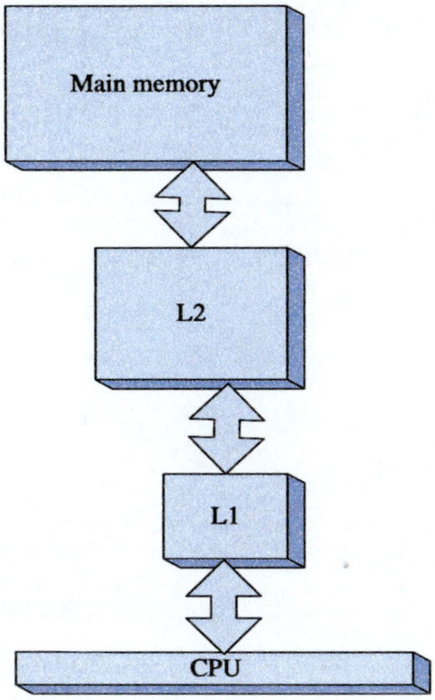

Fig. 12.5 Memory hierarchy description file example

```
//static unsigned char gauss_x_image[N][M];
//static unsigned char gauss_xy_image[N][M];
//static unsigned char comp_edge_image[N][M];
//static unsigned char out_compute[N][M][NB+1];
//static unsigned char max_compute[N][M];

int Dim1[2] = {N,M};
int Dim2[3] = {N,M,NB+1};
static CArrayWrapperT<unsigned char> gauss_x_image(2,Dim1,"gauss_x_image");
static CArrayWrapperT<unsigned char> gauss_xy_image(2,Dim1,"gauss_xy_image");
static CArrayWrapperT<unsigned char> comp_edge_image(2,Dim1,"comp_edge_image");
static CArrayWrapperT<unsigned char> out_compute(3,Dim2,"out_compute");
static CArrayWrapperT<unsigned char> max_compute(2,Dim1,"max_compute");
```

Fig. 12.6 XMSIM engagement example

compare equality between the XMSIM's objects and the C language's data types. Figure 12.6 depicts how XMSIM transforms C language data types to XMSIM data type declarations. The outcome of this process is the application's statements to remain untouched operating on XMSIM's data types.

In the second step the designer has the ability to map the XMSIM's arrays and scalars to specific memory locations. If this step is overridden for any data type then XMSIM places it automatically to main memory following a sequential placement on empty address space. Figure 12.7 illustrates the assignment of a specific memory address to an array while Fig. 12.8 depicts the application's code layout after XMSIM engagement. As it is obvious, nothing is changed in the code except for the data type declarations.

```
unsigned int base_address = 0xFFA3;

ATTACH_ARRAY_TO_MEMORY(array, main_memory);

ATTACH_ARRAY_TO_MEMORY(array, main_memory, base_address);
```

Fig. 12.7 Attachment of an array to memory example

Fig. 12.8 Example of code layout

Original Application Code

```
void cav_detect(){

/*C declarations */
static unsigned char gauss_x_image[N][M];
static unsigned short gauss_x_compute[N][M][(2*GB)+2];

/* C Statements */
for (x=GB; x<=N-1-GB; ++x){
 for (y=GB; y<=M-1-GB; ++y) {
          gauss_x_compute[x][y][0]=0;
           for (k=-GB; k<=GB; ++k)
                    gauss_x_compute[x][y][GB+k+1] =
gauss_x_compute[x][y][GB+k] +
(image_in_traf[x+k][y]*Gauss[abs(k)]);
                    gauss_x_image[x][y]=
           gauss_x_compute[x][y][(2*GB)+1]/tot;
 }
}
...
...
}
```

Application Code after XMSIM engagement

```
void cav_detect(){

/*C declarations */
/* Step 1 data type substitution */
//static unsigned char gauss_x_image[N][M]; // commented
//static unsigned short gauss_x_compute[N][M][(2*GB)+2]; // commented

int Dim1[2] = {N,M};
int Dim2[3] = {N,M,2*GB+2};
static CArrayWrapperT<unsigned char> gauss_x_image(2,Dim1,"gauss_x_image");
static CArrayWrapperT<unsigned short> gauss_x_image(2,Dim2,"gauss_x_compute");

/* Step 2 (optional) map arrays to specific memory segments */
unsigned int base_address = 0xFFA3;
ATTACH_ARRAY_TO_MEMORY(gauss_x_image, main_memory, base_address);

/* C Statements */
for (x=GB; x<=N-1-GB; ++x){
 for (y=GB; y<=M-1-GB; ++y) {
          gauss_x_compute[x][y][0]=0;
           for (k=-GB; k<=GB; ++k)
                    gauss_x_compute[x][y][GB+k+1] =
          gauss_x_compute[x][y][GB+k] + (image_in_traf[x+k][y]*Gauss[abs(k)]);
                    gauss_x_image[x][y]= gauss_x_compute[x][y][(2*GB)+1]/tot;
 }
}
...
...
}
```

The ultimate goal is to evaluate an application on a memory hierarchy in respect to power, area and speed metrics. For this reason XMSIM can be used in cooperation with existing power, time and area estimation kits developed for the different memory technologies supported by XMSIM. The CACTI tool [3] can be used for cache and scratch-pad [12] while for SDRAM exist power and speed estimation models such as the ones provided by Micron [4]. Before running the simulation, the aforementioned tools supply power, area and time metrics for each memory. The outcome of simulation provides the operational profile of the application in terms of power speed and area.

Finally, there are two spots in XMSIM where the designer can broaden its utilities. The first involves the Event Generation Engine while the second the Memory Model. XMSIM provides default functionality that refers to the analysis of the data structure access patterns and also the memory system performance. The designer can extended the analysis utilities following the principles described in Sect. 12.3 to record the frequency by which array elements are accessed, data locality etc.

12.5 XMSIM'S Graphical User Interface

In order to extend XMSIMS's usability, the development of a more friendly graphical interface is decided to assist the memory architecture specification described in Sect. 12.4. Tcl/Tk is considered to be the most appropriate environment to tackle with the task in mind, because this programming language and especially Tk has a wide variety of ready-to-use widgets. One can add events to this widgets and dictate graphical behavior easily and quickly. The graphical user interface (GUI) can be seamlessly glued on top of the existing application and do away with the command line. The Tcl/Tk extension is outlaid without limiting the core application's capabilities. So one can easily define different scenarios with different memory levels and architectures using the GUI, as one could do the same through the command line, writing code. Therefore the inputed data can be collected, turned into the appropriate form and be saved into an input file, ready to be processed by the XMSIM core program. Then the core program is invoked with the suitable input supplied and when the output is generated it is collected back to the graphic window and in order to be made available to the user.

The graphical extension initially asks for the total number of distinct memory units of all the scenarios to be used. Frames are made on-the-fly, one for every memory unit, (Fig. 12.9 rectangular regions), where the user can provide details about the architecture and functional behavior of each of them. These frames are placed horizontally on the window panel of the application. The upper memory frame represents a DDR or static RAM memory unit, which is served by the memory directly below it. Thus every unit on the pile serves the upper neighboring unit and is served by the one placed directly below it. When a cache memory unit is directly above a DDR or simple RAM, it signifies the end of a scenario and the

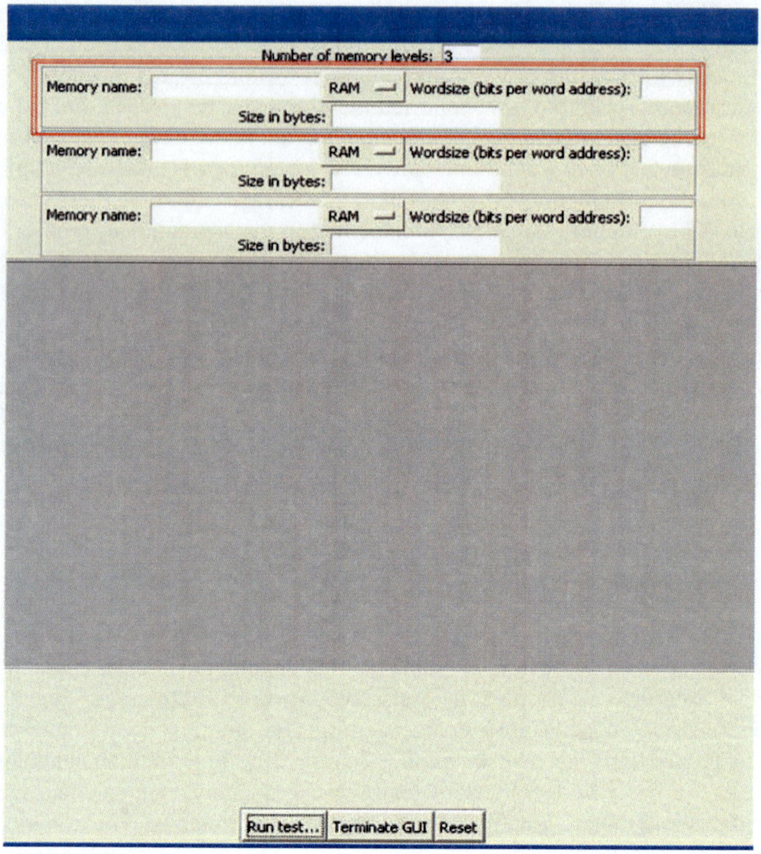

Fig. 12.9 GUI's main window

onset of a new one. In Fig. 12.10, the upper square signifies the first scenario of a RAM named M1 aided by the C1 cache. In the same figure the lower square describes a single RAM, named M2. Every scenario is discerned by a RAM/DDR memory on the top and some caches below it. The graphical interface will automatically differentiate among units of different scenarios and the input file can be generated. The necessary information for every memory unit vary according to the exact type of memory and thus every frame asks for different input.

Figure 12.11 depicts a case where two scenarios are about to be simultaneously studied. The first scenario is made of a DDR called Main memory and two levels of cache (L1 and L2). L1 serves Main memory, as it stands right below it and is served by L2. The second scenario describes another memory named DDR with one level of cache named LL2. The user is about to choose the cache type from the drop-down menu, for the memory named LL2, as this is shown in the square.

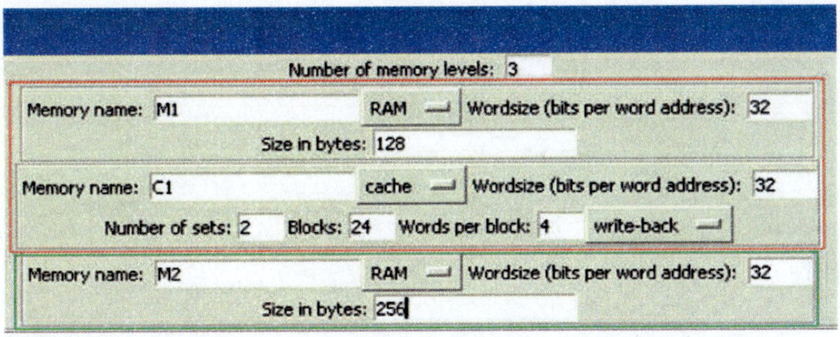

Fig. 12.10 Logical separation of two memory architecture scenarios

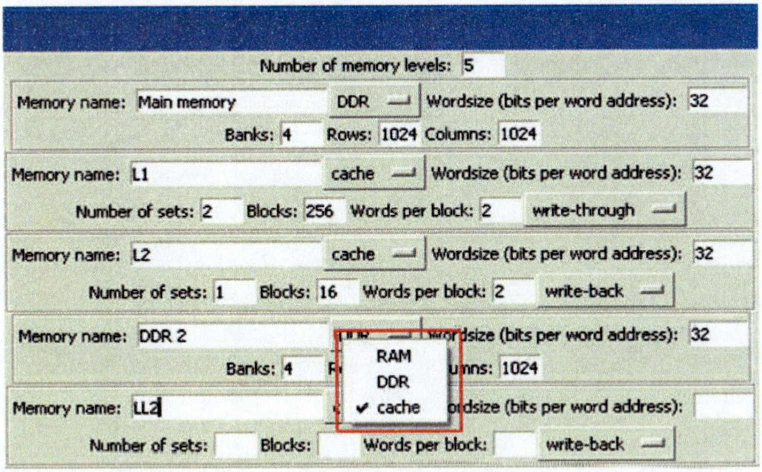

Fig. 12.11 An example with two memory architecture scenarios

In the same way, the user can define the cache memory policy using the drop down button labeled write-back to choose among a write-back or write-through policy.

After the input is given, the user can click on the start button. The input file is made and printed on the text frame, then XMSIM starts on the background using the input file. This is shown at Fig. 12.12. The input file is neatly made, since it includes remarks about scenarios and total number of memory units used. The data is accessible as the cursor suggests, and can be copy-pasted into another text application. The results are saved in a flat text file for later inspection. The name of the input file is given right before the results, as is shown in Fig. 12.13. The results are harvested from the command prompt and displayed in the text frame. The user can re-start the application with different scenarios.

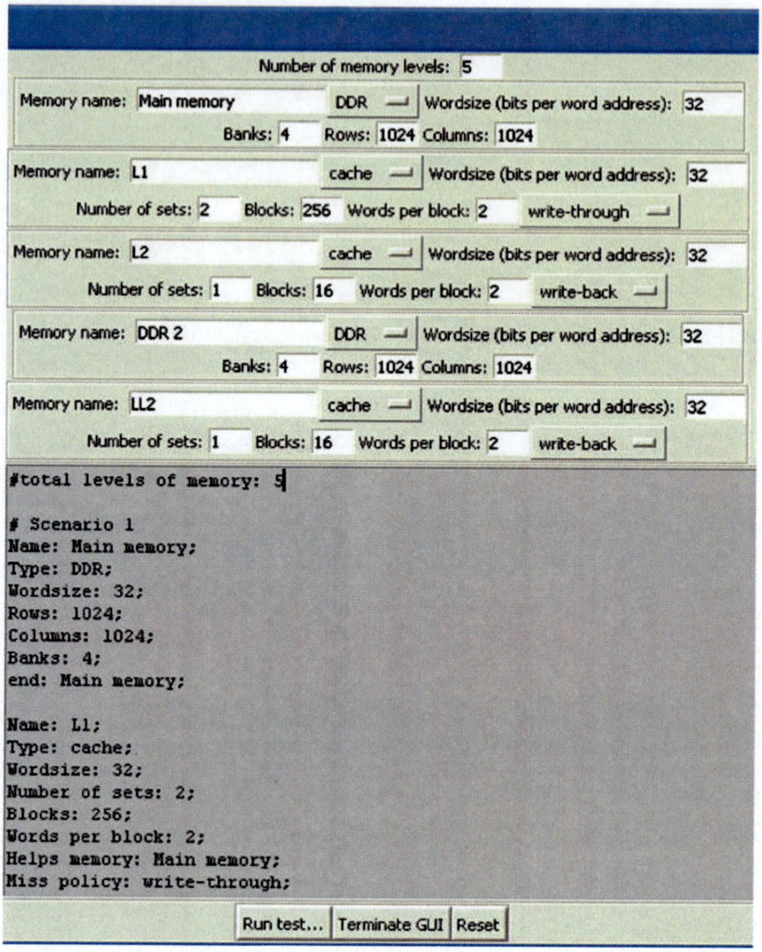

Fig. 12.12 Memory description in text form in the output frame

12.6 Experiments

Our experimental setup consists of the base XMSIM configuration and an image processing application called cavity detector [13]. Basically, cavity detector is a medical application consisting of three loop kernels. The first performs lowpass filtering to blur the input image, the second applies an edge detection algorithm and the third derives the negative of the image picture. For the experiments an optimized version of the cavity detection is also derived. A series of loop transformations [1] are applied to the original code focusing on data locality improvement. Subsequently XMSIM's utilities verified the correctness of the loop

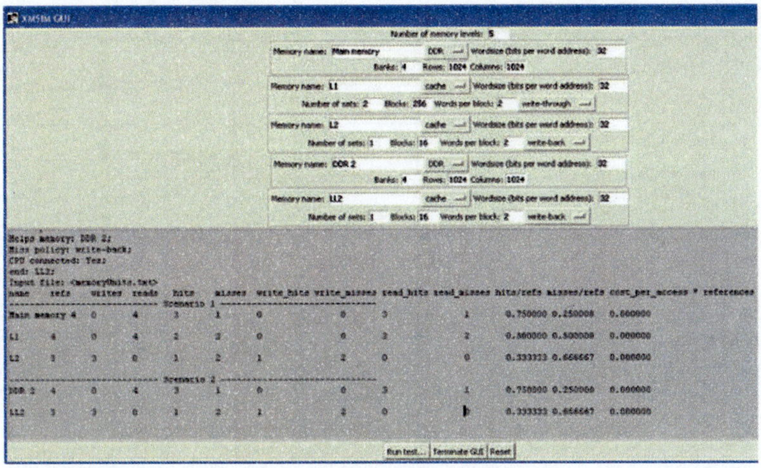

Fig. 12.13 Experimental results derived in the output frame

transformation, prior experimentation. The resulting code has one loop kernel corresponding to the union of the initial three.

Since there are many memory configuration parameters any many values to decide among them, a huge design space is created. The following experiments intent to give an overview of the feedback provided from XMSIM during the algorithm development or optimization phase and not how a fully optimized memory hierarchy is produced for the testbench application. In this context, the measurements presented concern the cache's and SDRAM's observed performance challenged over different major characteristics. In respect to the cache, these include the cache bank size, the number of blocks, the number of words per block and the number of cache hierarchy levels. Regarding the SDRAM these include various internal organizations concerning the cache rows and columns size. Finally, performance is evaluated with either of the following: the memory hits or misses, and the total memory references (read/write).

The graph in Fig. 12.14 depicts cache hits and misses for the initial and optimized (traf) cavity algorithm for one level direct-mapped write-through cache consisting in each case of 256 block of 4 words, 128 blocks of 8 words, 62 blocks of 16 words and 32 blocks of 32 words. As words per block increase we get a better performance for both algorithms while the optimized version of cavity has a lower cache activity because the sum of hits and misses is reduced in the case of the optimized version. Figure 12.15 illustrates cache performance for the initial and optimized (traf) cavity algorithm for one write through cache consisting of 256 block of 4 words, when sets are 1 (direct-mapped), 2(2-way associative), 4 and 8. Numbers are in thousands. We see a slightly better performance when sets are increased. Figure 12.16 depicts SDRAM row hits and misses [14] at various

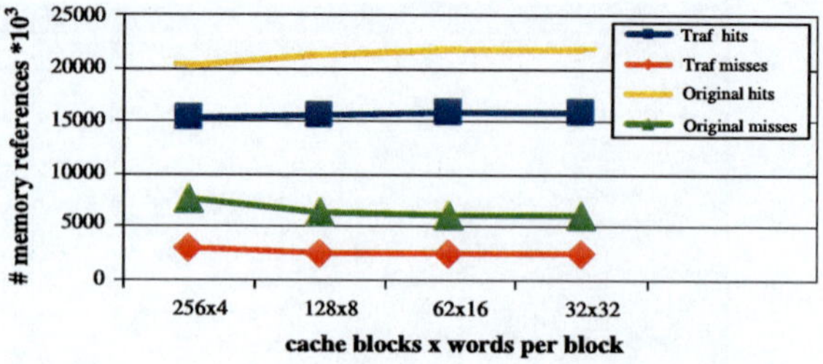

Fig. 12.14 Cache performance versus various cache block configurations

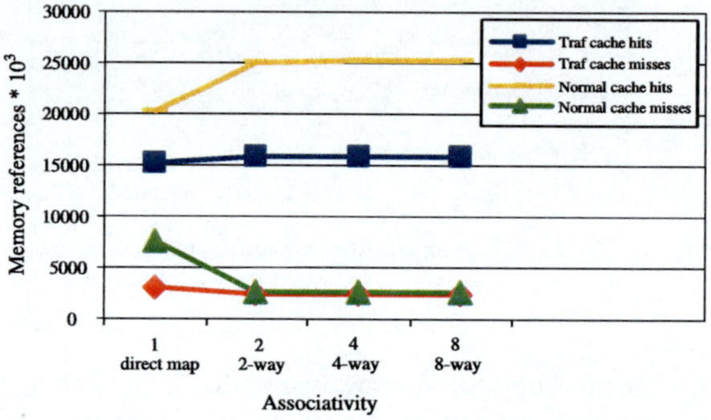

Fig. 12.15 Cache performance versus various cache associativities

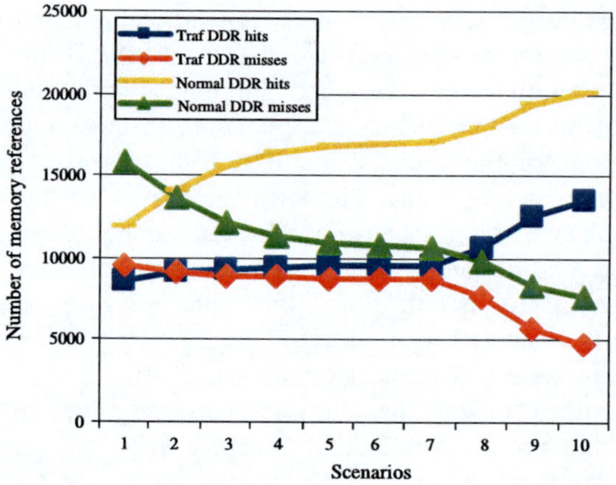

SDRAM 2MWords organized in

Scenario	Rows	Columns
1	524288	4
2	262144	8
3	131072	16
4	65536	32
5	32768	64
6	16384	128
7	8192	256
8	4096	512
9	2048	1024
10	1024	2048

Fig. 12.16 SDRAM performance figures

architectures for original and optimized cavity algorithm. From a range of 262,144 rows by 8 columns up to 2,048 by 1,024 columns we see that hits constantly rise when the number of columns increases.

Figures 12.17 and 12.18 present the explored memory hierarchy configurations for the two different versions of cavity detector. In all cases the main memory is an SDRAM memory chip with 4 banks each organized in 2,048 rows and 1,024

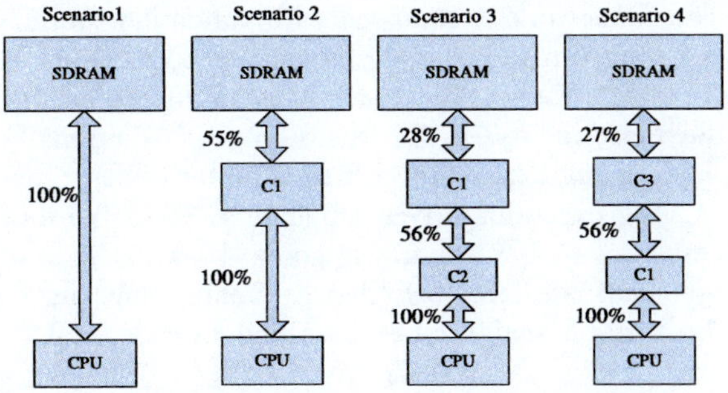

Fig. 12.17 Memory references for the optimized version of cavity

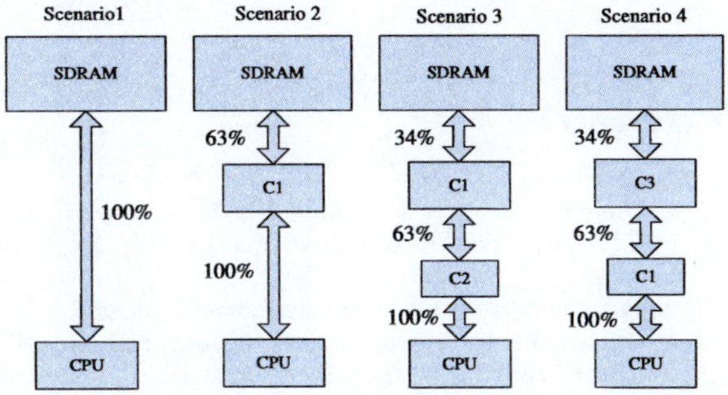

Fig. 12.18 Memory references for the initial version of cavity

columns. Totally, three different cache banks have been used in the various scenarios with the following characteristics: (1) Bank C1 is a 128 Kb 4-way write back cache with 128Kblocks and 1 word per block, (2) Bank C2 is a 64 Kb 4-way write back cache with 64Kblocks and 1 word per block and (3) Bank C3 is a 256 Kb 4-way write-back cache with 256Kblocks and 1 word per block.

Figures 12.17 and 12.18 exposes the experimental results over the aforementioned memory hierarchies for the optimized and initial version of cavity detector respectively. In detail the number of memory references among two adjacent levels of memory hierarchy has been recorded. The results have been normalized for clarity with the number 18,288,756 and 27,724,634 representing the 100% in Figs. 12.17 and 12.18 respectively. Finally, it must be noticed, that the overall experimental results have been derived in a very short amount of time which includes the editing of the memory configuration file and the execution of the testbench application. Hence, the proposed tool gives the opportunity to challenge a large number of scenarios in a direct and immediate way.

12.7 Conclusions

To conclude we believe that XMSIM, in its present form, is an evaluation framework that facilitates the software designer in the exploitation of existing memory hierarchies or the derivation of new ones for a given application.

The framework is equipped with validation routines to guarantee correctness of the transformations imposed on the application. The designer can amend the XMSIM's C++ native code, in order to extend the memory classes or enhance the functionality of the existing ones. Additionally, the tool provides platform independent exploration which is by far easier to understand and manipulate than existing execution-driven simulators that delve into the details of memory processor communication. On going work, is directed to the development of a graphical interface to further automate input and output, producing graphics for direct result analysis, best case finding and so on.

Acknowledgment The presented research work was co-funded by the European Union in the frame of the ENOSYS project (FP7-ICT-248821) (www.enosys-project.eu).

References

1. Catthoor F, Danckaert K, Kulkarni KK, Brockmeyer E, Kjeldsberg PG, Achteren T, Omnes T (2002) Data access and storage management for embedded programmable processors. Springer
2. http://msdn.microsoft.com/en-us/vstudio/default.aspx
3. Muralimanohar N, Balasubramonian R, Jouppi NP, CACTI 6.0: A tool to model large caches, technical Rep. HPL-2009-85 HP Laboratories
4. http://www.micron.com/support/part_info/design_analysis_kits, 2010
5. http://www.simplescalar.com/, 2010
6. Leticia P, Alejandro T, Julio S, José F (2007) Understanding cache hierarchy interactions with a program-driven simulator. Proceedings of the 2007 workshop on computer architecture education, pp 30–35
7. Edward ST, Jude AR, Gary ST, Edward SD (eds) (1998) Mlcache: a flexible multi-lateral cache simulator. Proceedings of MASCOTS'98
8. Edler J, Hill M, Dinero IV (2010) Trace-driven uniprocessor cache simulator, http://www.cs.wisc.edu/~markhill/DineroIV/
9. http://pages.cs.wisc.edu/~larus/spim.html, 2010
10. Sahuquillo J, Tomas N, Petit S, Pont A (2007) Spim-cache: a pedagogical tool for teaching cache memories through code-based exercises. Education, IEEE Transactions on pp 244–250 Aug 2007
11. Stroustrup B (2008) The C++ programming language (special edn). Addison-Wesley
12. Rajeshwari B, Stefan S, Bo-Sik L, Balakrishnan M, Peter M (2002) Scratchpad memory : a design alternative for cache on-chip memory in embedded systems. 10th International Symposium on Hardware/Software Codesign pp 73–78
13. Danckaert K, Catthoor F, De Man H (1999) Platform independent data transfer and storage exploration illustrated on a parallel cavity detection algorithm. CSREA Conference on parallel and distributed processing techniques and applications. pp 1669–1675
14. Jacob B, Spencer N, Wang D (2007) Memory systems: cache, DRAM, disk. Morgan Kaufmann

Chapter 13
Self-Freeze Linear Decompressors: Test Pattern Generators for Low Power Scan Testing

Vasileios Tenentes and Xrysovalantis Kavousianos

Abstract Even though linear decompressors constitute a very effective solution for compressing test data, they cause increased shift power dissipation during scan testing. Recently, new linear decompression architectures were proposed which offer reduced shift power at the expense however of increased test data volume and test sequence length. This chapter presents a linear encoding method which offers both high compression and low shift power dissipation at the same time. A low-cost, test-set-independent scheme is also described which can be combined with any linear decompressor for reducing the shift power during testing. Extensive experiments show that the new method offers reduced test power dissipation, test sequence length and test data volume at the same time, with very small area requirements.

Keywords LFSR · Scan testing · Low switching activity

13.1 Introduction

Currently, the most widely adopted test strategy is Test Resource Partitioning. According to this strategy, the test data are stored in a compressed form in the ATE (Automatic Test Equipment) memory, and they are downloaded on chip where they are decompressed by embedded decompressors before they are applied on the core under test (CUT). Many efficient test data compression techniques have been presented so far in the literature. Some of them utilize compression codes [2, 3, 6] while others utilize various broadcasting schemes [15]. However, the most widely

V. Tenentes (✉) · X. Kavousianos
Department of Computer Science, University of Ioannina, Ioannina, Greece
e-mail: tenentes@cs.uoi.gr

N. Voros et al. (eds.), *VLSI 2010 Annual Symposium*,
Lecture Notes in Electrical Engineering 105, DOI: 10.1007/978-94-007-1488-5_13,

adopted test data compression strategy in industry is based on linear decompressors [7, 8, 10, 12]. Linear decompressors constitute an effective means for exploiting the large volumes of unspecified ('X') values existing in test data in order to maximize the compression.

Although linear decompressors are very effective in compressing the test data, they elevate the power dissipation during testing above the functional power budget of the circuit. They fill the 'X' values pseudorandomly and they increase thus both the shift and capture power during scan testing. In particular, shift power dissipation is caused by successive complementary logic values shifted into the scan chains which generate transitions at the scan cells (and inevitably at the combinational part of the circuit) as they travel through the scan chains. Increased switching activity during the scan in–out process is responsible for increased average power dissipation and consequently increased heat dissipation which elevates the temperature of the chip beyond the acceptable limits.

Recently, various linear decompressors which reduce the switching activity of the CUT during scan testing were presented in [4, 5, 9, 11, 13]. The state-of-the-art low power dynamic reseeding method proposed in [5, 11] is very effective in reducing the switching activity during scan testing, but it requires additional test data compared to standard dynamic reseeding [12] for controlling the low power operation of the decompressor.

The method proposed in [13] offers low shift power dissipation and high compression efficiency at the same time. This technique exploits inherent properties of the test data to provide a fairly simple and low-cost weighted pseudorandom scheme which controls the decompression process and enables the power efficient encoding of test data, without the need of any additional control data. The major advantages of this scheme are: (a) it constitutes a generic test-set-independent architecture, and (b) it can be combined with any linear decompressor scheme for reducing shift power. Moreover, it offers a tradeoff between area overhead and shift power reduction. The combined use of this scheme with the test pattern generator proposed in [11] reduces the test data volume of [11] and achieves great power reductions with very small hardware cost.

13.2 Background

Hereafter "test cube" refers to a test pattern consisting of specified ('0' or '1') and unspecified values ('X'), while "test vector" refers to a completely specified test pattern.

Figure 13.1 presents the classical scan based architecture. The CUT consists of c scan chains of length r (for simplicity we assume that all scan chains are of equal length). The compressed test data are downloaded from the ATE, they are decompressed using the embedded decompressor and they are shifted into the scan chains. For applying a test vector to the CUT the decompressor first generates

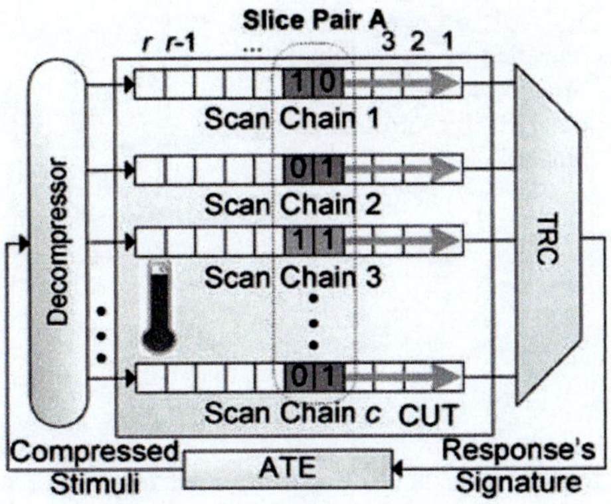

Fig. 13.1 Switching activity caused by successive slices

r successive test slices of size *c* which are shifted into the scan chains to reach their respective scan slices (hereafter, the term test slice t_j refers to the test bits of test cube *t* which correspond to scan slice *j* with $j \in [1, r]$). After the last test slice of *t* (i.e. t_r) is shifted into the scan chains, *t* is applied to the CUT and the response is shifted out concurrently with the loading of the next test vector. Linear decompressors fill 'X' values pseudorandomly, and thus they fail to control the number of incompatibilities between successive test slices.

In Fig. 13.1, every pair of successive test slices exhibits potential bitwise incompatibilities, i.e. pairs of successive complementary test bits loaded into the same scan chains. For example test slices denoted as "Slice Pair A" in Fig. 13.1 are incompatible in the bit positions corresponding to scan chains 1, 2, *c*. As the test slices travel through the scan chains during the scan-in process, every pair of complementary successive test bits causes transitions in the scan chains which propagate through the combinational logic and cause switching activity to the CUT. The number of incompatibilities between successive test slices can be reduced by exploiting the unspecified values which exist in large volumes in test sets. However, linear decompressors fill 'X' values pseudorandomly, and thus they fail to control the number of incompatibilities between successive test slices.

Recently, the authors of [11] proposed a linear based encoding method which exploits the 'X' values, wherever they exist, to reduce incompatibilities between successive test slices, and thus to reduce shift power. According to this method, whenever a group of *k* ($k > 1$) successive test slices of a test cube are compatible (i.e., every slice in this group exhibits no bitwise incompatibilities with any other slice in this group) one test slice S_k is computed which is compatible with all *k* test slices. This slice is encoded using the ring generator and it is loaded into the scan chains for *k* successive clock cycles. This is achieved by the use of a shadow register shown in Fig. 13.2 which can hold its contents if it is properly controlled. Specifically, instead of generating the first slice of this group, the ring generator generates slice S_k and it transfers this slice to the shadow register. This is called UPDATE operation. During the next *k* successive clock cycles, the shadow

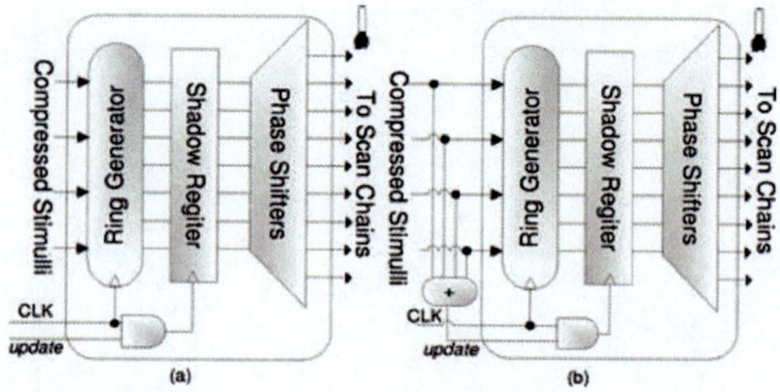

Fig. 13.2 **a** Low power EDT controlled by an additional "update" channel, **b** Low power EDT controlled by compressed stimuli

register holds its contents and loads the scan chains with slice S_k. This is called HOLD operation. The selection between these two operations of the shadow register requires additional control data which are either provided directly from the ATE (Fig. 13.2a) or they are encoded as compressed stimuli (Fig. 13.2b). In both cases the additional cost is considerable especially when the number of ATE channels is small and the number of slices per vector is large.

The additional control data can be completely eliminated by exploiting inherent properties of the test data. Specifically, during the generation of test slice t_j of any test cube t, the Update operation occurs with a unique probability. This probability depends solely on the test cubes and in particular on the probability a test slice t_j to be incompatible with the test slices corresponding to its predecessor test slices (i.e. t_{j-1}, t_{j-2},...) for any test cube t. By controlling the Update operation using predetermined weighted pseudorandom sequences generated by these probabilities, the additional control data are eliminated. Pseudorandomly controlled Update and Hold operations provide high power reduction and they can be easily implemented using embedded low-cost hardware modules. Let as see an example.

Example 1 An uncompacted test set for s9234 was encoded using the method proposed in [11], for $r = 16$, $c = 16$. X-axis in Fig. 13.3 shows the index of each scan slice. For each scan slice left y-axis shows the percentage of test cubes where the respective test slice was incompatible with its predecessor groups of $k \geq 1$ successive compatible test slices (line labeled "Probability of incompatibilities"). The right y-axis presents the percentage of test vectors which triggered an Update operation at scan slice i (line labeled "Update Operation"). The number of test vectors is smaller than the number of test cubes, as the ring generator encodes multiple test cubes on the same test vector (this elevates a slice's probability of incompatibility with its predecessors). The correlation between these two cases is clear.

To estimate the scan power dissipation we will use the metric proposed in [11], which counts the number of invoked transitions in successive scan cells, while taking into account their relative positions. Let t_j^i, t_{j+1}^i be two successive test bits of test vector t loaded into scan chain i, scan slice j. The average shift power dissipated is given by the formula:

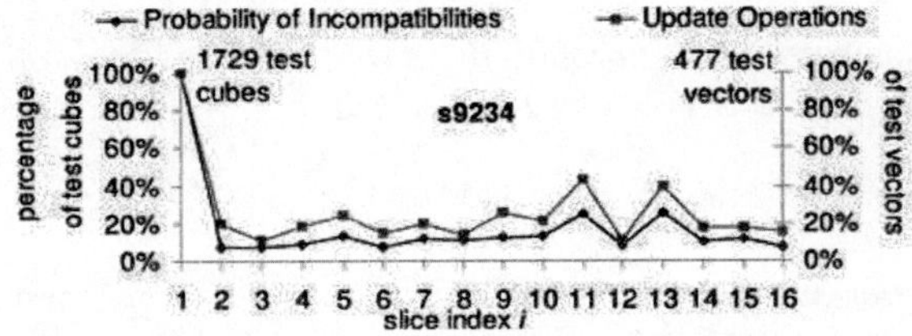

Fig. 13.3 Incompatibilities of the test set and UPDATES number per slices for deterministic freeze (DF)

$$S_{av}(t) = 2[cr(r-1)]^{-1} \sum_{i=1}^{c} \left[\sum_{j=1}^{r-1} (r-j)(t_j^i \oplus t_{j+1}^i) \right] \tag{13.1}$$

13.3 Power Aware Encoding

In this section we will first present the statistical analysis of test data and then we will present the encoding method.

13.3.1 Test Data Analysis

Let *TS* be a test set consisting of *N* test cubes for testing a CUT with *c* scan chains of length *r* (i.e., each test cube consists of *r* test slices of size *c* bits). Hereafter, we will refer to every scan cell using its location in the scan chain structure (for example, scan cell (j, i) is the cell located at the scan slice *j*, scan chain *i*). Let $N_0(j, i)$, $N_1(j, i)$ be the number of test cubes of *TS* with logic value 0, 1 respectively at the scan cell (j, i).

Definition 1 The *Zero* (*One*) *Fill Rate* of scan cell (j, i) is the probability scan cell (j, i) to be assigned to logic value '0' ('1') for any test cube of *TS*.

The *Zero*, *One Fill Rates* of scan cell (j, i) are denoted as $f_0(j, i), f_1(j, i)$ and they are computed as follows: $f_0(j, i) = N_0(j, i)/N$, $f_1(j, i) = N_1(j, i)/N$, with $j \in [1, r]$, $i \in [1, c]$.

Definition 2 The *Zero* (*One*) *Fill Rate* of scan slice *j* ($j \in [1, r]$) is the probability any scan cell of slice *j* to be assigned to logic '0' ('1') for any test cube of *TS*.

The *Zero*, *One Fill Rates* for slice *j* are denoted as $f_0(j), f_1(j)$ respectively and they are computed using formulas:

$$f_0(j) = \frac{1}{c} \cdot \sum_{i=1}^{c} f_0(j,i), \quad f_1(j) = \frac{1}{c} \cdot \sum_{i=1}^{c} f_1(j,i), \quad j \in [1,r]$$

Theorem 1 *The probability two test slices x, y of any test cube in TS to be compatible is given by the formula:*

$$P_{SC}(x,y) = (1 - f_0(x)f_1(y) - f_1(x)f_0(y))^c \tag{13.2}$$

Proof Let x_i, y_i be two bits of test slices x, y corresponding to scan chain i. If x_i, y_i are both specified and complementary then test slices x, y are incompatible. The probability x_i, y_i to be incompatible is equal to $P_{inc}(x_i, y_i) = f_0(x)f_1(y) + f_1(x)f_0(y)$ and thus the probability x_i, y_i to be compatible is equal to $P_c(x_i, y_i) = 1 - P_{inc}(x_i, y_i)$. Slices x, y are compatible when all bit pairs $(x_1, y_1), (x_2, y_2), \ldots, (x_c, y_c)$ are compatible. Thus, $P_{SC^{(x,y)}} = P_{C^{(x_1,y_1)}} \cdot P_{C^{(x_2,y_2)}} \cdot \ldots \cdot P_{C^{(x_C,y_C)}}$ which gives (13.2). □

Lemma 1 *The probability a group of k successive test slices j, j + 1, j + 2,..., j + k − 1 of any test cube in TS to be compatible is:*

$$P_{gc}(j, j+1, \ldots, j+k-1) = \prod_{a=j}^{j+k-2} \prod_{b=j+1}^{j+k-1} P_{sc}(a,b) \tag{13.3}$$

Proof A group of successive test slices is compatible if every two slices in this group are compatible. Thus the probability $P_{gc}(j, j+1, \ldots, j+k-1)$ is equal to the product of the probabilities $P_{sc}(a, b)$ of every possible slice pair a, b (a, $b \in [j, j+k-1]$). So,

$$P_{gc}(j, j+1, \ldots, j+k-1) = \prod_{a=j}^{j+k-2} \prod_{b=j+1}^{j+k-1} P_{sc}(a,b)$$

□

Let $u_j = 1$ ($u_j = 0$) denote the occurrence of an Update (Hold) operation during the generation of the test data loaded into scan slice j ($j \in [1, r]$). Then the *Update* vector $U = (u_1, u_2, \ldots, u_r)$ represents the Update-Hold operations occurring at the shadow register during the generation of a vector. Since the first scan slice of each vector has no predecessors we set $u_1 = 1$, that is an Update operation always occurs during the generation of it for every vector. Let t be a test cube consisting of r test slices i.e. $t = (t_1, t_2, \ldots, t_r)$.

Lemma 2 *Test cube t is encodable for Update vector* $U = (u_1, u_2, \ldots, u_r)$, *if for every* $j \in [1, r]$, $k < r$ *with* $u_j = 1$ *and* $u_{j+1} = u_{j+2} = \cdots = u_{j+k} = 0$ $(j + k \le r)$ test slices $t_j, t_{j+1}, \ldots, t_{j+k}$ *are compatible.*

Proof Since $u_j = 1$, during the generation of the test slice t_j the shadow register will be updated from the linear generator with a test slice s_j, and since $u_{j+1} = \cdots = u_{j+k} = 0$ then the same slice s_j will be loaded into scan slices $j, j+1, \ldots, j+k$. If test slices $t_j, t_{j+1}, \ldots, t_{j+k}$ are compatible then for every i, $i \in [1,c]$ the test bits of all test slices corresponding to scan chain i are either unspecified or exhibit the same logic value ('0' or '1'). Then, the test slice s_j can be computed as follows: for every $i \in [1, c]$ if any of the test slices $t_j, t_{j+1}, t_{j+2}, \ldots, t_{j+k}$ exhibit a logic value $v =$ '0' or $v =$ '1' the respective bit of s_j is set equal to v, else it is left unspecified. In that way s_j is compatible with all test cubes $t_j, t_{j+1}, t_{j+2}, \ldots, t_{j+k}$ and thus test cube t is encodable. □

The most power-efficient Update vector is $U = [1, 0, \ldots, 0]$ which can be used for encoding only those test cubes which have all their slices compatible. On the other hand, the most power consuming but in the same time highly efficient in respect to its encoding ability Update vector is $U = [1, 1, \ldots, 1]$. This vector can encode any test cube which is encodable by the decompressor. In order to maximize the power efficiency of linear decompressors without compromising their encoding efficiency, we need to maximize the volume of zeros in the Update vector of the decompressor and minimize at the same time the probability any test cube to become un-encodable. However, it is rather unlike that a single Update vector will suffice to encode all test cubes. We will show that multiple Update vectors achieving these goals can be generated in a weighted-pseudorandom fashion.

Let R_j be the probability of an Update operation during the generation of scan slice j ($1 - R_j$ is the probability of a Hold operation during the generation of scan slice j). We denote hereafter as *Pseudorandom-Configuration Vector* or simply as *Configuration*, the probability vector $R = [R_1, R_2, \ldots, R_r]$. Since $u_1 = 1$, we also set $R_1 = 1$.

Theorem 2 *The probability any test slice in TS corresponding to scan slice j to be encodable using configuration vector* $R = [R_1, R_2, \ldots, R_r]$ *is given by the formula.*

$$P_E(j) = \sum_{m=1}^{j} R_m \cdot P_{GC}(m, \ldots, j) \cdot \prod_{k=m+1}^{j} (1 - R_k) \tag{13.4}$$

Proof Any arbitrary test slice t_j corresponding to scan slice j is encodable if either the update operation occurs during the generation of this slice or if the update operation occurs during the generation of a predecessor slice t_k (of the same test cube) and all test slices $t_k, t_{k+1}, t_{k+2}, \ldots, t_j$ are bitwise compatible. Therefore, for slice j we have the following (also j in number) cases:

1. $P_1 = R_j$ is the probability of an update operation at slice j.
2. $P_2 = (1 - R_j)R_{j-1}P_{gc}(j - 1, j)$ is the probability the update operation to occur at slice $j - 1$ (and not at slice j) and at the same time test slices $j - 1, j$ to be compatible.

…

j. $P_j = (1 - R_j)(1 - R_{j-1})\ldots(1 - R_2)R_1P_{gc}(1,2,\ldots, j)$ is the probability the update operation to occur at slice 1 (and not at slices 2 … j) and test slices 1, 2, …, j to be compatible.

Thus $P_E(j) = P_1 + P_2 + \cdots + P_j$ which gives (13.4). □

Finally, since every test cube is encodable when all its test slices are encodable, we have that the overall probability P_{ET} for any test cube in *TS* to be encodable using configuration vector $R = [R_1, R_2, \ldots, R_r]$ is given by formula:

$$P_{ET}(R) = P_E(1) \cdot P_E(2) \cdots P_E(r) \tag{13.5}$$

Besides the encoding ability of the decompressor, the Configuration vector *R* affects also the switching activity during the scan-in process, which is calculated as follows.

Theorem 3 *The average scan-in switching activity SC_{av} for any test cube t in TS under Configuration $R = [R_1, R_2, \ldots, R_r]$ is:*

$$SC_{av}(R) = \frac{1}{r(r-1)} \sum_{j=1}^{r-1} (r-j) R_{j+1} \tag{13.6}$$

Proof Let t_j, t_{j+1} be two successive test slices, and let t_j^i, t_{j+1}^i be the test bits of these slices which correspond to scan chain *i*. Relation (13.1) gives the average switching activity for any test cube *t* in *TS*. The term $t_j^i \oplus t_{j+1}^i$ in (13.1) is equal to '1' if t_j^i, t_{j+1}^i are different else it is equal to '0'. Given a Configuration vector *R*, these bits can be different only if an update operation occurs during the generation of slice t_{j+1}. Since R_{j+1} is the probability of an update operation at slice t_{j+1} and 1/2 is the probability t_{j+1}^i to be generated complementary to t_j^i (assuming linear independent generation) the probability these test bits to be different is $P_{diff}(t_j^i \oplus t_{j+1}^i) = R_{j+1}/2$. Then, relation (13.1) becomes:

$$S_{av}(t) = 2[cr(r-1)]^{-1} \sum_{i=1}^{c} \left[\sum_{j=1}^{r-1} (r-j) P_{diff}(t_j^i, t_{j+1}^i) \right]$$

and provided that *t* is generated using configuration *R* we have:

$$SC_{av}(R) = 2[cr(r-1)]^{-1} \sum_{i=1}^{c} \left[\sum_{j=1}^{r-1} (r-j) \frac{R_{j+1}}{2} \right]$$ which gives (13.6). □

In the next Section we will give an algorithm to compute the configuration vector $R = [R_1, R_2, \ldots, R_r]$ for any given set of test cubes, which maximizes the switching activity reduction and does not violate a minimum encoding probability $P_{ET}(R)$.

13.3.2 Encoding Algorithm

The flowchart of the proposed encoding method is shown in Fig. 13.4 (Fig. 13.4a presents step E3 in details). The main target of the encoding method is to calculate the configuration *R* which offers the minimum average switching activity without

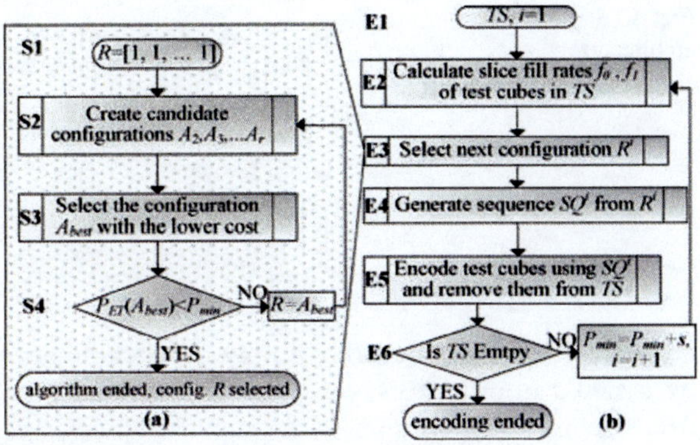

Fig. 13.4 **a** Configuration selection algorithm, **b** Test set encoding

compromising the encoding efficiency of the decompressor. This is shown in Fig. 13.4a. Specifically, $R = [R_1, R_2, \ldots, R_r]$ is initially set equal to [1, 1,...,1] which is the configuration offering the maximum encoding probability $P_{ET}(R) = 1$. Then, the values of R_j ($j \in [2, r]$) are iteratively decreased until $P_{ET}(R)$ drops below a pre-determined threshold P_{min} or when all $R_2, R_3, \ldots, R_r$ reach their minimum values and they cannot be further reduced. We remind that as the values of R_j decrease, both the average switching activity during scan-in and the encoding probability $P_{ET}(R)$ decrease too.

During every iteration, $r - 1$ candidate configurations $A_2, A_3, \ldots, A_{r-1}$ are generated based on R. Specifically, the candidate configuration A_j ($j \in [2, r]$) is derived from R by decreasing the probability R_j by a predetermined value p (all the other probabilities remain intact). Thus $A_j = [R_1, R_2, \ldots, R_j - p, \ldots, R_r]$, with $j \in [2, r]$ (note that R_1 is set always equal to 1). Next, candidate configurations are evaluated using the following formula

$$Cost(A_j) = \frac{\Delta P_{ET}(A_j)}{\Delta SC_{av}(A_j)} \quad \text{with} \quad \begin{array}{l} \Delta P_{ET}(A_j) = P_{ET}(A_j) - P_{ET}(R) \\ \Delta SC_{av}(A_j) = SC_{av}(A_j) - SC_{av}(R) \end{array} \tag{13.7}$$

$\Delta P_{ET}(A_j)$ is the reduction of the encoding probability and $\Delta SC_{av}(A_j)$ is the average switching activity reduction of A_j compared to R. The candidate A_{best} with the lower value of $Cost(A_{best})$ is selected and R is set equal to A_{best}.

Usually, one configuration does not suffice to encode all test cubes. Thus, multiple configurations must be generated using the algorithm shown in Fig. 13.4b. The algorithm begins with set TS of test cubes and it selects the first configuration, let say R^1 using the algorithm shown in Fig. 13.4a. Based on R^1, it generates a weighted pseudorandom bit sequence SQ^1, which controls the Update operation during the decompression process (the generation of this sequence is based on pseudorandom properties of simple hardware modules as we will show in the next section). Using SQ^1 the encoding process attempts to encode as many test cubes as possible and it drops the encoded test cubes from TS. This process is repeated and configurations R^2, R^3,... (and thus sequences SQ^2, SQ^3,...)

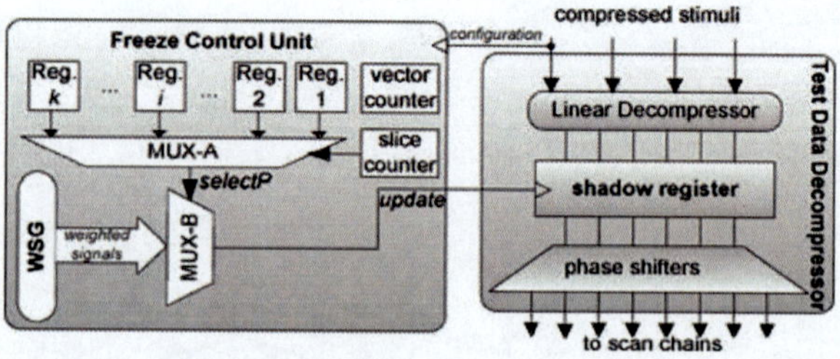

Fig. 13.5 Self-freeze architecture

are selected, until *TS* becomes empty. At each iteration, the value of P_{min} increases by a step *s* in order to favor the encoding ability of the next configurations and decrease thus their volume, at the expense however of an increase in the switching activity. Relations (13.2)–(13.7) are recomputed in each iteration using the remaining set of test cubes.

13.4 Architecture

The low power decompression architecture is shown in Fig. 13.5. It consists of the Test Data Decompression Unit (TDU) and the proposed Freeze Control Unit (FCU). TDU is a classical decompression architecture and it consists of the linear decompressor, the shadow register, and the phase shifter. Even though any linear decompressor can be used, ring generators [10] were used, as in the case of [11]. FCU generates the *update* signal which controls the shadow register based on the configuration *R* (when *update* = 1 the Update operation is applied). It consists of a set of *r* registers which store the configuration vector R_1, R_2, ..., R_r, the slice counter which selects the register for the next generated test slice, and the Weighted Signal Generation Unit (WSG) which generates a set of weighted pseudorandom signals with pre-determined weights. WSG unit generates a set of *n* pseudorandom signals WS_0, WS_1,..., WS_{n-1} with probabilities $W_0 < W_1 < \cdots < W_{n-1}$ respectively. Specifically, signal WS_i is assigned to logic value '1' with probability W_i and to logic value '0' with probability $1 - W_i$. Depending on the configuration *R*, register *j* is loaded from the ATE before the decompression begins with a value *d* in the range $[0, n-1]$. *d* selects the input of MUX-B which corresponds to signal WS_d with probability W_d equal to R_j. Slice counter counts from 1 to *j* and whenever it is equal to *j*, register *j* selects signal WS_d which is driven to the update input of the shadow register. Thus the Update operation is applied with probability R_j during the generation of the test data loaded into scan slice *j*.

Many techniques have been presented in the past for designing WSG units [1, 14]. A small LFSR which is loaded initially with a randomly selected seed is utilized, and a few *AND* gates of 2, 3 and 4 inputs driven by the LFSR cells (note that this small LFSR operates only as a pseudorandom generator and it does not participate in the decompression process). Since each LFSR cell is set to the logic

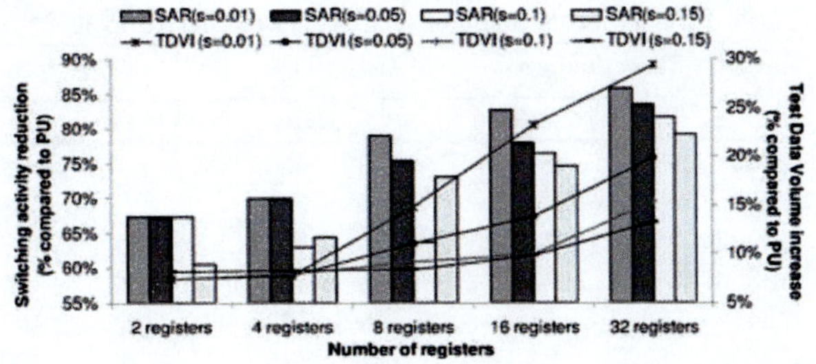

Fig. 13.6 Switching activity reduction, test data volume increase trade-off

value ‘1’ with probability $P_1 = 1/2$, every q-input *AND* gate produces a weighted pseudorandom signal at its output with probability P_1 equal to $(1/2)^q$. By using three *AND* gates of 2, 3 and 4 inputs driven by different LFSR cells, and by using both the normal and the inverted outputs of the *AND* gates, we generate signals with the following P_1 probabilities: 0.0625, 0.125, 0.25, 0.5, 0.75, 0.875, 0.9375. During the encoding process (Fig. 13.4a) the values R_j are selected among the P_1 probabilities only for the remaining test cubes. The encoding of the test cubes is done using the pseudorandom sequences generated at the outputs of the WSG unit. After the calculation of a configuration R^i, the WSG unit is simulated and it generates a pseudorandom sequence SQ^i using signals WS_0, WS_1,..., WS_{n-1}. The predetermined sequence SQ^i is used for encoding remaining test cubes.

The area overhead of this architecture (Fig. 13.5) increases as the number of slices (and thus the number of registers) increases. To overcome this problem an area-efficient alternative architecture will be described which reduces the number of register at the expense of a slight performance degradation. Specifically, k ($k < r$) registers are used and every register corresponds to more than one slices. The registers are assigned to scan slices in a modulo-k fashion. For example, register j is used for controlling the update signal during the generation of scan slices j, $j + k$, $j + 2k$,... etc. (note that scan cell 0 is excluded from this process because an update operation occurs always during the generation of this slice. In this case, the encoding method shown on Fig. 13.5a is modified accordingly in order to consider the reduced set R_1, R_2, ..., R_k. Thus, the process begins with set R where $R_j = R_{j \bmod k}$ and at each iteration k candidate configurations are generated.

13.5 Experiments

The proposed method was developed using the C programming language. We conducted experiments on test sets for complete stuck-at coverage generated using a commercial ATPG tool for the largest ISCAS’89. All the shift power estimations were done using formula (13.1).

Figure 13.6 presents the test data volume (TDV) increase (right y-axis) and the switching activity reduction (SAR at the left y-axis) of the proposed technique

Table 13.1 Proposed method results. TDV reported in Kbits

Circuit	SA reduction		TSL			TDV without repeat			TDV with repeat		
	DF (%)	Prop. (%)	PU	DF	Prop.	PU	DF	Prop.	PU	DF	Prop.
s5378	90	78	250	463	392	7	25	10	3	6.3	5.2
s9234	80	67	309	611	419	19	38	26	7	14	10
s13207	96	78	276	432	380	24	57	33	18	28	22
s15850	94	80	293	511	347	22	60	27	17	30	20
s38417	98	85	626	2374	924	65	493	96	33	124	48
s38584	98	82	267	1088	373	49	300	68	37	150	51

against the power unaware dynamic encoding (PU) method. In both cases the s13207 benchmark circuit was used assuming $c = 16$, $r = 44$ and the proposed method was applied for 2, 4, 8, 16 and 32 registers and various values of parameter s ($s = 0.01$, $s = 0.05$, $s = 0.1$ and $s = 0.15$). It is obvious that as the number of registers increase, the pseudorandom sequences reflect more accurately the specific requirements of the scan slices and thus the switching activity reduction improves. It is worth noting however, that even a relatively small number of registers suffices to achieve very high reduction of the switching activity. On the other hand, the TDV increases as the number of registers increase, because more data are required for loading the registers for every configuration. In respect with parameter s, it is obvious that small values of s improve the power reduction compared to PU but also increase the test data volume. The reason is that small values of s favor the switching activity reduction at the expense however of generation of more configurations.

Table 13.1 presents the results of (a) the proposed technique using 8 registers, (b) the power unaware dynamic encoding (PU) and (c) the deterministic freeze method (DF) presented in [11] and re-implemented here. We note that in the implemented DF method we assume that the control data are sent from the ATE to the CUT using an extra channel (Fig. 13.2a). In all cases 8 or 16 scan chains and 1 or 2 ATE channels were used (excluding the control channel for DF). Note that both DF and PU methods were implemented by omitting the fault simulation step in order to provide fair comparisons with the proposed method (the fault simulation step can be trivially included in all cases). The first column presents the circuit's name, while the next two columns present the average switching activity reduction of both DF and the proposed method against the power unaware method (PU). The next three columns in Table 13.1 present the test sequence length of the PU, the DF and the proposed technique. The results indicate that the proposed technique achieves a vast reduction of the average switching activity (67–85%). Note that, the proposed method is inferior compared to DF with respect to the switching activity. However this is attributed to the high TSL of the DF method which is a consequence of the trend of DF to minimize the volume of Update operations and to limit thus the ability of the decompressor to encode multiple test cubes on the same generated vector. As a result, the number of generated vectors (i.e. the TSL) increases considerably especially for large test sets. Nevertheless, the switching activity of the proposed technique remains significantly lower than PU and thus the probability to comply with the functional power budget of the CUT (which is the most important target of any low power testing technique) is still very high.

The next six columns present the test data volume (TDV) comparisons between the DF and the proposed method. The first three of these columns report the TDV results assuming that the repeat command is not supported by the ATE, while the next three report the TDV results assuming that the repeat command is supported by the ATE. As it has already been mentioned in [11] the use of the repeat command considerably reduces the TDV. The proposed method achieves very high TDV reduction against DF in both cases (in the range of [30–81%] whenever the repeat command is not supported and in the range of [17–66%] whenever the repeat command is supported).

Finally, we synthesized the proposed scheme for 8 registers. The hardware overhead of the proposed FCU unit is less than 100 gate equivalents (one gate equivalent corresponds to a 2-input nand gate). This overhead is less than the 25% of the overhead of the TDU unit. Additionally, we note that the same decompressor can be used for testing any number of cores, which makes its application very attractive to modern SoCs.

13.6 Conclusions

A new linear encoding method which exploits inherent properties of test data to reduce the scan-in switching activity during testing was presented. A low-cost embedded scheme was also presented which can be combined with any linear-decompressor architecture and achieves very high reduction of the switching activity at the expense of only a small increase on the test data volume. Compared to the state-of-the-art power aware linear encoding method, the method described in this chapter provides comparable shift power reduction with considerably lower test data volume.

Acknowledgments This work is co-financed by the European Union (European Social Fund – ESF) and Greek national funds through the Operational Program "Education and LifelongLearning" of the National Strategic Reference Framework (NSRF)—Research Funding Program: Heracleitus II. Investing inknowledge society through the European Social Fund.

References

1. Ahmed N, Tehranipoor M, Nourani M (2004) Low power pattern generation for BIST architecture. Proceedings IEEE ISCAS 2:689–692
2. Chandra A, Chakrabarty K (2001) System-on-a-chip test data compression and decompression architectures based on Golomb codes, IEEE Trans Comput Aided Des Integr Circuits Syst 20:355–368

3. Chandra A, Chakrabarty K (2003) Test data compression and test resource partitioning for system-on-chip using frequency-directed run-length (FDR) codes. IEEE Trans Comput 52(8):1076–1088
4. Czysz D, Mrugalski G, Rajski J, Tyszer J (2007) Low power embedded deterministic test. Proceedings of the IEEE VTS, pp 75–83
5. Czysz D et al (2009) Low-power scan operation in test compression environment. IEEE Trans CAD 28:1742–1755
6. Kavousianos X, Kalligeros E, Nikolos D (2007) Optimal selective Huffman coding for test-data compression. IEEE Trans Comput 56(8):1146–1152
7. Koenemann B (1991) LFSR-coded test patterns for scan designs. Proceedings ETS/ETC., VDE Verlag, pp 237–242
8. Krishna CV, Jas A, Touba NA (2001) Test vector encoding using partial LFSR reseeding. Proceedings of IEEE Int'l Test Conf. pp 885–893
9. Lee J, Touba NA (2007) LFSR-reseeding scheme achieving low-power dissipa-tion during test. IEEE Trans CAD 26(2):396–401
10. Mrugalski G, Rajski J, Tyszer J (2004) Ring generator—New devices for embedded test applications. IEEE Trans Comput Aided Des Integr Circuits Syst 23(9):1306–1320
11. Mrugalski G, Rajski J, Czysz D, Tyszer J (2007) New test data decompressor for low power applications. Proceedings of the ACM/IEEE Design Automation Conference. pp 539–544
12. Rajski J, Tyszer J, Kassab M, Mukherjee N (2004) Embedded deterministic test. IEEE Trans Comp Aided Design Integr Circuits Syst 23(5):776–792
13. Tenentes V, Kavousianos X (2010) Self-freeze linear decompressors for low power testing. IEEE Computer Society Annual Symposium on VLSI (ISVLSI). pp 63–68
14. Wang S, Gupta S (1999) LT-RTPG: a new test-per-scan BIST TPG for low heat dissipation. Proceedings in International Test Conference. pp 85–94
15. Wang LT et al (2005) UltraScan: Using time-division demultiplexing-multiplexing (TDDM/TDM) with VirtualScan for test cost reduction. Proceedings of the IEEE Int'l test conference. pp 946–953

Chapter 14
SUT-RNS Forward and Reverse Converters

E. Vassalos, D. Bakalis and H. T. Vergos

Abstract Stored Unibit Transfer (SUT) has recently been considered as a redundant high-radix encoding for the channels of a Residue Number System (RNS) that can improve the efficiency of conventional redundant RNS. In this work we propose modulo $2^n \pm 1$ forward and reverse converters for the SUT-RNS encoding. The proposed converters are based on parallel-prefix binary or modulo adders and are therefore highly efficient.

14.1 Introduction

The Residue Number System (RNS) [1, 2] is a number system commonly adopted for speeding up computations in digital signal processing [3–6], cryptography [7] and telecommunication applications [8, 9]. A non-positional RNS is defined by a set of L moduli, suppose $\{m_1, \ldots, m_L\}$ that are pair-wise relatively prime. Assume that $|A|_M$ denotes the modulo M residue of an integer A, that is, the least non-negative remainder of the division of A by M. A has a unique representation in the RNS, given by the set $\{a_1, \ldots, a_L\}$ of residues, where $a_i = |A|_{m_i}$. An operation $\otimes$

E. Vassalos · D. Bakalis
Electronics Laboratory, Department of Physics, University of Patras, Patras, Greece
e-mail: vassalos@upatras.gr

D. Bakalis
e-mail: bakalis@physics.upatras.gr

H. T. Vergos (✉)
Department of Computer Engineering and Informatics, University of Patras, Patras, Greece
e-mail: vergos@ceid.upatras.gr

N. Voros et al. (eds.), *VLSI 2010 Annual Symposium*,
Lecture Notes in Electrical Engineering 105, DOI: 10.1007/978-94-007-1488-5_14,

over an RNS is defined as $(z_1, \ldots, z_L) = (a_1, \ldots, a_L) \otimes (b_1, \ldots, b_L)$, where $z_i = |a_i \otimes b_i|_{m_i}$. The computation of each z_i depends only on a_i, b_i, and m_i and therefore all z_is can be computed in parallel in separate arithmetic units often called channels. Any carry propagation in an RNS is restricted inside each channel. Since each channel deals with narrow residues instead of the wide operands and since all channels operate in parallel, significant speedup over the binary may be achieved, provided that the arithmetic components required in each channel can be designed efficiently. RNSs built on moduli of the $2^n \pm 1$ forms have received significant attention due to the efficient architectures that have been proposed for the design of the respective arithmetic components.

The Binary Signed-Digit (BSD) [10] has been proposed as a redundant encoding, in which addition can be performed in constant time. The BSD encoding represents each number with a set of digits in $\{-1, 0, +1\}$. Each digit requires two bits for its representation leading to a significant overhead in storage, processing and interconnection requirements. Hybrid redundant number systems, such as those with weighted two-valued digit set encodings [11, 12], have been proposed as alternatives that limit the maximum length of carry propagation chains to any desired value and can lead to a wide representation range without the added costs of BSD. Furthermore, conventional components such as full/half adders can be utilized for the respective circuit implementations leading to highly efficient designs. Examples of such encodings are the Stored-Unibit Transfer (SUT) encoding [11, 12] and the Signed-LSB encoding [13].

Several attempts have been made to combine the parallel nature of RNS with the carry-free or carry-limited nature of redundant number systems. References [14–16] are among the most recent works that deal with the use of BSD inside each RNS channel for reducing the intra-channel carry propagation. They propose efficient arithmetic circuits, such as adders and multipliers, for the modulo $2^n \pm 1$ cases. The authors of [13] propose modulo $2^n \pm 1$ adders based on the Signed-LSB encoding. In order to trade-off the area overhead of the BSD with the delay, [17, 18] propose the use of the SUT encoding for the modulo $2^n \pm 1$ RNS channels and present SUT-RNS addition, subtraction and multiplication circuits. However, no architecture has been reported so far for converting a binary modulo $2^n \pm 1$ number from/to its corresponding SUT-RNS encoding, making the arithmetic circuits proposed for SUT-RNS in [17, 18] inapplicable.

In this work we fill this gap by presenting forward and reverse converters for modulo $2^n \pm 1$ SUT-RNS encoding. The proposed converters are based on parallel-prefix binary or modulo $2^n \pm 1$ adders and small extra logic and are very efficient.

The rest of this work is organized as follows. The next section presents an overview of the SUT-RNS encoding. Forward and reverse converters for modulo $2^n \pm 1$ SUT-RNS channels are given in Sects. 14.3 and 14.4, respectively. Section 14.5 evaluates the proposed circuits and presents some experimental results. Conclusions are drawn in the last section.

Table 14.1 Dot notation, symbolic notation and binary encoding of twits

Twit	Dot notation	Symbolic notation	Lower value	Upper value
Negabit	○	Y_i	0 (−1)	1 (0)
Posibit	●	y_i	0 (0)	1 (+1)
Unibit	■	y'_i	0 (−1)	1 (+1)

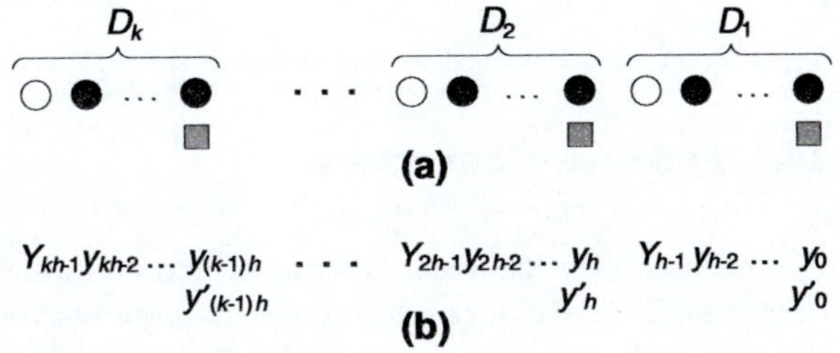

Fig. 14.1 **a** Dot and **b** symbolic notation of a k-digit radix-2^h SUT-RNS number Y

14.2 Redundant High-Radix SUT-RNS

Every SUT-encoded number is composed of SUT digits. Each SUT digit consists of two-valued digits (twits) of three types: posibits {0, +1}, negabits {−1, 0}, and unibits {−1, +1}. A posibit has a lower value equal to 0 whereas a negabit and a unibit have a lower value equal to −1. Furthermore, a posibit and a negabit use a gap size equal to 1 whereas a unibit uses a gap size equal to 2 [11]. All three twits require one bit for their representation. and use bias encoding, that is, their lower value is encoded in binary with logical 0 whereas their upper value is encoded in binary with logical 1. The commonly used dot notations, symbolic notations and binary encodings of the three different twits are presented in Table 14.1.

SUT-RNS has been proposed as a redundant, high-radix, encoding for modulo $2^n \pm 1$ numbers [17, 18]. Every SUT-RNS encoded number consists of k radix-2^h SUT digits ($n = k \times h$), where each SUT digit consists of (h + 1) twits, that is, (h−1) posibits, 1 negabit and 1 unibit. The negabit along with the posibits represent the radix-2^h main part $[-2^{h-1}, +2^{h-1}-1]$ whereas the unibit represents the transfer part of the SUT digit. The symbolic and dot notation of a k-digit ($D_k \ldots D_1$) SUT-RNS number Y are shown in Fig. 14.1. The maximum representable number is equal to $+2^{h-1} \cdot R$ whereas the minimum representable number is equal to $(-2^{h-1}-1) \cdot R$, where $R = (2^{kh}-1)/(2^h-1)$. For example, when $n = 6$ ($k = 2$ and $h = 3$), the range of representable numbers is from −45 up to +36. A modulo $2^n + 1$ ($2^n - 1$) number $X, X \in [0, 2^n]$ ($X \in [0, 2^n - 1]$ assuming a double representation of zero), can be encoded in SUT-RNS by utilizing the positive value range of the SUT-RNS encoding for some values of X and the negative value range for the remaining values. Assuming the values of n, k and h of the previous example, 7 can be encoded as +7 in both moduli cases whereas 37 can be encoded as −28 in modulo $2^6 + 1$ since $|37|_{65} = |-28|_{65}$ and can be encoded as −26 in modulo 2^6-1 since $|37|_{63} = |-26|_{63}$.

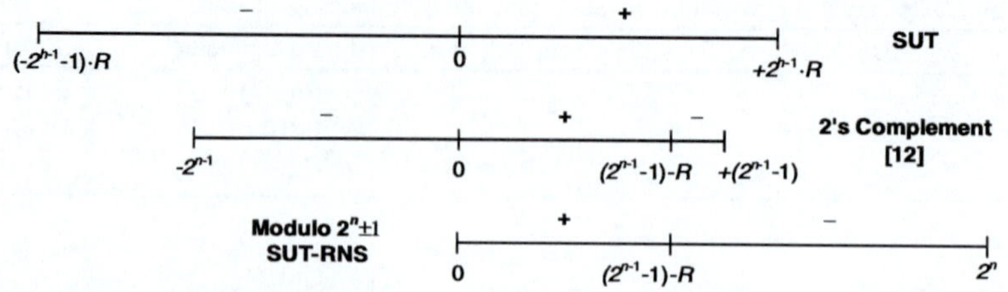

Fig. 14.2 Positive and negative value range of the SUT-RNS encoding

14.3 Forward Converters

In order to utilize the adder, subtractor and multiplier circuits that were proposed in [18] for SUT-RNS, one has to use forward converters to derive the SUT-RNS encodings of the input operands. In [18] no such circuits have been presented. Jaberipur and Parhami [12] reported an algorithm for converting a signed two's complement number to an SUT representation. However this algorithm cannot be directly applied to the SUT-RNS encoding since in this case the input is a modulo $2^n + 1$ or a modulo $2^n - 1$ unsigned number. We present in this section efficient forward converters of a modulo $2^n \pm 1$ number to its corresponding SUT-RNS encoding. We consider the two cases of moduli separately.

14.3.1 Modulo $2^n - 1$

Consider an n-bit modulo $2^n - 1$ number $X = x_{n-1} .. x_0 \in [0, 2^n - 1]$. The algorithm for forward conversion presented in [12] can correctly encode in SUT-RNS every value of X that lies in the positive value range of the SUT-RNS encoding, as shown in Fig. 14.2. However, for all values of X that are encoded in SUT-RNS in the negative value range, a value decreased by one compared to the correct modulo $2^n - 1$ value is produced since modulo 2^n arithmetic is used instead and $|X - 2^n|_{2^n-1} = |X - (2^n - 1) - 1|_{2^n-1} = |X - 1|_{2^n-1}$. Hence, for all the SUT-RNS encoded values of X that lie in the negative value range, we have to increase the corresponding value of X by one in order to get the correct modulo $2^n - 1$ value.

The following two-step algorithm is a modification of the forward conversion algorithm of [12] that deals with the above-mentioned problem and performs correct modulo $2^n - 1$ forward SUT-RNS conversion:

Step I: Compute $z = X + R + s = z' + s$, where z and $R = (2^{kh}-1)/(2^h-1)$ denote n-bit operands, $z' = X + R$, and s denotes a sign indication bit whose value is equal to 0 when the value of X lies in the positive value range of the SUT-RNS encoding and is equal to 1 when the value of X lies in the negative value range. According to Fig. 14.2, $s = \begin{cases} 0, & X + R < 2^{n-1} \\ 1, & X + R \geq 2^{n-1} \end{cases}$. Hence, sign s can be derived by

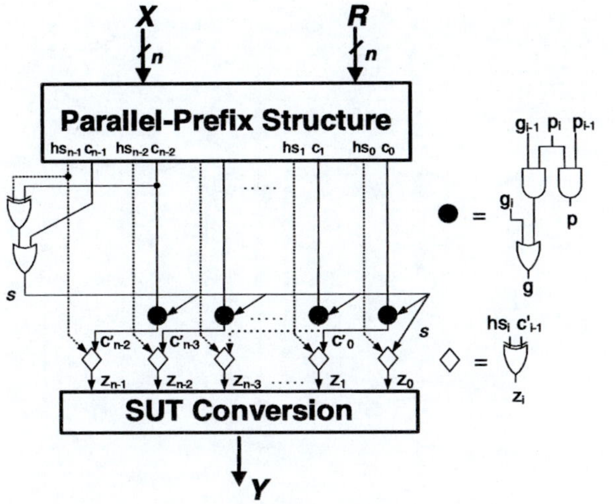

Fig. 14.3 Proposed modulo $2^n - 1$ SUT-RNS forward converter

the logical equation $s = z'_{n-1} \vee c_{n-1}$, where z'_{n-1} and c_{n-1} are the most significant bit and carry out of the $(X + R)$ addition and $\vee$ denotes the logical OR operation. A straightforward solution for deriving the bits of z uses a binary adder for deriving z' and a controllable incrementer for incorporating s. However, those two operations can be efficiently merged in a parallel-prefix-based adder, as shown in Fig. 14.3. The X and R operands are driven to a parallel-prefix structure that derives the n carries $(c_{n-1}, \ldots, c_0)$ of $X + R$ in $\log_2 n$ levels. Then, s can be derived by an XOR and an OR gate, as $s = (hs_{n-1} \oplus c_{n-2}) \vee c_{n-1}$, where $hs_{n-1} = x_{n-1} \oplus R_{n-1}$ is the half-sum bit of the most significant bit position. An extra prefix level can then be used for adding the value of s and for producing the required set of carries $(c'_{n-2}, \ldots c'_0)$.

Finally, the n-bits of z can be derived by 2-input XOR gates. We have to note that since R is a constant that has a value equal to 1 in all bit positions with weights 2^{ih}, $0 \leq i < k$, and a value equal to 0 in all other bit positions, the parallel-prefix structure can be significantly simplified.

Step II: Use the following logic equations [12] to transform the bits of z to the corresponding SUT-RNS encoding of X, denoted as Y, assuming that $z_{-1} = 0$ and that $\wedge$ denotes the logical AND operation, while $\bar{w}$ denotes the complement of bit w:

a. negabits : $Y_{ih-1} = \bar{z}_{ih-1}$, for $1 \leq i \leq k$
b. posibit : $y_{ih} = z_{ih} \oplus z_{ih-1}$, for $0 \leq i \leq k-1$
c. posibits : $y_{ih-j} = z_{ih-j}$, for $1 \leq i \leq k$ and $2 \leq j < h$
d. unibits : $y'_{ih} = z_{ih} \wedge z_{ih-1}$, for $0 \leq i \leq k-1$

Step II implies that the unibit y'_0 is always equal to 0. The complete circuit structure that realizes the above algorithm is given in Fig. 14.3 and is capable of

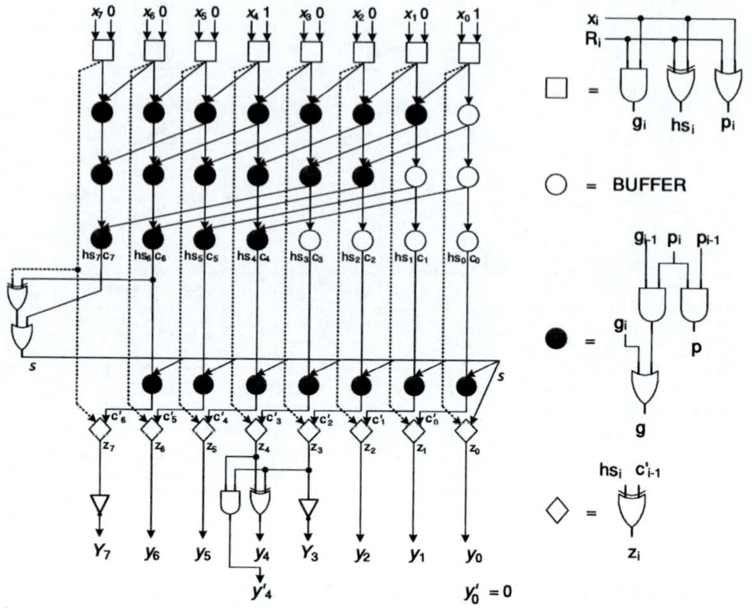

Fig. 14.4 Proposed modulo 2^8-1 SUT-RNS forward converter ($k = 2$, $h = 4$)

dealing with both representations of zero in modulo $2^n - 1$ arithmetic, that is, 0 and $2^n - 1$.

Example 1: Suppose that $n = k \times h = 2 \times 4 = 8$ and $X = 153$. Then $R = 17$, $z' = 170$, $s = 1$ and $z = 153 + 17 + 1 = 171 = 10101011_2$. According to Step II, $Y_7 = \bar{z}_7 = 0$, $y_6 = z_6 = 0$, $y_5 = z_5 = 1$, $y_4 = z_4 \oplus z_3 = 1$, $y'_4 = z_4 \wedge z_3 = 0$, $Y_3 = \bar{z}_3 = 0$, $y_2 = z_2 = 0$, $y_1 = z_1 = 1$, $y_0 = z_0 \oplus z_{-1} = 1$, and $y'_0 = z_0 \wedge z_{-1} = 0$, thus $Y = \left\langle \begin{matrix} Y_7y_6y_5y_4 & Y_3y_2y_1y_0 \\ y'_4 & y'_0 \end{matrix} \right\rangle = \left\langle \begin{matrix} 0011 & 0011 \\ 0 & 0 \end{matrix} \right\rangle = (-6) \times 2^4 + (-6) \times 2^0 =$ $|-102|_{255} = 153$. The architecture of the forward converter for $n = 8$, $k = 2$ and $h = 4$, assuming a Kogge-Stone parallel-prefix structure [19], is shown in Fig. 14.4. It is obvious that since operand R is actually a constant, several simplifications are possible not only in the pre-processing level but in the prefix levels of the parallel prefix structure as well.

14.3.2 Modulo $2^n + 1$

Consider now a $(n + 1)$-bit modulo $2^n + 1$ number $X = x_n x_{n-1} \cdots x_0 \in [0, 2^n]$. The forward converter of [12] could be used to encode X in SUT-RNS. However, for all values of X that lie in the negative value range of the SUT-RNS encoding, an increased by one value compared to the correct one would be produced since $|X - 2^n|_{2^n+1} = |X - (2^n + 1) + 1|_{2^n+1} = |X + 1|_{2^n+1}$. Hence, for all these values of X we have to decrease by one in order to get the correct modulo $2^n + 1$ SUT-RNS encoding.

The following algorithm is similar to the one previously presented for modulo $2^n - 1$ and performs modulo $2^n + 1$ SUT-RNS forward conversion.

Step I: Compute $z = X + R - s = X + (R - 1) + \bar{s} = x_n 2^n + z' + \bar{s}$ where z and $R = (2^{kh}-1)/(2^h-1)$ denote n-bit operands, $z' = |X|_{2^n} + (R - 1)$, and s denotes the sign indication bit. According to Fig. 14.2,

$$s = \begin{cases} 0, & X + R < 2^{n-1} \\ 1, & X + R \geq 2^{n-1} \end{cases} = \begin{cases} 0, & X + (R-1) < 2^{n-1} - 1 \\ 1, & X + (R-1) \geq 2^{n-1} - 1 \end{cases}$$
$$= \begin{cases} 0, & X + (R-1) < 2^{n-1} - 1 \\ 1, & X + (R-1) \geq 2^{n-1} \\ 1, & X + (R-1) = 2^{n-1} - 1 \end{cases}.$$

The first two conditions can be identified by the logical equation $s = z'_{n-1} \vee c_{n-1} \vee x_n$, where z'_{n-1} and c_{n-1} are the most significant bit and carry out of the $|X|_{2^n} + (R - 1)$ addition, respectively. Note that s also incorporates the most significant bit of X, x_n, in order to add the value $x_n 2^n$. Hence, the n least significant bits of X and $(R - 1)$ are driven to an n-bit parallel-prefix structure. Then $\bar{s}$ is derived by an XOR and a NOR gate while an extra parallel prefix level is used to add the value of $\bar{s}$ and produce z. Similarly to the previous modulo case, simplifications inside the parallel-prefix structure are possible, due to the fact that one of its inputs $(R - 1)$ consists of constant bits.

Step II: Use the same logic equations as in the modulo $2^n - 1$ case to transform the bits of z to the corresponding SUT-RNS encoding Y. The above algorithm produces the correct SUT-RNS encoding for all values of X except when $X + (R - 1) = 2^{n-1} - 1$ (third condition of the previous equation). In this case we still have to subtract one. This can be easily achieved by deriving a signal m indicating the case where $X + (R - 1) = 2^{n-1} - 1$ and correcting in this case the SUT-RNS encoding only of the least significant SUT digit. The logic equation for m is: $m = \overline{hs_{n-1}} \wedge hs_{n-2} \wedge \ldots \wedge hs_0$, where $hs_{n-1}, \ldots, hs_0$ denote the half-sum bits of the parallel-prefix structure. m can be derived as fast as the carries of the parallel-prefix structure and hence it doesn't increase the delay of the forward converter. The twits of the least significant SUT digit are then derived by the following logic equations:

a. negabit : $Y_{h-1} = m \oplus \bar{z}_{h-1}$
b. posibit : $y_1 = z_1$
c. posibits : $y_{h-j} = m \oplus z_{h-j}$, for $2 \leq j \leq h, j \neq h - 1$
d. unibit : $y'_0 = m$

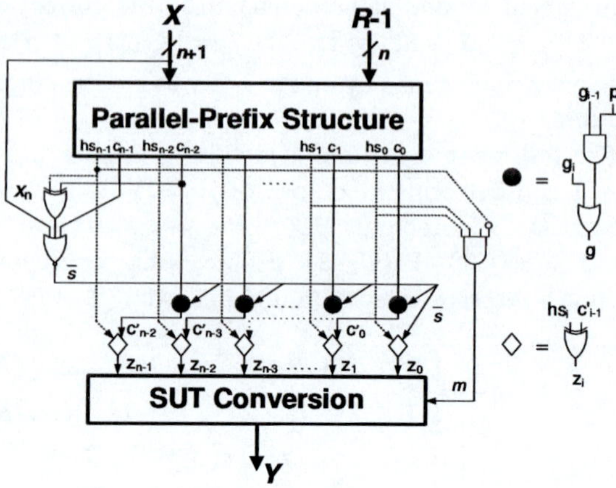

Fig. 14.5 Proposed modulo $2^n + 1$ SUT-RNS forward converter

Hence, unibit y'_0 is equal to 1 only when $X + (R - 1) = 2^{n-1} - 1$. The complete circuit structure that realizes the above algorithm is given in Fig. 14.5.

Example 2: Suppose that $n = k{\times}h = 2{\times}4 = 8$ and $X = 153$. Then $R = 17$, $\bar{s} = 0$, $m = 0$ and $z = 153 + 16 + 0 = 169 = 10101001_2$. According to Step II, $Y_7 = \bar{z}_7 = 0$, $y_6 = z_6 = 0$, $y_5 = z_5 = 1$, $y_4 = z_4 \oplus z_3 = 1$, $y'_4 = z_4 \wedge z_3 = 0$, $Y_3 = m \oplus \bar{z}_3 = 0$, $y_2 = m \oplus z_2 = 0$, $y_1 = z_1 = 0$, $y_0 = m \oplus z_0 = 1$, and $y'_0 = m = 0$, thus $Y = \left\langle \begin{matrix} Y_7y_6y_5y_4 & Y_3y_2y_1y_0 \\ y'_4 & y'_0 \end{matrix} \right\rangle = \left\langle \begin{matrix} 0011 & 0001 \\ 0 & 0 \end{matrix} \right\rangle = (-6) \times 2^4 + (-8) \times 2^0$ $= |-104|_{257} = 153$. Figure 14.6 presents a detailed view of the modulo $2^8 + 1$ forward converter, assuming a Kogge-Stone parallel-prefix structure.

14.4 Reverse Converters

We present in this section efficient reverse converters of an SUT-RNS encoded modulo $2^n \pm 1$ number to its corresponding binary encoding.

14.4.1 Modulo $2^n - 1$

In order to get the binary encoding X of an SUT-RNS encoded modulo $2^n - 1$ number Y, we need to add in modulo $2^n - 1$ the following 4 n-bit vectors, as shown in dot notation in Fig. 14.7:

- The $P = 0y_{kh-2}\dots y_{(k-1)h}\ 0y_{(k-1)h-2}\dots y_{(k-2)h}\dots\ 0y_{h-2}\dots y_0$ posibits vector.

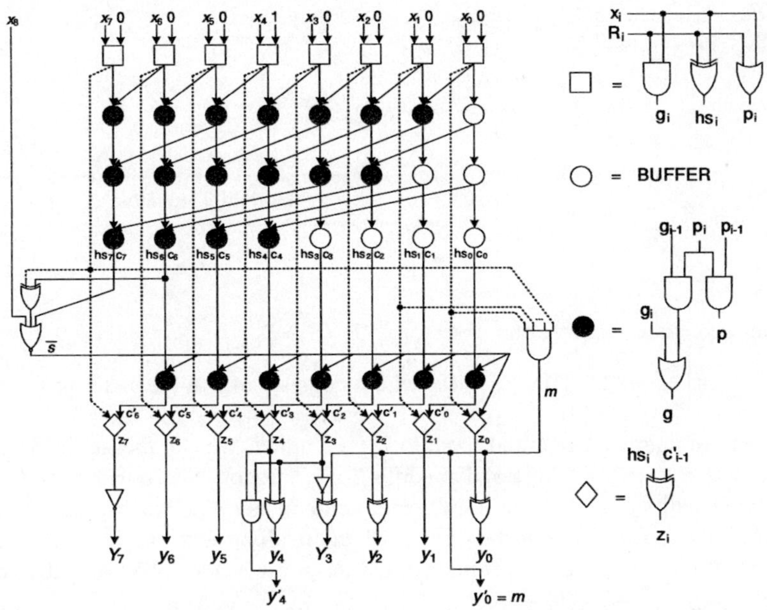

Fig. 14.6 Proposed modulo $2^8 + 1$ SUT-RNS forward converter ($k = 2$, $h = 4$)

Fig. 14.7 Vector formation for modulo 2^n-1 SUT-RNS reverse conversion

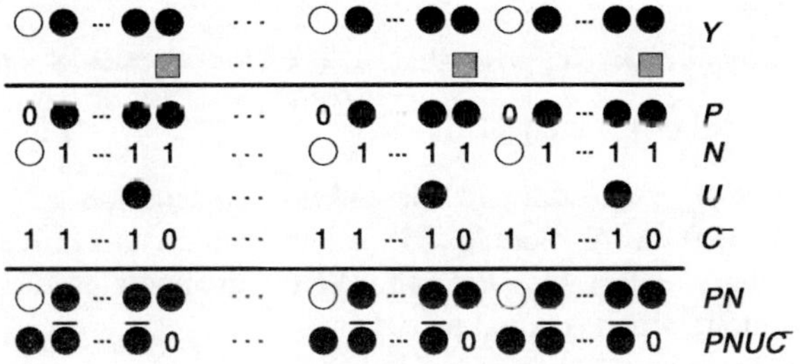

- The negabits vector denoted as N. Vector N in modulo $2^n - 1$ arithmetic is equal to $N = Y_{kh-1}1\ldots1\ Y_{(k-1)h-1}1\ldots1 \ldots Y_{h-1}1\ldots1$. This is justified as follows: Due to the bias encoding, a negabit n_i with a weight equal to 2^i represents a value equal to $-2^i\bar{n}_i$. Hence, $N = \left|\sum_{i=1}^{k}\left(-2^{ih-1}\bar{Y}_{ih-1}\right)\right|_{2^n-1} = \left|(2^n-1) - \sum_{i=1}^{k}\left(2^{ih-1}\bar{Y}_{ih-1}\right)\right|_{2^n-1}$. Since, for every bit w it holds that $1-\bar{w}=w$, we conclude that $N = Y_{kh-1}1\ldots1\ Y_{(k-1)h-1}1\ldots1 \ \ldots\ Y_{h-1}1\ldots1$, that is, it consists of k h-bit patterns $Y_{ih-1}1\ldots1$, $1 \le i \le k$.
- The unibits vector denoted as U. Unibits can be treated as doublebits or equivalently as posibits in the next higher bit position, as long as we also consider a correction equal to $-R$. Hence, $U = 0\ldots0y'_{(k-1)h}0\ 0\ldots0y'_{(k-2)h}\,0 \ldots 0\ldots0y'_00$.

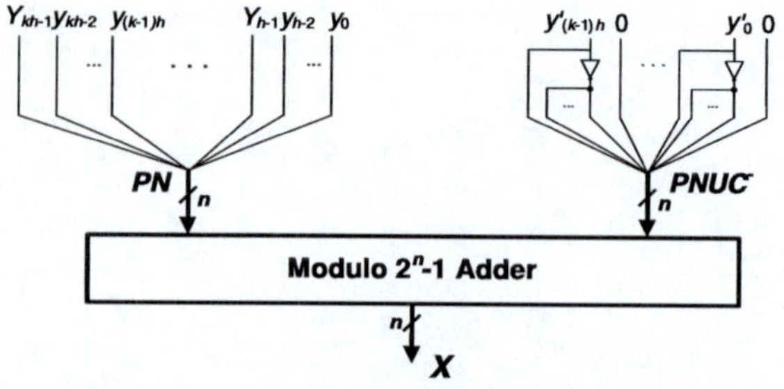

Fig. 14.8 Proposed modulo 2^n-1 SUT-RNS reverse converter

- The constant correction vector $C^- = \left|-R\right|_{2^n-1} = \left|(2^n-1)-\left(2^{kh}-1\right)/\left(2^h-1\right)\right|_{2^n-1} = 1\ldots10 \ldots 1\ldots10$, which consists of k h-bit patterns $1\ldots10$.

Instead of using a 4-operand modulo $2^n - 1$ adder, we can merge the 4 vectors in two and use only a 2-operand modulo $2^n - 1$ adder. The posibits of P along with the negabits of N form an n-bit vector denoted as PN. PN is actually the main part of Y. The remaining constant bits of P and N along with the U vector and the constant vector C^- can be replaced by an n-bit vector $PNUC^-$ defined as $PNUC^- = b_{n-1}\ldots b_0$, where $b_{(i+1)h-1}\ldots b_{ih} = y'_{ih}\overline{y'_{ih}}\ldots\overline{y'_{ih}}0$, $0 \le i \le k-1$, that is, $PNUC^-$ consists of repeating h-bit patterns of $y'_{ih}\overline{y'_{ih}}\ldots\overline{y'_{ih}}0$, with $0 \le i \le k-1$. As a result, only the PN and $PNUC^-$ vectors need to be added; these are driven in a modulo $2^n - 1$ adder in order to derive the binary encoding X, as shown in Fig. 14.8.

Example 3: Suppose that $n = k \times h = 2 \times 4 = 8$ and $X = 104$. The SUT-RNS encoding of X is equal to $Y = \left\langle \begin{matrix} Y_7 y_6 y_5 y_4 & Y_3 y_2 y_1 y_0 \\ y'_4 & y'_0 \end{matrix} \right\rangle = \left\langle \begin{matrix} 1110 & 0001 \\ 1 & 0 \end{matrix} \right\rangle = 7 \times 2^4 - 8 \times 2^0$. According to the previous discussion $PN = 11100001$ and $PNUC^- = 10000110$. A modulo 255 adder (which is equivalent to an end-around-carry binary adder) with PN and $PNUC^-$ as inputs produces the value 01101000 at the output which is equal to 104.

14.4.2 Modulo $2^n + 1$

A similar approach can be used in the modulo $2^n + 1$ case as well. In order to get the binary encoding X of an SUT-RNS encoded modulo $2^n + 1$ number Y, we need to add in modulo $2^n + 1$ arithmetic 4 n-bit vectors, as shown in Fig. 14.9: vectors P, N and U for the posibits, negabits and unibits, respectively, which are equal to those in the modulo $2^n - 1$ case and a constant correction vector C^+ which in the case of modulo $2^n + 1$ is equal to $C^+ = \left|2 - R\right|_{2^n+1}$. The constant term 2 is justified by the fact that the negabits vector in modulo $2^n - 1$ and the corresponding negabits vector in modulo $2^n + 1$ always differ by 2.

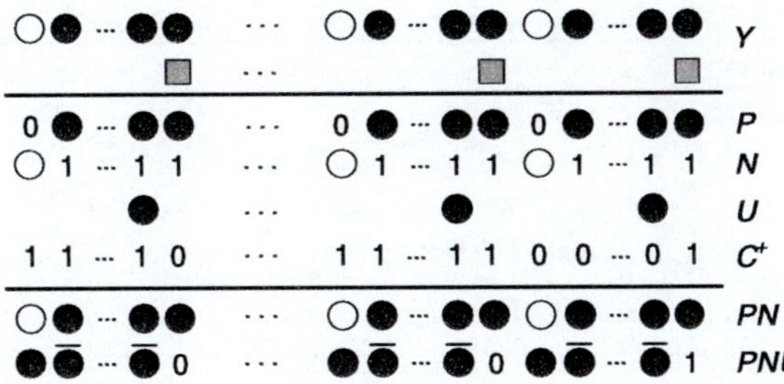

Fig. 14.9 Vector formation for modulo $2^n + 1$ SUT-RNS reverse conversion

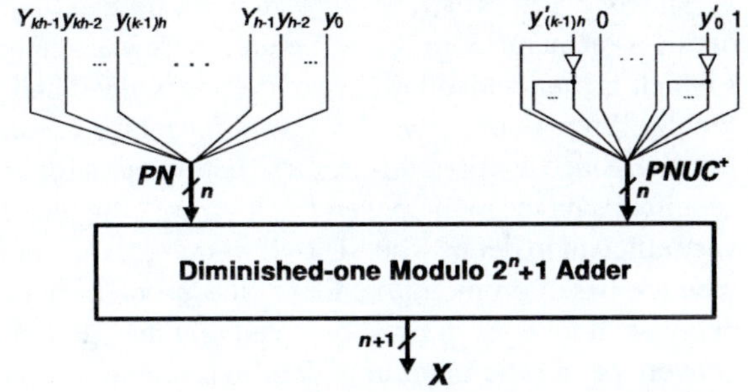

Fig. 14.10 Proposed modulo $2^n + 1$ SUT-RNS reverse converter

The 4 vectors can be merged in two: the *PN* vector which is the main part of the SUT-RNS encoded number *Y* and the $PNUC^+$ vector which depends on the transfer part of *Y* and is equal to $PNUC^+ = b_{n-1}\ldots b_0$, where $b_{h-1}\ldots b_0 = y'_{ih}\overline{y'_{ih}}\ldots\overline{y'_{ih}}1$ and $b_{(i+1)h-1}\ldots b_{ih} = y'_{ih}\overline{y'_{ih}}\ldots\overline{y'_{ih}}0, 1 \leq i \leq k-1$. The two *n*-bit vectors *PN* and $PNUC^+$ are then driven to an enhanced diminished-one modulo $2^n + 1$ adder [20] that produces the $(n + 1)$-bit binary encoding *X*, as shown in Fig. 14.10. We have to note that in $PNUC^+$ a constant correction term equal to -1 is also taken into account since a diminished-one adder always increases the sum of its two input operands by one.

Example 4: Let *n*, *h*, *k* and *X* have the same values as in the previous example. The SUT-RNS encoding of *X* is equal to $Y = \left\langle \begin{matrix} Y_7y_6y_5y_4 & Y_3y_2y_1y_0 \\ y'_4 & y'_0 \end{matrix} \right\rangle = \left\langle \begin{matrix} 1110 & 0001 \\ 1 & 0 \end{matrix} \right\rangle$. Then $PN = 11100001$ and $PNUC^+ = 10000111$. An enhanced diminished-one modulo 257 adder sums *PN* and $PNUC^+$ and produces the value 001101000 at its output which is equal to 104.

14.5 Evaluation and Experimental Results

In this section we evaluate the forward and reverse converters that were proposed in Sects. 14.3 and 14.4, respectively, and we present some experimental results based on CMOS VLSI circuit implementations.

Table 14.2 Unit-gate area and delay requirements of the proposed circuits

Architecture	Delay	Area
SUT-RNS Forward Converters		
Modulo $2^n - 1$	$2\log n + 8$	$3n\log n + n + (7k-3)$
Modulo $2^n + 1$	$2\log n + 9$	$3n\log n + 2n + (7k + 1)$
SUT-RNS Reverse Converters		
Modulo $2^n - 1$	$2\log n + 3$	$3n\log n + 4n$
Modulo $2^n + 1$	$2\log n + 3$	$(9/2)n\log n + (1/2)n + 5$

The SUT-RNS forward converters for both modulo $2^n - 1$ and $2^n + 1$ are based on an n-bit parallel-prefix structure. A few gates are used to derive the sign bit s which is then added with an extra prefix level and a level of 2-input XOR gates. Finally, Step II of forward conversion requires some extra gates that operate in parallel. Since the parallel-prefix structure has a logarithmic delay and all remaining subcircuits have small constant delays, we conclude that the forward converters are very efficient in delay. The SUT-RNS reverse converters are also very efficient since they are based on modulo adders whose input operands are formed at a minimum delay of an inverter. It must be noted that the parallel-prefix structure in the forward converters and the modulo adders in the reverse converters can be designed using any desirable architecture, while several simplifications are possible.

Table 14.2 summarizes the area and delay requirements, in gate equivalents, of the proposed architectures, according to the unit gate model [21]. In the forward converters case, we assume a Kogge-Stone parallel-prefix structure, whereas in the reverse converters case we assume the modulo $2^n - 1$ and diminished-one modulo $2^n + 1$ adders architectures of [22] and [23], respectively.

For including in the results all possible logic simplifications, we described in HDL forward and reverse converters for both moduli cases and for several values of n, k and h. After validating the correct operation of the HDL descriptions via simulation, we synthesized them in a power-characterized 90 nm CMOS technology, using a standard delay optimization script, and derived estimates for area, delay and average power dissipation. The attained results, given in Table 14.3, indicate that the proposed converters are very fast and require small area and power dissipation. Since we are not aware of any other work on forward and reverse modulo $2^n \pm 1$ SUT-RNS converters, no comparison with other proposals is possible.

14.6 Conclusions

Redundant number systems can be used to reduce the carry propagation inside each channel of an RNS. SUT has been proposed as a redundant high-radix encoding for RNS that can improve the efficiency of BSD-based RNS since it can utilize conventional arithmetic components such as full/half adders. We have

Table 14.3 CMOS experimental results

n	*k*	*h*	SUT-RNS forward converters			SUT-RNS reverse converters		
			Area (um^2)	Delay (ns)	Power (mW)	Area (um^2)	Delay (ns)	Power (mW)
Modulo $2^n - 1$								
8	2	4	864	0.235	0.25	974	0.172	0.37
12	4	3	1462	0.270	0.50	1932	0.212	0.69
12	3	4	1457	0.269	0.43	1813	0.215	0.67
16	4	4	2218	0.278	0.62	2592	0.211	0.97
20	5	4	2554	0.313	0.70	3757	0.259	1.32
20	4	5	2485	0.307	0.61	3809	0.252	1.34
Modulo $2^n + 1$								
8	2	4	1191	0.250	0.29	1533	0.170	0.62
12	4	3	1559	0.285	0.47	2601	0.209	0.99
12	3	4	1467	0.287	0.37	2842	0.209	1.14
16	4	4	2345	0.292	0.59	3426	0.217	1.36
20	5	4	2777	0.331	0.73	5129	0.252	2.04
20	4	5	2706	0.320	0.61	5386	0.251	2.17

presented in this work efficient forward and reverse converters for the SUT-RNS encoding for the two most commonly used moduli cases, that is, modulo $2^n \pm 1$. The forward converters are based on parallel-prefix binary adders and simple logic gates whereas the reverse converters are based on modulo $2^n \pm 1$ adders and simple inverters.

References

1. Ananda Mohan PV (2002) Residue number systems: algorithms and architectures. Kluwer, Netherlands
2. Omondi A, Premkumar B (2007) Residue number systems: theory and implementation. Imperial College Press, London
3. Chaves R, Sousa L (2003) RDSP: a RISC DSP based on residue number system. In: Proceedings of 6th Euromicro symposium on digital system design, pp 128–135. doi: 10.1109/dsd.2003.1231911
4. Fernandez PG, Lloris A (2003) RNS-based implementation of 8x8 point 2D-DCT over field-programmable devices. Electron Lett 39:21–23. doi:10.1049/el:20030084
5. Liu Y, Lai E (2004) Moduli set selection and cost estimation for RNS-based FIR filter and filter bank design. Des Autom Embed Syst 9:123–139. doi:10.1007/s10617-005-1186-4
6. Cardarilli G, Nannarelli A, Re M (2007) Residue number system for low-power DSP applications. In: Proceedings of asilomar conference on signals, systems and computers, pp 1412–1416. doi:10.1109/acssc.2007.4487461
7. Bajard JC, Imbert L (2004) A full RNS implementation of RSA. IEEE Trans Comput 53:769–774. doi:10.1109/tc.2004.2
8. Meyer-Baese U, Garcia A, Taylor F (2001) Implementation of a communications channelizer using FPGAs and RNS arithmetic. J VLSI Signal Process 28:115–128. doi:10.1023/a:1008167323437

9. Madhukumar AS, Chin F (2004) Enhanced architecture for residue number system-based CDMA for high-rate data transmission. IEEE Trans Wireless Commun 3:1363–1368. doi: 10.1109/twc.2004.833509
10. Avizienis A (1961) Signed-digit representation for fast parallel arithmetic. IRE Trans Electron Comput EC-10:389–400. doi:10.1109/tec.1961.5219227
11. Jaberipur G, Parhami B, Ghodsi M (2005) Weighted two-valued digit-set encodings: unifying efficient hardware representation schemes for redundant number systems. IEEE Trans Circuits Syst I 52:1348–1357. doi:10.1109/tcsi2005.851679
12. Jaberipur G, Parhami B (2007) Stored-transfer representations with weighted digit-set encodings for ultrahigh-speed arithmetic. IET Circuits Devices Syst 1:102–110. doi:10.1049/iet-cds:20050228
13. Jaberipur G, Parhami B (2009) Unified approach to the design of modulo-($2^n \pm 1$) adders based on signed-LSB representation of residues. In: Proceedings of IEEE international symposium on computer arithmetic, pp 57–64. doi:10.1109/arith.2009.14
14. Lindstrom A, Nordseth M, Bengtsson L, Omondi A (2003) Arithmetic circuits combining residue and signed-digit representations. In: Proceedings of 8th Asia-Pacific computer systems architecture conference, pp 246–257. doi:10.1007/978-3-540-39864-6_20
15. Wei S (2008) A new residue adder with redundant binary number representation. In: Proceedings of 6th international IEEE north-east workshop on circuits and systems, pp 157–160. doi:10.1109/newcas.2008.4606345
16. Persson A, Bengtsson L (2009) Forward and reverse converters and moduli set selection in signed-digit residue number systems. J Signal Process Syst 56:1–15. doi:10.1007/s11265-008-0249-8
17. Timarchi S, Navi K (2007) Efficient class of redundant residue number system. In: Proceedings of IEEE international symposium on intelligent signal processing, pp 475–780. doi:10.1109/wisp.2007.4447506
18. Timarchi S, Navi K (2009) Arithmetic circuits of redundant SUT-RNS. IEEE Trans Instrum Meas 58:2959–2968. doi:10.1109/tim.2009.2016793
19. Kogge PM, Stone HS (1973) A parallel algorithm for the efficient solution of a general class of recurrence equations. IEEE Trans Comput 22:786–792. doi:10.1109/tc.1973.5009159
20. Vergos HT, Bakalis D, Efstathiou C (2010) Fast modulo $2^n + 1$ multi-operand adders and residue generators. Integr VLSI J 43:42–48. doi:10.1016/j.vlsi.2009.04.002
21. Tyagi A (1993) A reduced-area scheme for carry-select adders. IEEE Trans Comput 42: 1163–1170. doi:10.1109/12.257703
22. Kalampoukas L, Nikolos D, Efstathiou C, Vergos HT, Kalamatianos J (2000) High-speed parallel prefix modulo 2^n–1 adders. IEEE Trans Comput 49:673–680. doi:10.1109/12.863036
23. Vergos HT, Efstathiou C, Nikolos D (2002) Diminished-one modulo $2^n + 1$ adder design. IEEE Trans Comput 51:1389–1399. doi:10.1109/tc.2002.1146705

Chapter 15
Off-Chip SDRAM Access Through Spidergon STNoC

Khaldon Hassan and Marcello Coppola

Abstract External memory access in MPSoCs becomes more challenging with the growing requirements for high bandwidth and low latency. We propose a novel method for optimizing external memory access in term of latency for NoC-based MPSoCs. Our approach considers the off-chip memory access within a system approach: from the initiators to the memory modules through the NoC-based interconnect. We couple QoS of both NoC and memory scheduler in order to guarantee continued services throughout the request and the response paths, between the masters and the SDRAM modules. We study the influence of low-priority requests over high-priority requests. We also analyze the influence of the number of the conflict points inside the NoC over high-priority requests latency. We compare the use of virtual channels with the physical direct connection to map latency-sensitive IPs requests towards the memory subsystem, and demonstrate that both solutions are equivalent in term of memory access latency.

15.1 Introduction

Off-chip DRAM access bottleneck becomes more challenging with the growing gap of MPSoCs[1] requirements for high bandwidth and low latency. The multi-threading technique used nowadays in multimedia SoCs[2] with heterogeneous cores

[1] Multi Processor System on Chip.

[2] System on Chip.

K. Hassan · M. Coppola (✉)
STMicroelectronics, Grenoble, France
e-mail: marcello.coppola@st.com

K. Hassan
e-mail: khaldon.hassan@st.com

N. Voros et al. (eds.), *VLSI 2010 Annual Symposium*,
Lecture Notes in Electrical Engineering 105, DOI: 10.1007/978-94-007-1488-5_15,

increases the contention on the main memory system and demands memory systems with more complex architecture and higher performance [1].

We first analyze the classical CPU-DDR case. The frequency gap between the CPU and the main memory eventually offsets most performance gains from further improvements on the CPU speed. For instance, a cache miss is equivalent to hundreds cycles for today's CPUs, a time long enough for the processor to execute hundreds of instructions. While the DDR SDRAM[3] IO frequency has been improving by 37% per year since 2001, the CAS[4] Latency of SDRAM has been only improving by 5% per year [2–4]. This shows that the general trend of SDRAMs evolution is bandwidth-oriented rather than latency-oriented. Hennessy and Patterson showed that microprocessor performance has been improving by 55% per year since 1987, which emphasizes the growing gap between CPUs speed and SDRAM access time [5].

Schumann reported that 30–60% of memory latency is attributable to system overhead rather than to latency DRAM components in Alpha workstations [6]. These rather CPU oriented trends also do exist for the MPSoCs designs, in which the performance of memory system is even more important due to the share of the interconnect and the memory bus between different heterogeneous cores. Figure 15.1 shows an example of a MPSoC architecture with heterogeneous cores.

A large number of paths has been taken by researchers to reduce the primary memory system overhead. These paths have been divided into two main approaches. The first one focuses on the memory components and their scheduler, whereas the second one takes into consideration the interconnect architecture [7]. There are very few studies that match both approaches. Recent study shows that memory-oriented approaches can reduce application time execution [8]. However, focusing on memory access alone is not enough. Even with zero latency SDRAM access, the overhead of primary memory system would not be eliminated, because the transactions through a shared on-chip communication system still require time.

The goal of this study is to provide a system vision of the off-chip memory access considering globally the interconnect system and the memory scheduler. We focus on the configuration of the interconnect and the memory scheduler in order to exploit at best their QoS. Finding a theoretical solution to this problem is hardly feasible [9], therefore we use an experimental approach based on traffic classes whose support has been implemented on the Spidergon STNoC. Our goal with these experiments is to optimize the external memory accesses for high-priority transactions in term of latency.

The rest of this article is organized as follows. Section 15.2 introduces previous studies made on memory access optimization, these studies concern memory controllers and complex interconnection architectures. Section 15.3 mentions the

[3] Double Date Rate Synchronous Dynamic Random Access Memory

[4] Column Access Strobe.

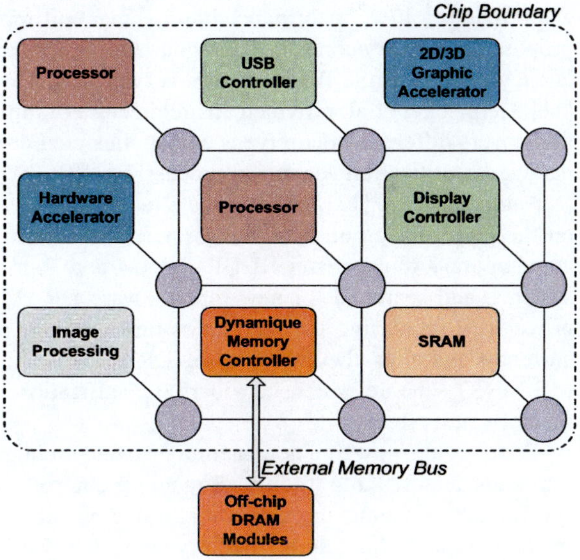

Fig. 15.1 Example of a simplified architecture of MPSoC with heterogeneous Cores

problem of shared resources in MPSoC systems, and the problem of high latency in modern interconnect structures such as NoCs. In Sect. 15.4 we propose our solution to minimize high-priority read requests through the NoC and the memory controller. In Sect. 15.5 we discuss our experiments results. Finally, we present the conclusion and the future work in Sect. 15.6.

15.2 Related Work

Recognizing the importance of high performance off-chip SDRAM communication as a key to a successful system design, several memory controllers and on-chip interconnection systems have been proposed. As we study the QoS extension through the interconnection and the memory controller within a system approach, we separate the related work in three subsections:

15.2.1 Memory Schedulers

Several DRAM controllers and schedulers have been proposed to make the most efficient use of the off-chip memory subsystem. Rixner et al. show that none of the fixed policies studied provide the best performance for all workloads and under all circumstances. However, the FR FCFS (first-ready first-come-first-served) policy exploits the locality within the 3-D memory structure (bank/row/column), and provides a 17% performance improvement on applications [10]. Akesson et al. proposed a memory controller that guarantees minimum bandwidth and maximum latency bounds to the IPs using a novel approach to predictable SDRAM sharing [11]. Heithecker et al. provided an architecture of multi-stream SDRAM controller that covers different stream types and applies memory scheduling to achieve high-bandwidth utilization for image processing [12].

Natarajan et al. [13] examine the effect of different memory controller policies on the performance of multiprocessor server systems. Zhu and Zhang evaluated contemporary multi-channel DDR DRAM and Rambus DRAM systems in SMT systems, and searched for new thread-aware DRAM techniques [1]. Their study proves that increasing the number of threads tends to increase the memory concurrency and thus the pressure on DRAM system. Zheng et al. proposed the ME-LREQ scheme which considers the utilization of both processor cores and memory subsystem.

Carter et al. showed a new memory system architecture, Impulse, that adds two important features to a traditional memory controller [14]. It supports application specific optimizations through configurable physical address remapping; and does intelligent prefetching at the memory controller. Zhu et al. provided the fine grain memory scheduler [15]. It issues multiple DRAM requests for a single cache miss, where each request fetches a sub-block of a cache line.

Lee et al. presented a multi-layer quality aware memory controller [16] that contains partitioned functionality layers to achieve high SDRAM utilization and meets requirements for bandwidth and latency. They prove through digital set-top-box emulation that accesses delay of latency-sensitive data flows can be reduced by 37–65%.

Another memory scheduler providing QoS to improve the system performance is proposed by Nesbit et al. [17]. It is based on concepts developed for network fair queuing scheduling algorithms and targets high performance Chip Multi Processors (CMPs). Mutlu and Moscibroda studied also the shared memory problem in CMPs [18]. They analyzed the interference between threads sharing memory system and proposed a memory access scheduler called Stall-Time Fair Memory. The goal of the scheduler is to equalize the DRAM-related slowdown experienced by each thread due to the interference from other threads, without hurting overall system performance.

All of these previous studies focus only on the architecture and QoS provided by memory schedulers and do not tackle neither the interconnect services nor the manner in which the masters requests are brought to the memory system.

15.2.2 On-chip Interconnection

In the mid 90's Hosseini-Khayat and Bovopoulos presented an efficient bus management scheme which allows the bus to support both continuous media transfer and regular random transactions. The algorithm ensures that continuous streams can meet their real-time constraints independently of random traffic, and random traffic is not delayed significantly by continuous traffic except when the traffic load is very high [19]. In the early 2000 the applications needs in term of throughput have led the interconnection systems to the NoC idea [20], in which the support for these traffic classes is still required.

Grot et al. propose a QoS scheme (*Preemptive Virtual Clock*) specifically designed for cost and performance sensitive on-chip interconnects [21]. PVC requires neither per flow buffering in the router nor large queue in the source nodes. Instead, it provides fairness guarantees by tracking each flow's bandwidth consumption over a time interval and prioritizing packets based on the consumed bandwidth.

Lee et al. present a new scheme called GSF (*Globally Synchronized Frames*) to implement QoS for multi-hop on-chip networks [22]. GSF provides guaranteed and differentiated bandwidth as well as bounded network delay without increasing the complexity of the on-chip routers. They quantize the time into frames and the system only tracks a few frames into the future to reduce time management costs. Each QoS packet from a source is tagged with a frame number indicating the desired time of future delivery to the destination. At any point in time, packets in the earliest extant frame are routed with highest priority but sources are prevented from inserting new packet into this frame.

QNoC has four types of service level. *Signaling* for urgent short packets that have the highest priority; *Real-Time* that guarantees bandwidth and latency for streamed audio and video; *Read/Write* for short memory and register accesses; and *Block-Transfer* for long messages such as DMA transfers. It combines multiple service levels (SL) with multiple equal-priority virtual channels (VC) within each level. The VCs are assigned dynamically per each link. A different number of VCs may be assigned to each SL and per each link [23].

The DSPIN network-on-chip provides guaranteed service traffic by using VCs technique with a buffer per virtual channel [24]. The advantage of this technique is a full separation of the traffic classes. Two traffic classes are defined, Best Effort (BE) and Guaranteed Service (GS) packets. Thus, when one traffic class is blocked the other is neither suspended or blocked. Consequently, the deadlock situations can be avoided.

MANGO stands for message passing asynchronous NoC providing Guaranteed Service traffic over a virtual channel approach [25]. MANGO routers are the nodes of 2D mesh. They has five ports where one is a local port. The router consists of a BE router, a GS router and a link arbiter. The GS router is implemented as a non-blocking switching module. Each output port has seven GS communications and one BE communication. The GS communications are multiplexed using virtual channels within a buffer per channel approach.

Spidergon STNoC is a customizable on-chip communication platform that addresses heterogeneous, application specific requirements of MPSoCs. It allows customizable pseudo-regular or hierarchical topologies. As a programmable distributed hardware/software component, Spidergon STNoC offers a set of services to design advanced application features such as quality of service, security, and exception handling [26]. Two virtual channels can be used to map traffic classes. In addition, two arbitration stages are implemented in the building components . The first one is an intra-channel arbitration, which arbitrates the packets going though the same channel. The second one is an inter-channel arbitration, which arbitrates between channels going through the same physical link.

15.2.3 Combined "Interconnect-Memory Scheduler" Solutions

Few studies treat the off-chip memory system as a system matter.

Burchard et al. presented a design of a real-time streaming memory controller (SMC) that supports off-chip network services [27]. The SMC has been designed to allow external SDRAM to be accessed from a PCI Express network. They proposed a fully parametrized credit-based arbitration algorithm. They also proposed the extension of the virtual channels provided by the PCIEx inside the SMC. As they map one stream by VC, the maximum number of parallel streams accessing the SMC is limited by the number of PCIEx VCs (eight VCs).

Sonics has developed algorithms for memory load balancing in a multi-channel memory system with an advanced memory scheduler to optimize SDRAM access (Interleaved Memory Technology) [28]. The global address space, covered by an address region of SonicsSX SMART Interconnect, may be partitioned into a set of channels. The channels are non-overlapping and collectively cover the whole region. The number of channels for a region is a static value derived from the number of individual targets associated with the region. The memory load balancing unit distributes application workloads over memory channels through the interconnect.

Jang and Pan presented a NoC router with an SDRAM-aware flow control. It improves the SDRAM utilization and latency, and decouples the Noc design cost from the number of SDRAMs [29]. The router arbiter schedules the packets to access SDRAM efficiently. The packets arrive at the memory subsystem into the order that is more friendly to SDRAM operations. In consequence, the complexity of the memory decreases while the memory performance is more improved.

15.3 Problem Statement

Shared resources pose a significant resource management problem in designing MPSoC systems. Different threads can interfere with each other while accessing the shared resources. If thread interference is not controlled, some threads could be unfairly prioritized over others while other threads, perhaps having higher priority, could be starved for long time [18]. In addition, todays SoCs need to have efficient interconnect structures to respond to the increasing number of on-chip IPs and their needs in term of bandwidth and latency. Nowadays, networks-on-chip seem to be the adequate interconnect system for complex SoCs. A NoC is scalable in the sense that adding more routers results in more bandwidth. However, increasing the number of routers in a NoC may statistically increase the transactions latency through the interconnect.

The Dynamic memory controller (DMC) plays a principal role in memory access optimization process. Unfortunately, the optimization of a DMC is hard because its task is complicated. This complication is due to two reasons. First, the DMC needs to obey all SDRAM timing constraints to provide correct functionality. Second, the controller must intelligently prioritize SDRAM commands from different memory requests to optimize system performance [30].

The interconnect latency between a master and the memory subsystem becomes trickier for latency-sensitive masters, e.g. a cache controller. Moreover, most of memory controllers store requests before sending them to the SDRAM, what adds more delay to the transactions latency. That makes sense to optimize the combination of external-memory controller and interconnect, and shows the importance of a *system approach* for minimizing the overall latency when using a complex interconnection system such as a network-on-chip.

15.4 Proposed Solution

Realizing the importance of a system approach for optimizing external memory access in SoCs, we propose in this study a way to optimally interface the memory controller and the network-on-chip. In Sect. 15.4.2.4 we are going to show how to homogenize the services offered by the Spidergon STNoC and the memory controller in order to make shorten the Load/Store operations of latency-sensitive IPs. *Within the frame of this study, we focus on the latency of high priority requests and consider the other classes of transaction with a best-effort approach.*

We give an overview of the Spidergon STNoC and the memory subsystem we use in our simulations.

The Spidergon STNoC contains four building blocks:

- The network interface (NI), which provides a hardware access point to external IP or processor cores and the necessary hardware to implement a set of communication primitives and low-level platform services.
- The router, responsible for implementing the network layer of Spidergon STNoC protocol stack. It must ensure a reliable packet transfer trough the on-chip network, according to a proper QoS policy. From a very high-level perspective, a router is based on a crossbar switch with a given number of input and output ports.
- The network plug switch (NPS), used to aggregate several NIs for accessing the network. This component enables the connection of several network interfaces to the *NI port* of a router.

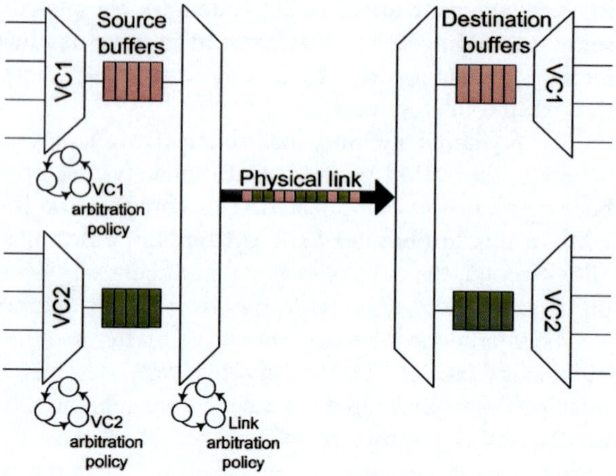

Fig. 15.2 Spidergon STNoC virtual channels implementation

- The physical link implements the physical layer of the Spidergon STNoC protocol. It is responsible for connecting routers to each other, and also router to NIs. There are several possible ways of implementing physical links, including combinations of synchronous / asynchronous and serial / parallel links.

The QoS in Spidergon STNoC indicates the ways to manage bandwidth and latency to ensure the requirements for each traffic flow. Arbitration is a critical part of the router, since it determines the level of QoS support of the network and impacts router performance in terms of critical path delay. Two factors affect performance, the number of request ports of the arbiter and the complexity of the arbitration scheme. Another important issue is the capability of the router to be flexible enough to allow a certain degree of configurability of the global network arbitration policy.

As far as bandwidth is concerned, Spidergon STNoC supports the Fair Bandwidth Allocator (FBA) QoS mechanism. It is an end-to-end service that guarantees fair and programmable weighted bandwidth allocation on the top of a distributed network, just by tuning the injection point [26].

The Network Plug Switch and the Router can implement two virtual channels through one physical link with the necessary logical blocks for arbitration within a given channel, and between two channels. The main advantages of the virtual channels (VCs) technique is a low wire area overhead per additional virtual channel compared to the duplication of the physical link. This stems from the fact that the traffic classes are multiplexed over the same long wires. Figure 15.2 shows a simplified scheme of virtual channels implementation with a buffer per channel.

We use a single-port memory controller which offers QoS in term of latency for read transactions. Entries are arbitrated with an algorithm that optimizes the efficiency of the memory data bus. The algorithm can be modified to meet any programmed QoS requirement. To achieve optimum memory bus efficiency entries might be arbitrated out of order from their arrival time. QoS is defined as a method of increasing the arbitration priority of a read access that requires low-latency read data. The QoS for read access is determined when the arbiter receives it. No QoS exists for write accesses.

In order to provide an extended QoS to high-priority transactions such as *cache controller* transactions, we separate high-priority transactions from other transactions by mapping them on a dedicated virtual channel (VC). In addition, we give this VC the highest priority in NIs and Routers through the NoC. Moreover, we program the memory controller so as to minimize the stall time of transactions coming through the high-priority VC. The next subsection provides more details about the transactions mapping and the platform configuration.

15.4.1 Platform Composition

Figure 15.3 shows a simplified architecture of the simulation platform. It is made up of:

- 4 traffic generators representing: 2 cache controller ports, a DMA, and an ARM processor.
- 4 SRAMs and 1 ROM.
- 2 SDRAM DDR subsystems, made up of memory controller and Micorn DDR SDRAM modules [31].
- A Spidergon STNoC, composed of two separated and symmetric networks, one for requests and one for responses. Both networks contain 6 routers. This interconnect is the backbone part of an STMicroelectronics design. Its role is to connect several clusters with the memory subsystems.

The next subsection shows the configuration of each block of this platform especially the implementation of the virtual channels inside the NoC.

15.4.2 Platform Configuration

Several configuration of each component are proposed in order to be able to evaluate the performance of the previous NoC architecture and calculate the SDRAM access speedup when implementing virtual channels.

15.4.2.1 Traffic Generators

Each traffic generator has an AMBA AXI interface with 2 separated channels for read and write requests. The number of outstanding transactions for each channel is configurable. It has the capability of generating constrained-random traffic in accordance with a statistical distribution which determines the inter-transaction

Table 15.1 Traffic generator characteristics

IP Name	Data Rate	Latency	Jitter	Burst Length[a]	Issuing Capabilities
$ Ctrl port0	Low	Low	Low	32[b]	2 reads, 2 writes
$ Ctrl port1	Low	Low	Low	32	2 reads, 2 writes
DMA	High	Tolerant	Tolerant	16 → 128[a,c]	2 reads, 2 writes
Streaming IP	High	Tolerant	High	16 → 128	2 reads, 2 writes

[a] In Bytes
[b] Corresponds to the cache line width
[c] Allowed burst sizes are : 16, 32, 64, 96, 128 bytes

time. We have a full control over AMBA AXI bus parameters such as address range; transaction ID and burst length. A preview of traffic generators characteristics is shown in Table 15.1.

The traffic balancing of all these generators is defined as 50% towards on-chip SRAMs, and 50% towards off-chip DDRs.

15.4.2.2 Memory Subsystem

Two similar memory subsystems are connected to the NoC. Each one is made up of the combination of a single port dynamic memory controller (DMC) and two 16-bit DDR SDRAMs. The DMC is programed after the reset signal. During this period we configure the DMC in specifying the maximal admissible latency value for each initiator (identified by its unique source ID). Thus the DMC will be able to schedule the requests towards the memory according to these latency values.

The QoS in the DMC indicates the ways to manage bandwidth and latency. The latency guarantee of a flow is based on the flow ID, while the minimal bandwidth of a flow is based on the flow ID and the memory bank status (page hit/miss).

15.4.2.3 Interconnect

We use in our simulation platform the Spidergon STNoC interconnect system. We implement two separated NoCs, one for requests and one for responses. In order to minimize the number of buffers and thus the interconnect area, we only implement two channels on the path between *cache controller* ports and *memory subsystems* (see Fig. 15.3). *Channel splitters* aim to separate *cache controller* transactions on two channels. The *channel splitter* can be enabled or not:

- When enabled, it separates cache controller transactions towards off-chip memory subsystems on one channel (ch2 that provides the highest QoS in the NoC) and all other transactions on the other channel (ch1).
- When disabled, it forwards all cache controller transactions on one channel (ch1). DMA and Streaming IP transactions are mapped on channel 1 with low

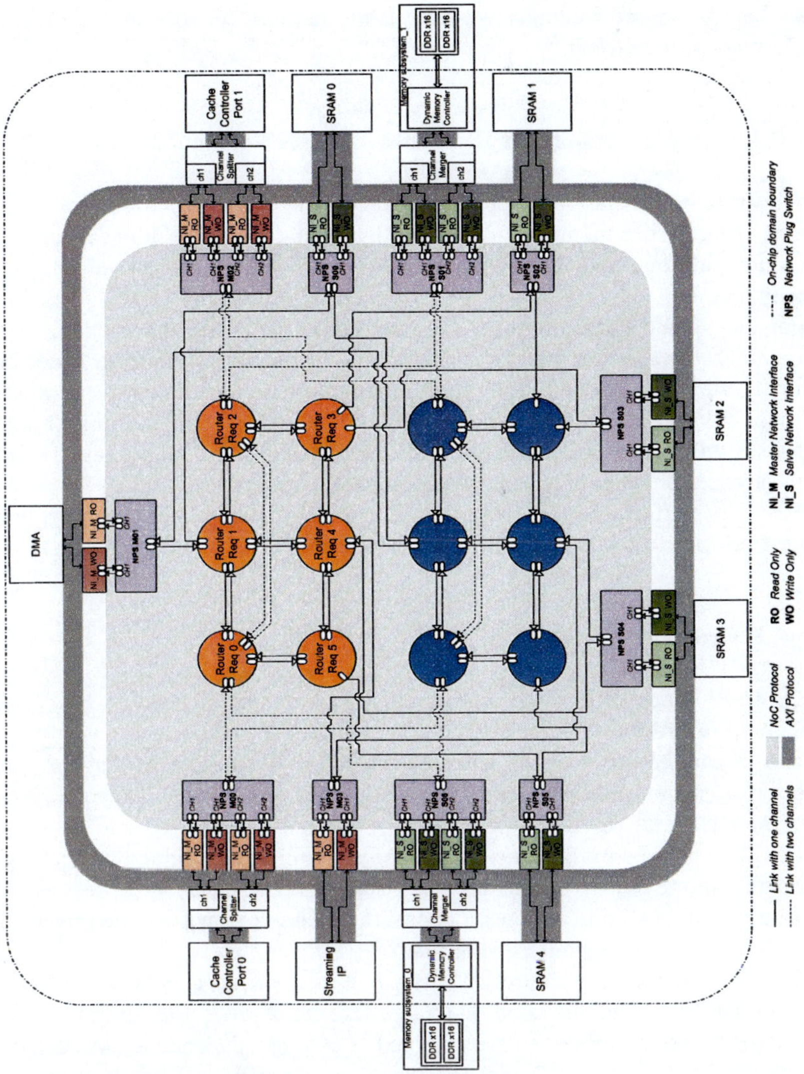

Fig. 15.3 Simplified architecture of the simulation platform

priority, whereas the cache controller transactions on this channel are given the highest priority.

Therefore, we are able to make a fair comparison of the external memory access latencies when we use a separated channel for cache controller transactions.

As we use a single port memory controller, we need a *channel merger* block to merge both channels in one AXI bus when the channel splitter is enabled.

Note that the source routing algorithm in the request network and the response network are symmetrical.

15.4.2.4 Services Coupling of Both Spidergon STNoC and DMC

For the first configuration (1 channel), we prioritize the cache controller transactions towards the memory subsystems by giving them the highest priority in the router arbiters, and by choosing an arbitration algorithm based on packets priority.

For the second configuration (2 channels), we prioritize the cache controller transactions towards the memory subsystems by mapping them on a dedicated channel (ch2), and by configuring the router arbiters in such a way as to prioritize ch2 over ch1 without locking packet on ch1. In this way, the high priority requests/responses on ch2 do not stall behind the other requests/responses on ch1 on the same physical channel between routers (see Fig. 15.3).

For both platform configurations (one channel or two channels), we give the highest priority to the cache controller transactions.

15.4.2.5 Ordering Aspects

There are 3 kinds of components in our platform which need to be carefully configured in order to guarantee data consistency for the masters.

The *Channel Splitter* which is connected to the *Cache Controller* port has an arbiter that lets multiple IDs route to the same slave, but a given ID can only router to a single slave at any particular time.

The *Master Network Interface* implements a similar technique by checking the destination of the request and its ID. For instance, if two consecutive requests have the same ID and two different destinations, the second one will not be granted until the reception of the response of the first one.

The dynamic memory controller is aware of the requests dependency when they have the same type and the same ID (read after read & write after write). In case of hazard detection, the arbiter entry is flagged as having a dependency. As the arbiter entries are invalidated, so the dependencies are reduced until finally there are no outstanding dependencies and the entry is free to start.

15.5 Simulation Environment and Results

We measure in this section the latency access of off-chip memory for cache controller transactions. We change the mapping of their transactions through the interconnect and we measure the difference of latency by means of several transactions spies.

Fig. 15.4 Off-chip memory access speedup of cache controller read transactions when the number of conflict points increases inside the NoC

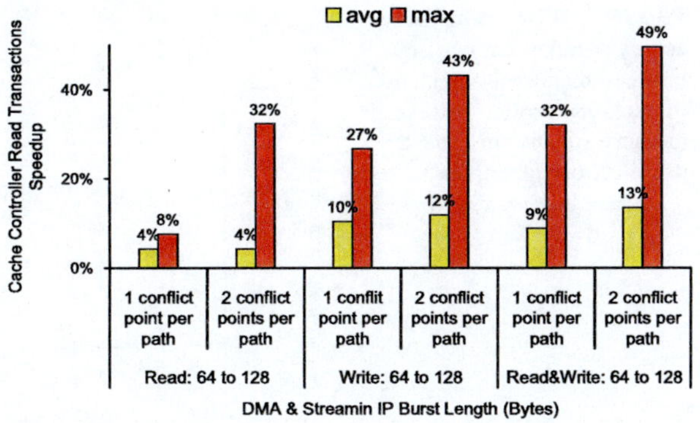

15.5.1 Number of Conflict Points Influence Over Cache Controller Read Transactions Latency

A conflict point is defined as a physical link shared between low-priority master(s) and high-priority master(s). To show how the NoC routing could influence high-priority transactions latency, we calculate the gain of memory access latency when we use two channels by running simulations with two different routing configurations. The first one has one conflict point per path, and the second one has two conflict points per path. For both configurations, the QoS in the dynamic memory controller for read requests is guaranteed. Thus, we can ensure the extension of QoS provided by the interconnect into memory subsystems. Three low-priority traffic patterns are in use in order to evaluate the influence of each of them over high-priority read transactions. Traffic patterns are: reads, writes, and mixed read&write. All of them have a random burst length between 64 and 128 bytes. The memory speedup is calculated using the Eq. 15.1 for both cases. We apply this formula on maximum latency and average latency value.

$$Speedup = 100 \cdot \frac{Lat_{2chs}^{DMC\_QoS_{on}} - Lat_{1ch}^{DMC\_QoS_{on}}}{Lat_{1ch}^{DMC\_QoS_{on}}} \tag{15.1}$$

Figure 15.4 provides an overview of the memory access speedup evolution with the DMA and Streaming IP transactions. In general, the use of two channels when there are two conflict points per path is more efficient in comparison with the case of one conflict point per path.

We notice that the gain obtained by implementing two channels when low-priority IPs issue write requests is more important than the gain obtained when they issue read requests. This is expected because the DMC provides no QoS for write requests. In addition, cache controller read requests are the only read requests in the arbitration queue of DMC, therefore they are scheduled immediately. In normal operation mode of low-priority IPs (read&write requests), the implementation of two channels speeds up the high-priority reads of 13% in case of two

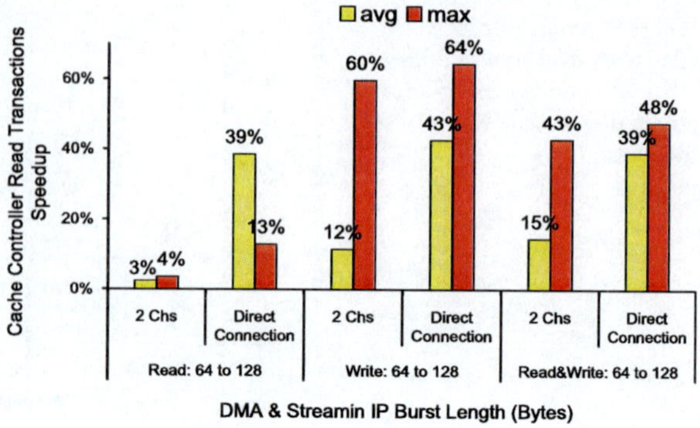

Fig. 15.5 Off-chip memory access speedup:comparison between the direct connection of cache controller ports to memory subsystems and the use of virtual channels

conflict points per path, and by 9% in case of one conflict point per path. The read accesses speed up based on maximum latency with two conflict points per path reaches 49% in a normal operation mode for low priority IPs versus 32% for one conflict point per path.

We do not show in this experiment the results for cache controller write requests, because the DMC model used inside the platform can not provide QoS for write requests.

15.5.2 Comparison Between Cache Controller Direct Connection to Memory Subsystem and the Connection Through Virtual Channels

We change the routing table in order to fairly compare the direct connection of cache controller ports to memory subsystems and the connection trough virtual channels. We delete the path between *cache controller port0* and *memory subsystem0*. We also delete the path between *cache controller port1* and *memory subsystem1* (see Fig. 15.3). Therefore, *port0* of cache controller can only communication with *memory subsystem1*, and *port1* of cache controller with *memory subsystem0*. We run simulations with this mapping using one channel, and then two channels. We disconnect then the *cache controller port0/1* from the NoC to connect them through a direct bus to *memory subsystem1/0*. Figure 15.5 shows the simulations results.

$$Speedup = 100 \cdot \frac{Lat^{DMC\_QoS_{on}}_{direct\_connection} - Lat^{DMC\_QoS_{on}}_{1ch}}{Lat^{DMC\_QoS_{on}}_{1ch}} \qquad (15.2)$$

We use Eq. 15.1 to compute memory access speedup for cache controller read requests when we use the virtual channels, and Eq. 15.2 to calculate memory speed up when we use a physical direct connections.

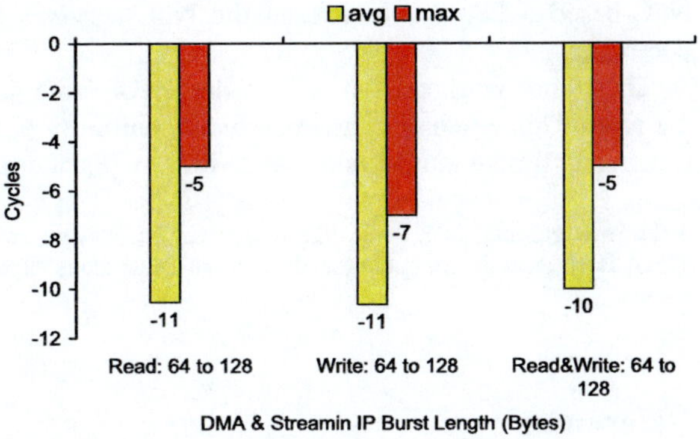

Fig. 15.6 Latency difference of cache controller read transactions when we use direct connections instead of virtual channels

Memory speedup access obtained through two channels competes with physical direct connections in term of maximum latency. Actually, the average gap between speedup values based on maximum latency is 6%. However the gap between speedup values based on average latency is still big. This is due to the way in which we compute the memory speedup. Equations 15.1 and 15.2 hide the real difference of latency between direct connection mode and virtual channels mode. For this reason, we use Eq. 15.3 to calculate the difference of latency between the two connection modes for cache controller read transactions.

$$\Delta Lat = Lat^{DMC\_QoS_{on}}_{direct\_connection} - Lat^{DMC\_QoS_{on}}_{2ch} \quad (15.3)$$

Figure 15.6 shows both maximum latency difference and average latency difference for cache controller read transactions. We see that the difference of average latency is almost invariable (11 clock cycles), which corresponds to the delay created by the pipeline inside the Spidergon STNoC at the request and response paths.

15.6 Conclusion and Future Work

We show the importance of the external memory access optimization in MPSoCs. We highlight the need of coupling the services provided by the network-on-chip with the services provided by the memory scheduler in order to improve the overall system performance. We studied the influence of the number of conflict points on the memory subsystems paths, and proved that the use of two virtual channels can speedup the memory access by 13% on average (up to 49% for maximum latency) compared to the case of one channel. We made a comparison between the connection of latency-sensitive IPs through a reserved virtual channel in the NoC and the direct connection of these IPs to the memory subsystems. We find that both solutions are equivalent in term of access latency. This study lead us to a conclusion that the best way to improve the memory subsystem in a

NoC-based MPSoC is to extend the NoC services inside the dynamic memory controller.

The future work consists of the design of a memory scheduler integrated into the NoC. This scheduler will ensure the continuity of QoS provided by the NoC in order to optimize latency and bandwidth of IPs accessing the memory subsystem.

Acknowledgments We would like to express our sincere gratitude to Prof. Frédéric Pétrot of TIMA Laboratory in Grenoble for offering his tremendous experience in the field to promote this work.

References

1. Zhu Z, Zhang Z (2005) A performance comparison of dram memory system optimizations for smt processors. In: Proceedings HPCA-11, pp 213–224
2. Double data rate (ddr) SDRAM specification, May 2002. URL http://www.jedec.org/download/search/JESD79F.pdf
3. Double data rate (ddr2) SDRAM specification, January 2005. URL http://www.jedec.org/download/search/JESD79-2E.pdf
4. Double data rate (ddr3) SDRAM specification, April 2008. URL http://www.jedec.org/download/search/JESD79-3B.pdf
5. Hennessy JL, Patterson DA (2006) Computer architecture: a quantitative approach. 4th edn. Morgan Kaufmann Publishers Inc., San Francisco, CA
6. Schumann RC (1997) Design of the 21174 memory controller for digital personal workstations. Digital Tech J 9(2):57–70
7. Cuppu V, Jacob B (2001) Concurrency, latency, or system overhead: Which has the largest impact on uniprocessor dram-system performance? In: Proceedings 28th Annual international symposium on computer architecture, pp 62–71
8. Cuppu V, Jacob B, Davis B, Mudge T (1999) A performance comparison of contemporary dram architectures. In: Proceedings of the 26th International symposium on computer architecture, pp 222–233
9. Wang Z, Crowcroft J (1996) Quality-of-service routing for supporting multimedia applications. IEEE J Sel Areas Commun 14(7):1228–1234
10. Rixner S, Dally WJ, Kapasi UJ, Mattson P, Owens JD (2000) Memory access scheduling. In: Proceedings ISCA '00, pp 128–138
11. Akesson B, Goossens K, Ringhofer M (2007) Predator: a predictable SDRAM memory controller. In: Proceedings CODES+ISSS '07, pp 251–256
12. Heithecker S, do Carmo Lucas A, Ernst R (2003) A mixed qos SDRAM controller for fpga-based high-end image processing. In: Proceedings SIPS 2003, pp 322–327
13. Natarajan C, Christenson B, Briggs F (2004) A study of performance impact of memory controller features in multi-processor server environment. In: Proceedings WMPI '04, pp 80–87
14. Carter J, Hsieh W, Stoller L, Swanson M, Zhang L, Brunvand E, Davis A, Kuo C-C, Kuramkote R, Parker M, Schaelicke L, Tateyama T(1999) Impulse: building a smarter memory controller. Fifth international symposium on high-performance computer architecture, Proceedings, pp 70–79
15. Zhu Z, Zhang Z, Zhang X (2002) Fine-grain priority scheduling on multi-channel memory systems. In: Proceedings HPCA'02, pp 107–116
16. Lee K-B, Lin T-C, Jen C-W (2005) An efficient quality-aware memory controller for multimedia platform soc. IEEE Trans Circ Syst Video Technol 15(5):620–633 May 2005
17. Nesbit KJ, Aggarwal N, Laudon J, Smith JE (2006) Fair queuing memory systems. In: Proceedings MICRO-39, pp 208–222

18. Mutlu O, Moscibroda T. Stall-time fair memory access scheduling for chip multiprocessors. In: Proceedings MICRO-40, pp 146–160
19. Hosseini-Khayat S, Bovopoulos AD (1995) A simple and efficient bus management scheme that supports continuous streams. ACM Trans Comput Syst 13(2):122–140
20. Guerrier P, Greiner A (2000) A generic architecture for on-chip packet-switched interconnections. In: Proceedings DATE '00, pp 250–256
21. Grot B, Keckler SW, Mutlu O (2009) Preemptive virtual clock: a flexible, efficient, and cost-effective qos scheme for networks-on-chip. In: Proceedings Micro-42, pp 268–279
22. Lee JW, Ng MC, Asanovic K (2008) Globally-synchronized frames for guaranteed quality-of-service in on-chip networks. In: SIGARCH Comput. Archit. News, vol 36, pp 89–100
23. Dobkin R, Ginosar R, Cidon I (2007) Qnoc asynchronous router with dynamic virtual channel allocation. In: Proceedings NoCS'07, pp 218–218, May 2007
24. Panades IM (2008) Design and Implementation of a Network-on-Chip with Guaranteed Service. PhD thesis, Pierre etMarie Curie University-Paris VI, May 2008
25. Bjerregaard T, Sparso J (2006) Implementation of guaranteed services in the mango clockless network-on-chip. In: Computer Digital Techniques, IEE Proceedings, vol 153, pp 217–229, July 2006
26. Coppola M, Grammatikakis MD, Locatelli R, Maruccia G, Pieralisi L (2008) Design of Cost-Efficient Interconnect Processing Units: Spidergon STNoC. CRC Press, Inc., Boca Raton, FL, USA, 2008. ISBN 1420044710, 9781420044713
27. Burchard A, Hekstra-Nowacka E, Chauhan A (2005) A real-time streaming memory controller. In: Proceedings DATE '05, pp 20–25
28. Sonics. Sonics sx smart interconnect solution. Datasheet, 2010. URL http://www.sonicsinc.com/uploads/pdfs/sonicssx_DS_021610.pdf
29. Jang W, Pan DZ (2009) An SDRAM-aware router for networks-on-chip. In: Proceedings DAC '09, pp 800–805
30. Ipek E, Mutlu O, Martinez JF, Caruana R. Self-optimizing memory controllers: A reinforcement learning approach. In: Proceedings ISCA'08, pp 39–50
31. Micron. 1gb x4, x8, x16 double date rate SDRAM. Datasheet, 2003. URL http://download.micron.com/pdf/datasheets/dram/ddr/1GbDDRx4x8x16.pdf

Chapter 16
Digital Microfluidic Biochips: A Vision for Functional Diversity and More than Moore

Krishnendu Chakrabarty and Yang Zhao

Abstract Microfluidics-based biochips are revolutionizing high-throughput sequencing, parallel immunoassays, blood chemistry for clinical diagnostics, and drug discovery. These emerging devices enable the precise control of nanoliter volumes of biochemical samples and reagents. They combine electronics with biology, and they integrate various bioassay operations, such as sample preparation, analysis, separation, and detection. Compared to conventional laboratory procedures, which are cumbersome and expensive, miniaturized biochips offer the advantages of higher sensitivity, lower cost due to smaller sample and reagent volumes, system integration, and less likelihood of human error. This chapter provides an overview of droplet-based "digital" microfluidic biochips. It describes emerging computer-aided design (CAD) tools for the automated synthesis and optimization of biochips from bioassay protocols. Recent advances in fluidic-operation scheduling, module placement, droplet routing, pin-constrained chip design, and testing are presented. These CAD techniques allow biochip users to concentrate on the development of nanoscale bioassays, leaving chip optimization and implementation details to design-automation tools.

16.1 Introduction

Advances in digital microfluidics have led to the promise of miniaturized biochips for applications such as immunoassays for point-of-care medical diagnostics, DNA sequencing, and the detection of airborne particulate matter [1–8].

K. Chakrabarty (✉) · Y. Zhao
Department of Electrical and Computer Engineering, Duke University, Durham, NC 27708, USA
e-mail: krish@ee.duke.edu

N. Voros et al. (eds.), *VLSI 2010 Annual Symposium*,
Lecture Notes in Electrical Engineering 105, DOI: 10.1007/978-94-007-1488-5_16,

These devices enable the precise control of nanoliter droplets of biochemical samples and reagents, and integrated circuit (IC) technology can be used to transport and process "biochemical payload" in the form of tiny droplets. Biochips facilitate the convergence of electronics with the life sciences, and they integrate on-chip various bioassay operations, such as sample preparation, analysis, separation, and detection [1]. Compared to conventional laboratory procedures, which are cumbersome and expensive, miniaturized biochips offer the advantages of higher sensitivity, lower cost due to smaller sample and reagent volumes, higher levels of system integration, and less likelihood of human error. As a result, non-traditional biomedical applications and markets are opening up fundamentally new uses for ICs. For example, the worldwide market for in vitro diagnostics in 2007 was estimated at $38 billion [9], and 1.5 billion diagnostic tests/year worldwide has been predicted for malaria alone [10].

However, continued growth in this emerging field depends on advances in chip/system integration. In particular, design methods are needed to ensure that biochips are as versatile as the macro-labs that they are intended to replace. The few commercial biochips available today (e.g., from Agilent, Fluidigm, Caliper, I-Stat, BioSite, etc.) are specific to an application and they offer no flexibility to the user.

This chapter is focused on droplet-based "digital" microfluidic biochips. The digital microfluidics platform offers the flexibility of dynamic reconfigurability and software-based control of multifunctional biochips. Next the paper describes emerging computer-aided design (CAD) tools for the automated synthesis and optimization of biochips from bioassay protocols. Recent advances on fluidic-operation scheduling, module placement, droplet routing, testing, and dynamic reconfiguration are also presented. These CAD techniques allow biochip users to concentrate on the development of nanoscale bioassays, leaving chip optimization and implementation details to design-automation tools.

It is expected that an automated design flow will transform biochip research and use, in the same way as design automation revolutionized IC design in the 80 s and 90 s. This approach is therefore especially aligned with the vision of functional diversification and "More than Moore", as articulated in the International Technology Roadmap for Semiconductors (ITRS) 2007, which highlights "Medical" as being a "System Driver" for the future [11]. Biochip users will adapt more easily to emerging technology if appropriate design methods/tools and in-system automation methods are available.

The rest of this chapter is organized as follows. Section 16.2 describes the digital microfluidic platform. Section 16.3 presents synthesis techniques, including solutions published in the literature for operation scheduling, module placement, and droplet routing. Section 16.4 describes pin-constrained chip methods. Section 16.5 presents advances in testing, diagnosis, and dynamic reconfiguration. Finally, Sect. 16.6 concludes the chapter.

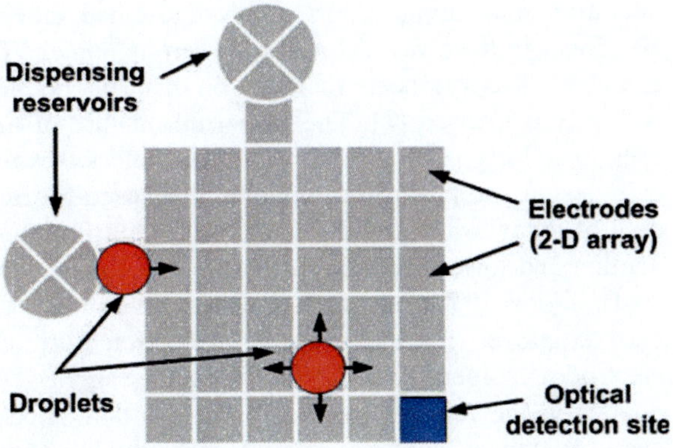

Fig. 16.1 Conceptual view of a digital microfluidic biochip

16.2 Technology Platform

The basic idea of a microfluidic biochip is to integrate all necessary functions for biochemical analysis using microfluidics technology. These micro-total-analysis-systems are more versatile than microarrays. Integrated functions include assay operations, detection, and sample preparation.

A digital microfluidic biochip utilizes electrowetting on dielectric (EWOD) to manipulate and move microliter or nanoliter droplets containing biological samples on a two-dimensional electrode array [1–4, 12–21]. A unit cell in the array includes a pair of electrodes that acts as two parallel plates. The bottom plate contains a patterned array of individually controlled electrodes, and the top plate is coated with a continuous ground electrode. A conceptual view of a digital microfluidic biochip is shown in Fig. 16.1. A droplet is moved by applying a control voltage to an electrode adjacent to the droplet and, at the same time, deactivating the electrode just under the droplet. Using interfacial tension gradients, droplets can be moved to any location on a two-dimensional array. A film of silicone oil is used as a filler medium to prevent cross contamination and evaporation [7]. Recent work has demonstrated chips with an electrode pitch of 100 μm and the gap height is 20 μm. The insulator thickness is 200 nm. To dispense a 300 pl droplet, an actuation voltage of 11.4 V is required [4].

The division of a volume of fluid into discrete, independently controllable "packets" or droplets, provides several advantages over continuous-flow. The reduction of microfluidics to a set of basic repeated operations (i.e., "move one unit of fluid one distance unit") allows a hierarchical and cell-based design approach to be utilized. By varying the patterns of control-voltage activation (a clock signal with logic-high and logic-low values), many fluid-handling operations such as droplet merging, splitting, mixing, and dispensing can be easily executed. The digital microfluidic platform offers dynamic reconfigurability, since fluidic operations can be performed anywhere on the array. Droplet routes and operation scheduling results are programmed into a microcontroller that drives electrodes in the array. As a result, there is no need for dedicated on-chip reaction chambers. Reservoirs are included on the array boundary, from which droplets can be easily dispensed [1]. The disposable nature of these chips precludes multiple uses over long periods of time; nevertheless, reconfigurability allows the same chip design and fabrication method to be used for multiple applications. As in the case of today's integrated circuit, such multifunctional chips facilitate mass production and lower product cost.

To address the need for low-cost, printed circuit board (PCB) technology has been employed for inexpensive fabrication [22]. Using a copper layer for the electrodes, solder mask as the insulator, and a Teflon AF coating for hydrophobicity, the microfluidic array can be fabricated using an existing PCB process. A typical coplanar digital microfluidic chip has an electrode pitch of 1.5 mm, with a gap of 90 μm [8]. Actuation voltages of around 220 V are applied for fluidic operation. Power supplies are therefore external to the microfluidic chip.

Demonstrated applications of digital microfluidics include the on-chip detection of explosives such as commercial-grade 2,4,6-trinitrotoluene (TNT) and pure 2,4-dinitrotoluene [6], automated on-chip measurement of airborne particulate matter [16, 17], and colorimetric assays [7]. Measured performance metrics for such colorimetric assays have been reported in prior work, e.g., [2, 7]. Digital microfluidic biochips are being designed for on-chip gene sequencing through synthesis [1], and integrated hematology, pathology, molecular diagnostics, cytology, microbiology, and serology on the same platform [13]. A prototype has been developed for pyrosequencing [1], which targets the simultaneous execution of 106 fluidic operations and the processing of a large number of droplets. Other lab-on-chip/biochip systems are being designed for protein crystallization, which requires the concurrent execution of hundreds of operations [23, 24].

16.3 Synthesis Methods

In this section, we examine a progression of CAD problems related to biochip synthesis.

16.3.1 Scheduling and Module Placement

Recent years have seen growing interest in the automated design and synthesis of microfluidic biochips [25–44]. Optimization goals here include the minimization of assay completion time, minimization of chip area, and higher defect tolerance. The minimization of the assay completion time, i.e., the maximization of throughput, is essential for environmental monitoring applications where sensors can provide early warning. Real-time response is also necessary for surgery and neonatal clinical diagnostics. Finally, biological samples are sensitive to the environment and to temperature variations, and it is difficult to maintain an optimal clinical or laboratory environment on chip. To ensure the integrity of assay results, it is therefore desirable to minimize the time that samples spend on-chip before assay results are obtained.

Increased throughput also improves operational reliability. Long assay durations imply that high actuation voltages need to be maintained on some electrodes, which accelerate insulator degradation and dielectric breakdown, reducing the number of assays that can be performed on a chip during its lifetime.

Architectural-level synthesis for microfluidic biochips can be viewed as the problem of scheduling assay functions and binding them to a given number of resources so as to maximize parallelism, thereby decreasing response time [35, 42]. A behavioral model for a set of bioassays is first obtained from their laboratory protocols. Architectural-level synthesis is then used to generate a macroscopic structure of the biochip; this is analogous to a structural register-transfer level (RTL) model in electronic CAD [45]. On the other hand, geometry-level synthesis (physical design) addresses the placement of resources and the routing of droplets to satisfy objectives such as area or throughput. It creates the final layout of the biochip, consisting of the placement of microfluidic modules such as mixers and storage units, the routes that droplets take between different modules, and other geometrical details [41].

As in the case of high-level synthesis for integrated circuits, resource binding in the biochip synthesis flow refers to the mapping from bioassay operations to available functional resources. Note that there may be several types of resources for any given bioassay operation. For example, a 2×2-array mixer, a 2×3-array mixer and a 2×4-array mixer can be used for a droplet mixing operation, but with different mixing times. In such cases, a resource selection procedure must be used. On the other hand, resource binding may associate one functional resource with several assay operations; this necessitates resource sharing. Once resource binding is carried out, the time duration for each bioassay operation can be easily determined. Scheduling determines the start times and stop times of all assay operations, subject to the precedence and resource-sharing constraints.

A key problem in the geometry-level synthesis of biochips is the placement of microfluidic modules such as different types of mixers and storage units. Since digital microfluidics-based biochips enable dynamic reconfiguration of the microfluidic array during run-time, they allow the placement of different modules on the same location during different time intervals.

A simulated annealing-based heuristic approach has been developed to solve the NP-complete problem in a computationally efficient manner [41]. Solutions for the placement problem can provide the designer with guidelines on the size of the array to be manufactured. If module placement is carried out for a fabricated array, area minimization frees up more unit cells for sample collection and preparation.

Architectural synthesis is based on rough estimates for placement costs such as the area of the microfluidic modules. These estimates provide lower bounds on the exact biochip area, since the overheads due to spare cells and cells used for droplet transportation are not known a priori. However, it cannot be accurately predicted if the biochip design meets system specifications, e.g., maximum allowable array area and upper limits on assay completion times, until both high-level synthesis and physical design are carried out. [36] proposed a unified system-level synthesis method for microfluidic biochips based on parallel recombinative simulated annealing (PRSA), which offers a link between these two steps. This method allows users to describe bioassays at a high level of abstraction, and it automatically maps behavioral descriptions to the underlying microfluidic array.

Efficient reconfiguration techniques have been developed to bypass faulty unit cells in the microfluidic array [43]. A microfluidic module containing a faulty unit cell can easily be relocated to another part of the microfluidic array by changing the control voltages applied to the corresponding electrodes [38]. Defect tolerance can also be achieved by including redundant elements in the system; these elements can be used to replace faulty elements through reconfiguration techniques [37]. Another method is based on graceful degradation, in which all elements in the system are treated in a uniform manner, and no element is designated as a spare [39]. In the presence of defects, a subsystem with no faulty element is first determined from the faulty system. This subsystem provides the desired functionality, but with a gracefully-degraded level of performance (e.g., longer execution times). Due to the dynamic reconfigurability of digital microfluidics-based biochips, microfluidic components (e.g., mixers) can be viewed as reconfigurable virtual devices. For example, a 2×4 array mixer (implemented using a rectangular array of control electrodes—two in the X-direction and four in Y-direction) can easily be reconfigured to a 2×3 array mixer or a 2×2 array mixer.

Figure 16.2a shows the module placement results and the microfluidic array design for a representative protein assay [36]. The XY-plane refers to the placement of modules on the chip real estate. The Z-axis refers to time. As shown in Fig. 16.2b, we can further integrate optical detectors as well as on-chip reservoirs/dispensing ports into the microfluidic array to form a complete digital microfluidic biochip for the protein assay. Figure 16.2c shows the corresponding results when some of the unit cells in the array are faulty, and reconfiguration is used in a unified manner with synthesis. The solution obtained for the fault-free array yields a biochip design with a 9×9 microfluidic array and the completion time for the protein assay is 363 s. The design for the faulty array allows the protein assay to operate with an increase of only 6% in the completion time, i.e., the completion time is now 385 s.

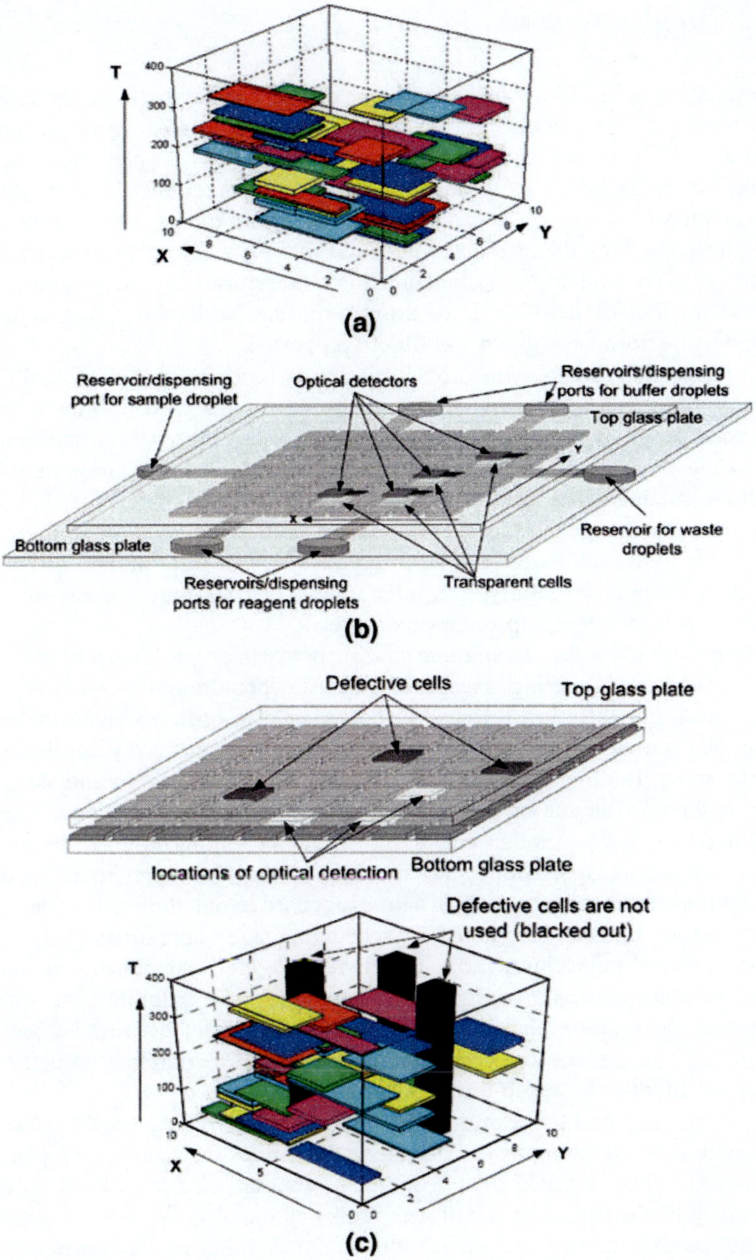

Fig. 16.2 **a** A 3-D model illustrating the synthesis results. **b** A digital microfluidic biochip for a protein assay. **c** A defective array and module placement for the protein assay on this array

16.3.2 Droplet Routing

A key problem in biochip physical design is droplet routing between modules, and between modules and I/O ports (i.e., on-chip reservoirs). The dynamic reconfigurability inherent in digital microfluidics allows different droplet routes to share cells on the microfluidic array during different time intervals. In this sense, the routes in microfluidic biochips can be viewed as virtual routes, which make droplet routing different from the classical wire VLSI routing problem. Systematic routing method for digital microfluidic biochips have therefore been developed to minimize the number of cells used for droplet routing, while satisfying constraints imposed by performance goals and fluidic properties.

One of the first methods for droplet routing in biochips was published in [40]. The main objective in routing is to find droplet routes with minimum lengths, where route length is measured by the number of cells in the path from the starting point to the destination. For a microfluidic array of fixed size, minimum-length droplet routes lead to the minimization of the total number of cells used in droplet routing, thus freeing up more spare cells for fault tolerance. As in the case of electronic circuits, the fluidic ports on the boundary of microfluidic modules are referred to as pins. Similarly, we refer to the droplet routes between pins of different modules or on-chip reservoirs as nets.

During droplet routing, a minimum spacing between droplets must be maintained to prevent accidental mixing, except for the case when droplet merging is desired. Fluidic constraint rules in [40] need to be satisfied in order to avoid undesirable mixing. We view the microfluidic modules placed on the array as obstacles in droplet routing. In order to avoid conflicts between droplet routes and assay operations, a segregation region is added to wrap around the functional region of microfluidic modules. Another constraint in droplet routing is given by an upper limit on droplet transportation time. The delay for each droplet route should not exceed some maximum, e.g., 10% of a time-slot used in scheduling, in order that the droplet-routing time can be ignored for scheduling assay operations [40].

Since a digital microfluidic array can be reconfigured dynamically at run-time, a series of 2-D placement configurations of modules in different time spans are obtained in the module placement phase [37]. Therefore, the droplet routing is decomposed into a series of sub-problems. We obtain a complete droplet-routing solution by solving these sub-problems sequentially.

Based on this problem formulation, a two-stage routing method has been proposed in [40]. In the first stage, M alternative routes for each net are generated. In the second stage, a single route from the M alternatives for each net is selected independent of the routing order of nets. This method also exploits the features of dynamic reconfigurability and independent controllability of electrodes to modify droplet pathways to override potential violation of fluidic constraints.

Droplet routing should be considered in the synthesis flow for digital microfluidics, in order to generate a routable synthesized design for the availability of routing paths. [33] proposed a method to incorporate droplet-routability in the

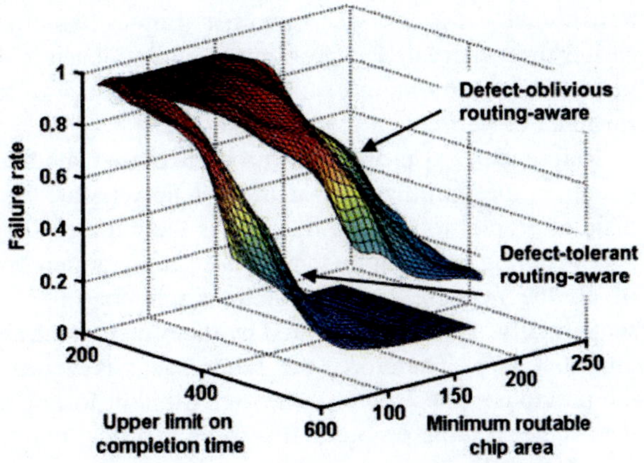

Fig. 16.3 Feasibility frontier surface and feasible design region for defect-tolerant and defect-oblivious routing-aware synthesis methods

PRSA-based synthesis flow. This method estimates the droplet-routability using two metrics. It adopts the average module distance (over all interdependent modules) as the first design metric to guarantee the routability of modules in the synthesized biochip. It also adopts the maximum module distance as the second design metric to approximate the maximum length of droplet manipulation. Since synthesis results with high routability values are more likely to lead to simple and efficient droplet pathways, this method incorporates the above two metrics into the fitness function by a factor that can be fine-tuned according to different design specifications to control the PRSA-based procedure.

We ran the defect-tolerant routing-aware and defect-oblivious routing-aware algorithms under a set of combinations of weights in the fitness function for the protein assay example. We carried out random defect injection into each design and obtain its failure rate. We mapped each design G to a 3D point (T_G, A_G, F_G), where T_G, A_G, F_G are completion time, chip area, and failure rate of the design, respectively. A point (T_G, A_G, F_G) is referred to as a feasibility boundary point if there are no other points (T_m, A_m, F_m) such that $T_m < T_G$, $A_m < A_G$, and $F_m < F_G$. A feasibility frontier surface is obtained by connecting all the feasibility boundary points, as shown in Fig. 16.3. The feasible design region corresponds to the space above the feasible surface. Any design specification can be met whose corresponding is point located in this region; otherwise, no feasible design exists for this specification. As shown in Fig. 16.3, defect-tolerant routing-aware synthesis leads to a lower-feasibility frontier surface and a larger feasible design space as compared to the defect-oblivious method.

Cross-contamination is likely to occur when multiple droplet routes intersect or overlap with each other. At the intersection site of two droplet routes, a droplet that arrives at a later clock cycle can be contaminated by the residue left behind by another droplet that passed through at an earlier clock cycle. Disjoint droplet routes are utilized to avoid the cross-contamination. In a set of disjoint routes, a droplet route does not share any cell in its path with each of the other droplet routes in that set. Such routes eliminate the possibility of a droplet being transported via a cell when another droplet has already passed through it in an earlier time interval. The problem of finding feasible disjoint routes can be directly mapped to the problem of finding disjoint paths (vertex-disjoint or edge-disjoint) for pairs of vertices in a graph [46].

Reference [25] proposed a droplet-routing method that avoids cross-contamination in the optimization of droplet flow paths. The proposed approach targets disjoint droplet routes and minimizes the number of cells used for droplet routing. A net-routing ordering method is used to obtain an optimized order for the routing of all the nets in a sub-problem. Each net in the net-routing order is routed sequentially. The cells occupied by the routed paths are marked as obstacles for the unrouted nets. Therefore, the latter route is vertex-disjoint with respect to all the previous routes. Since it is often difficult to find vertex-disjoint paths to solve the droplet-routing problem, it is more practical to relax the overlap restriction and search for edge-disjoint routes in such cases.

16.4 Pin-Constrained Chip Design

Electrode addressing is an important problem in biochip design. It refers to the manner in which electrodes are connected to and controlled by input pins. Early design-automation techniques relied on the availability of a direct-addressing scheme. For large arrays, direct-addressing schemes lead to a large number of control pins, and the associated interconnect routing problem significantly adds to the product cost. Thus, the design of pin-constrained digital microfluidic arrays is of great practical importance for the emerging marketplace. In this section, we describe a number of pin-constrained biochip design methods.

16.4.1 Droplet-Trace-Based Array Partitioning

An array-partitioning-based pin-constrained design method of digital microfluidic biochips proposed in [44]. This method uses array partitioning and careful pin assignment to reduce the number of control pins. The key idea is to "virtually" partition the array into regions. The partitioning criterion here is to ensure at most one droplet is included in each partition. The droplet trace, defined as the set of cells traversed by a single droplet, serves as the basis for generating the array partitions. The droplet trace can be easily extracted from the droplet routing information and the placement of the modules to which it is routed. If droplets traces intersect on the array, the partitions derived by this method overlap in some regions. Sets of pins from an "overlapping" partition cannot be used in the overlapped region since the reuse of the pins may lead to droplet interference. The solution to this problem is to make the overlapping region a new partition, referred to as the overlapping partition, and use direct addressing (one-to-one mapping) for it.

A Connect-5 algorithm is used to address the problem of how to map control pins to the electrodes in a partition, which can be easily implemented using a 3-layer-PCB. The Connect-5 algorithm succeeds in avoiding droplet interference while moving a single droplet inside the partition. It can be integrated into the droplet-trace-based array partitioning method to generate droplet-interference-free layouts with a minimum number of pins. However, this method requires detailed information about the scheduling of assay operations, microfluidic module placement, and droplet routing pathways. Thus, the array design in such cases is specific to a target biofluidic application.

16.4.2 Cross-Referencing-Based Droplet Manipulation

An alternative method based on a cross-reference driving scheme is presented in [32]. This method allows control of an $N \times M$ grid array with only $N + M$ control pins. The electrode rows are patterned on both the top and bottom plates, and placed orthogonally. In order to drive a droplet along the X-direction, electrode rows on the bottom plate serve as driving electrodes, while electrode rows on the top serve as reference ground electrodes. The roles are reversed for movement along the Y-direction. This cross-reference method facilitates the reduction of control pins. However, due to electrode interference, this design cannot handle the simultaneous movement of more than two droplets. For the concurrent manipulation of multiple droplets on a cross-referencing-based biochip, multiple row and column pins must be selected to activate the destination cells, i.e., cells to which the droplets are supposed to move. However, the selected row and column pins may also result in the activation of cells other than the intended droplet destinations.

A solution based on destination-cell categorization has been proposed to tackle the above problem. The key idea is to group the droplet movements according to their destination cells. A group consists of droplets whose destination cells share the same column or row. In this way, the manipulation of multiple droplets is ordered in time; droplets in the same group can be moved simultaneously without electrode interference, but the movements for the different groups must be sequential. The problem of finding the minimum number of groups can be directly mapped to the problem of determining a minimal clique partition from graph theory [47]. A linear-time heuristic algorithm based on row-scanning and column-scanning has been used to derive the clique partitions.

16.4.3 Broadcast-Addressing Method

One drawback of the cross-reference driving scheme is that this design requires a special electrode structure (i.e., both top and bottom plates contain electrode rows), which results in increased manufacturing cost. Thereby, a broadcast-addressing based design technique for pin-constrained and multi-functional biochips has been developed in [31].

To execute a specific bioassay, routing and scheduling information must be stored in the form of electrode activation sequences, where each bit representing the status of the electrode at a specific time-step. The status can be either "1" (activate), "0" (deactivate) or "x" (don't-care). Each electrode activation sequence contains several don't-care terms, which can be replaced by "1" or "0". If two sequences can be made identical by careful replacing these don't-care terms with "0" or "1", they are referred to as compatible sequences. Compatible sequences can be generated from a single signal source.

The number of control pins can be reduced by connecting together electrodes with mutually-compatible activation sequences, and addressing them using a single control pin. Therefore, the resulting electrode-access method is referred to as a broadcast addressing. The first step here is to partition the electrodes into groups. For all the electrodes in any group, the corresponding activation sequences must be pairwise-compatible. The problem of finding an optimal partition that leads to the minimum number of groups can be easily mapped to the problem of determining a minimal clique partition from graph theory [47]. The minimum number of groups yields the minimum number of control pins.

16.5 Testing and Diagnosis

In this section, we describe recent advances in the testing of digital microfluidic biochips and fault localization techniques.

16.5.1 Fault Modeling

As in microelectronic circuits, a defective microfluidic biochip is said to have a failure if its operation does not match its specified behavior [48]. In order to facilitate the detection of defects, fault models that efficiently represent the effect of physical defects at some level of abstraction are required. These models can be used to capture the effect of physical defects that produce incorrect behaviors in the electrical or fluidic domain. As described in [49], faults in digital microfluidic systems can be classified as being either catastrophic or parametric. Catastrophic faults lead to a complete malfunction of the system, while parametric faults cause degradation in the system performance. A parametric fault is detectable only if this deviation exceeds the tolerance in system performance.

Catastrophic may be caused by a number of physical defects, for example:

- Dielectric breakdown: The breakdown of the dielectric at high voltage levels creates a short between the droplet and the electrode. When this happens, the droplet undergoes electrolysis, thereby preventing further transportation.

- Short between the adjacent electrodes: If a short occurs between two adjacent electrodes, the two electrodes effectively form one longer electrode. When a droplet resides on this electrode, it is no longer large enough to overlap the gap between adjacent electrodes. As a result, the actuation of the droplet can no longer be achieved.
- Degradation of the insulator: This degradation effect is unpredictable and may become apparent gradually during the operation of the microfluidic system. A consequence is that droplets often fragment and their motion is prevented because of the unwanted variation of surface tension forces along their flow path.
- Open in the metal connection between the electrode and the control source: This defect results in a failure in activating the electrode for transport.

Table 16.1 lists some common failure sources, defects and the corresponding fault models for catastrophic faults in digital microfluidic lab-on-chip. Examples of some common parametric faults include the following:

- Geometrical parameter deviation: The deviation in insulator thickness, electrode length and height between parallel plates may exceed their tolerance value.
- Change in viscosity of droplet and filler medium. These can occur during operation due to an unexpected biochemical reaction, or changes in operational environment, e.g., temperature variation.

16.5.2 Structural Test Techniques

A unified test methodology for digital microfluidic biochips has recently been presented, whereby faults can be detected by controlling and tracking droplet motion electrically [50]. Test stimuli droplets containing a conductive fluid (e.g., KCL solution) are dispensed from the droplet source. These droplets are guided through the unit cells following the test plan towards the droplet sink, which is connected to an integrated capacitive detection circuit. Details of the capacitive readout circuit are presented in [51]. Most catastrophic faults result in a complete cessation of droplet transportation. Therefore, we can determine the fault-free or faulty status of the system by simply observing the arrival of test stimuli droplets at selected ports. An efficient test plan ensures that testing does not conflict with the normal bioassay, and it guides test stimuli droplets to cover all the unit cells available for testing. The microfluidic array can be modeled as an undirected graph, and the pathway for the test droplet can be determined by solving the Hamiltonian path problem [52]. With negligible hardware overhead, this method also offers an opportunity to implement self-test for microfluidic systems and therefore eliminate the need for costly, bulky, and expensive external test equipment. Furthermore, after detection, droplet flow paths for bioassays can be

Table 16.1 Examples of fault models for digital microfluidic biochip [54]

Cause of defect	Defect type	Number of cells	Fault model	Observable error
Excessive actuation voltage applied to an electrode	Dielectric breakdown	1	Droplet-electrode short (a short between the droplet and the electrode)	Droplet undergoes electrolysis, which prevents its further transportation
Electrode actuation for excessive duration	Irreversible charge concentration on an electrode	1	Electrode-stuck-on (the electrode remains constantly activated)	Unintentional droplet operations or stuck droplets
Excessive mechanical force applied to the chip	Misalignment of parallel plates (electrodes and ground plane)	1	Pressure gradient (net static pressure in some direction)	Droplet transportation without activation voltage
Coating failure	Non-uniform dielectric layer	1	Dielectric islands (islands of Teflon coating)	Fragmentation of droplets and their motion is prevented
Abnormal metal layer deposition and etch variation during fabrication	Grounding Failure	1	Floating droplets (droplet are not anchored)	Failure of droplet transportation
	Broken wire to control source	1	Electrode open (electrode actuation is not possible)	Failure to activate the electrode for droplet transportation
	Metal connection between two adjacent electrodes	2	Electrode short (short between electrodes)	A droplet resides in the middle of the two shorted electrodes, and its transport along one or more directions cannot be achieved
Particle contamination or liquid residue	A particle that connect two adjacent electrodes	2	Electrode short	
Protein adsorption during bioassay [12]	Sample residue on electrode surface	1	Resistive open at electrode	Droplet transportation is impeded
			Contamination	Assay results are outside the range of possible outcomes

reconfigured dynamically such that faulty unit cells are bypassed without interrupting the normal operation.

Even though most catastrophic faults lead to a complete cessation of droplet transportation, there exist differences between their corresponding erroneous behaviors. For instance, to test for the electrode-open fault, it is sufficient to move a test droplet from any adjacent cell to the faulty cell. The droplet will always be stuck during its motion due to the failure in charging the control electrode. On the other hand, if we move a test droplet across the faulty cells affected by an electrode-short fault, the test droplet may or may not be stuck depending on its flow direction. Therefore, to detect such faults, it is not enough to solve only the Hamiltonian path problem. In [53], a solution based on Euler paths in graphs is described for detecting electrode shorts.

Despite its effectiveness for detecting electrode shorts, testing based on an Euler path suffers from long test application time. This approach uses only one droplet to traverse the microfluidic array, irrespectively of the array size. Fault diagnosis is carried out by using multiple test application steps and adaptive Euler paths. Such a diagnosis method is inefficient since defect-free cells are tested multiple times. Moreover, the test method leads to a test plan that is specific to a target biochip. If the array dimensions are changed, the test plan must be completely altered. In addition, to facilitate chip testing in the field, test plans need to be programmed into a microcontroller. However, the hardware implementations of test plans from [50] are expensive, especially for low cost, disposable biochips. More recently, a cost-effective testing methodology referred to as "parallel scan-like test" has been proposed [51]. The method is named thus because it manipulates multiple test droplets in parallel to traverse the target microfluidic array, just as test stimuli can be applied in parallel to the different scan chains in an integrated circuit.

A drawback of the above "structural" test methods is that they focus only on physical defects, and they overlook module functionality. Therefore, these methods can only guarantee that a biochip is defect-free. However, a defect-free microfluidic array can also malfunction in many ways. For example, a defect-free reservoir may result in large volume variations when droplets are dispensed from it. A splitter composed of three defect-free electrodes may split a big droplet into two droplets with significantly unbalanced volumes. These phenomena, referred to as malfunctions, are not the result of electrode defects. Instead, they are activated only for certain patterns of droplet movement or fluidic operations. Such malfunctions can have serious consequences on the integrity of bioassay results.

16.5.3 Functional Testing Techniques

Functional testing involves test procedures to check whether groups of cells can be used to perform certain operations, e.g., droplet mixing and splitting. For the test of a specific operation, the corresponding patterns of droplet movement are carried out on the target cluster of cells. If a target cell cluster fails the test, e.g., the mixing test, we label it as a malfunctioning cluster. As in the case of structural testing, fault models must be developed for functional testing. Malfunctions in fluidic operations are identified and included in the list of faults; see Table 16.2.

Functional test methods to detect the defects and malfunctions have recently been developed. In particular, dispensing test, mixing test, splitting test, and capacitive sensing test have been described in [54] to address the corresponding malfunctions.

Table 16.2 Functional fault models [54]

Cause of malfunction	Malfunction type	Number of cells	Fault model	Observable error
Electrode actuation for excessive duration	Irreversible charge concentration on the dispensing electrode	3	Dispensing-stuck-on (droplet is dispensed by not fully cut off from the reservoir)	No droplet can be dispensed from the reservoir
Electrode shape variation in fabrication	Deformity of electrodes	3	No overlap between droplets to be mixed and center electrode	Mixing failure
Electrode electrostatic property variation in fabrication	Unequal actuation voltages	3	Pressure gradient net static pressure in some direction)	Unbalanced volumes of split droplets
Bad soldcring	Parasitic capacitance in the capacitive sensing circuit	1	Oversensitive or insensitive capacitive sensing	False positive/ negative in detection

Fig. 16.4 Fabricated lab-on-chip used for PCR

Functional test methods were applied to a PCB microfluidic platform for the Polymerase Chain Reaction (PCR), as shown in Fig. 16.4. The platform consists of two columns and two rows of electrodes, three reservoirs, and routing electrodes that connect the reservoirs to the array. A dispensing malfunction is shown in Fig. 16.5.

An illustration of the mixing and splitting test is shown in Fig. 16.6. The bottom row was first targeted and five test droplets were dispensed to the odd electrodes, as shown in Fig. 16.6a. Next, splitting test for the even electrodes was carried out. Droplets were split and merged on the even electrodes. In Fig. 16.6b, we see a series of droplets of the same volume resting on the even electrodes, which means that all the odd electrodes passed the splitting test, and merging at the even electrodes worked well. However, when the splitting test was carried out on the

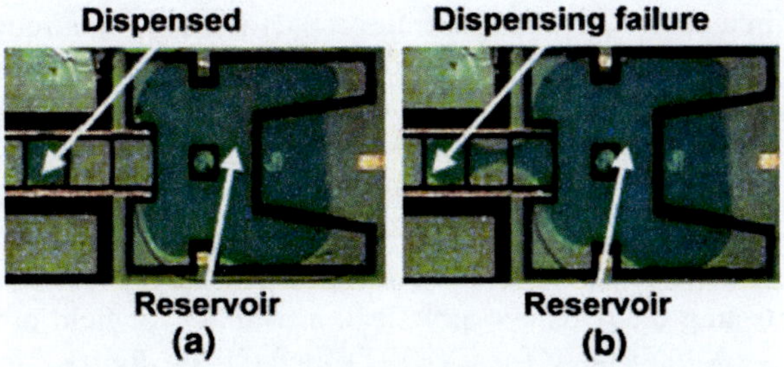

Fig. 16.5 Illustration of **a**, normal dispensing and **b**, dispensing failure, for a fabricated lab-on-chip

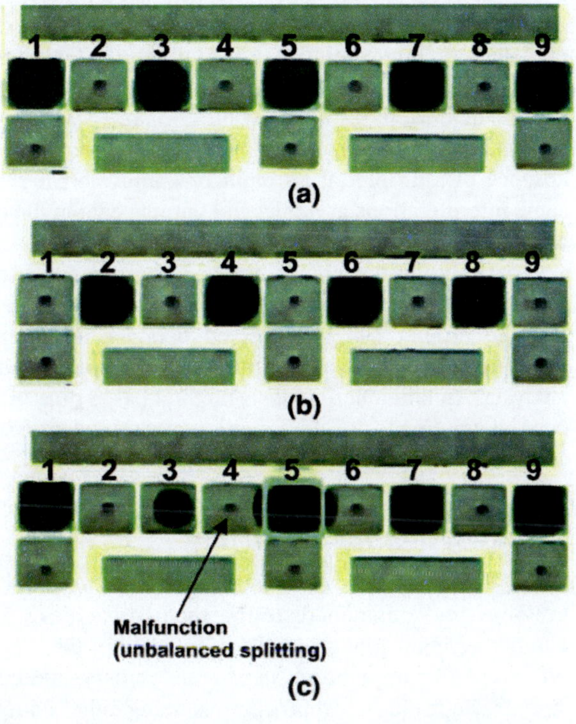

Fig. 16.6 Mixing and splitting test for a fabricated PCR chip

even electrodes, a large variation in droplet volume was observed on the 3rd and 5th electrodes; see Fig. 16.6c. This variation implied a malfunction, leading to unbalanced splitting on the 4th electrode. The malfunction was detected when the droplets were routed to the capacitive sensing circuit. The 4th electrode on the bottom row was marked as an unqualified splitting site.

16.5.4 Built-In Self-Test Techniques

Previous test methods for digital microfluidic platforms use capacitive sensing circuits to read and analyze test outcomes. After reading the test-outcome droplets in a consecutive manner, the capacitive sensing circuit generates a pulse-sequence corresponding to the detection of these droplets. This approach requires an additional step to analyze the pulse sequence to determine whether the microfluidic array under-test is defective. The reading of test outcomes and the analysis of pulse sequences increase test time; the latter procedure is especially prone to errors arising from inaccuracies in sensor calibration. The complexity of the capacitive-sensing circuit and the need for pulse-sequence analysis make previously proposed testing methods less practical, especially for field operation.

A built-in self-test (BIST) method for digital microfluidic lab-on-chip is proposed in [55]. This method utilizes microfluidic logic gates to implement the "compactor" in a BIST architecture. Using the principle of electrowetting-ondielectric, microfluidic AND, OR and NOT gates are implemented through basic droplet-handling operations such as transportation, merging, and splitting.

The definitions of logic values '0' or '1' are as follows: the presence of a droplet of 1x volume at an input or output port indicates a logic value of '1'. The absence of a droplet at an input or output port indicates the logic value '0'. The same interpretations at inputs and outputs enable the output of one gate to be fed as an input to another gate, thus logic gates can be easily cascaded.

The microfluidic compactor can compress the test-outcome droplets of both structural test and functional test into one droplet in a very short amount of time, and the droplet can be detected using a simple photodiode detector, thereby avoiding the need for a capacitive-sensing circuit and complicated pulse-sequence analysis. An efficient diagnosis method based on a "microfluidic encoder" has also been developed to locate a single defective electrode in a microfluidic array.

16.5.5 Design for Testability

Previous pin-constrained design methods achieve a significant reduction in the number of input pins needed for controlling the electrodes. However, as a trade-off, droplet manipulation steps must satisfy additional constraints. These constraints can result in test procedures being either completely ineffective or effective only for a small part of the chip. As a result, chip testability is significantly reduced. Note that the reduction in testability is due to the conflicts between the fluidic operation steps required by functional test and the constraints on droplet manipulations introduced by the mapping of pins to electrodes. Figure 16.7 shows a pin-constrained chip design for a representative protein-dilution assay. The functional test procedure requires a splitting operation to be executed on the highlighted functional unit. To do this, we first activate Pin 13 to hold a test droplet at E_2. Next, we deactivate Pin 13 and activate Pin 12 and Pin 14 to split the test droplet into two small droplets seated on E_1 and E_3. However, E_4 is also charged by activating Pin 12. As a result, the split droplet that is supposed to be seated on E_2 will be moved unintentionally to the boundary of E_4 and E_3. Note that different

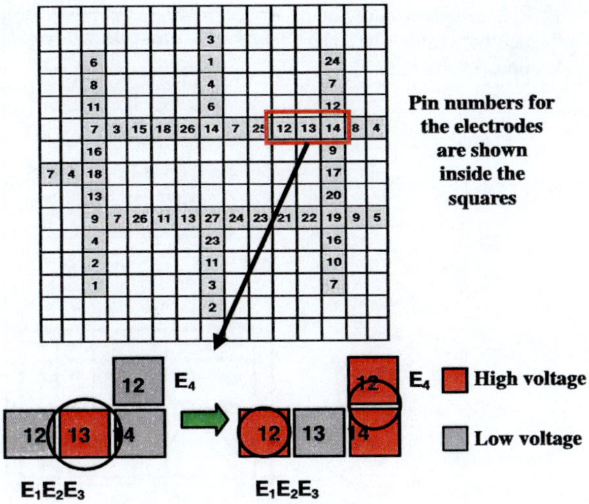

Fig. 16.7 An example of an untestable functional unit on a pin-constrained chip for multiplexed assay

mappings for a pin-constrained chip lead to different untestable functional units, thereby different levels of chip testability.

The authors in [56] proposed a design-for-testability (DFT) solution to facilitate the testing in pin-constrained design. A test-aware pin-constrained design method that incorporates test procedures into the fluidic manipulation steps in the target bioassay protocol is proposed. First, the fluidic operations required by the test procedure are derived from the scheduling and routing steps related to the test droplets. Next, we merge these fluidic operations with the droplet manipulation steps needed for the target bioassay. The merging can be carried out by attaching the electrode-activation sequences for the test procedure to the electrode-activation sequences for the target bioassays. For each electrode in the array, its activation sequence during the test procedure is added to that for the target bioassay to form a longer sequence. These merged longer electrode-activation sequences are provided as input to the broadcast-addressing method, thereby the resulted chip design will support not only the target bioassay but also the test operations. Figure 16.8 shows a linear array consisting of four electrodes. A simple "routing assay" is mapped to the array, where a droplet is to be routed from E_4 to E_1, one electrode per step. We first list the activation sequence for each electrode (Table T_1) in Fig. 16.8. Next we add a splitting test on E_3. The electrode-activation sequences for the splitting test are shown in Table T_2 of Fig. 16.8. These activation sequences are then combined with the activation sequences in T_1. The resulted longer activation sequences are listed in table T_3. The broadcast-addressing method is then applied to generate the eventual pin assignment according to sequences in T_3.

By applying pin-constrained design to the testability-aware bioassay protocol, the proposed method ensures that the resulting chip layout supports the effective execution of test-related droplet operations for the entire chip. Therefore, the

Fig. 16.8 Illustration of the influence by adding test operations to the bioassay

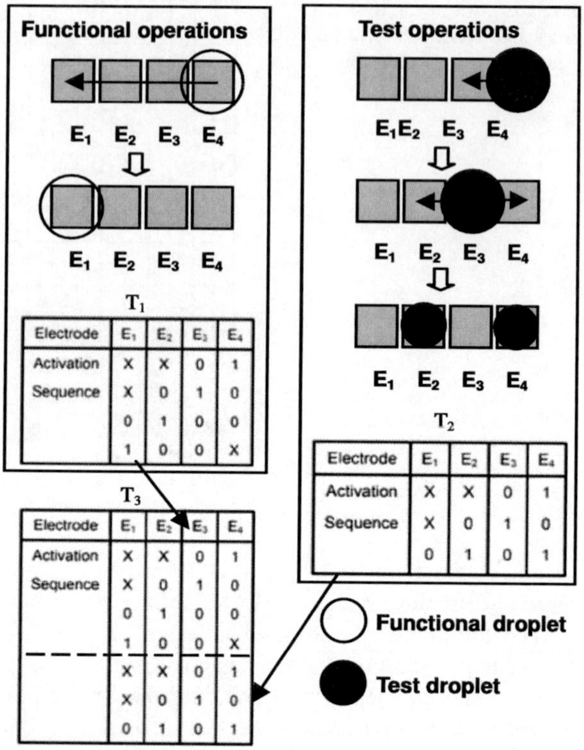

proposed DFT method allows design of pin-constrained biochips with a high level of testability with negligible overhead in terms of the number of control pins.

16.6 Chapter Summary and Conclusions

We have presented a survey of research on design automation and test techniques for digital microfluidic biochips. We first provided an overview of the digital microfluidic platform, and then highlighted advances in synthesis and droplet routing techniques. Practical design techniques for achieving high throughout with a small number of control pins have been presented. Testing and design-for-testability techniques have also been presented. Common defects have been identified and related to logical fault models. Based on these fault models, test techniques for emerging lab-on-chip devices and digital microfluidic modules have been presented. The use of these test techniques for fabricated devices has been highlighted. These design techniques are expected to pave the way for the deployment and use of biochips in the emerging marketplace.

References

1. Fair RB, Khlystov A, Tailor TD, Ivanov V, Evans RD, Griffin PB, Srinivasan V, Pamula VK, Pollack MG, Zhou J (2007) Chemical and biological applications of digital-microfluidic devices. IEEE Des Test Comput 24:10–24
2. Srinvasan V, Pamula VK, Pollack MG, Fair RB (2003) Clinical diagnostics on human whole blood, plasma, serum, urine, saliva, sweat, and tears on a digital microfluidic platform. In: Proceedings of MicroTAS, pp 1287–1290
3. Guiseppi-Elie A, Brahim S, Slaughter G, Ward KR (2005) Design of a subcutaneous implantable biochip for monitoring of glucose and lactate. IEEE Sens J 5:345–355
4. Lin Y-Y, Evans RD, Welch E, Hsu B-N, Madison AC, Fair RB (2010) Low voltage electrowetting-on-dielectric platform using multi-layer insulators. Sens Actuators B 150: 465–470
5. Ottesen EA, Hong JW, Quake SR, Leadbetter JR (2006) Microfluidic digital PCR enables multigene analysis of individual environmental bacteria. Science 314:1464–1467
6. Zhao Y, Cho SK (2006) Microparticle sampling by electrowetting actuated droplet sweeping. Lab Chip 6:137–144
7. Srinivasan V, Pamula VK, Fair RB (2004) An integrated digital microfluidic lab-on-a-chip for clinical diagnostics on human physiological fluids. Lab Chip 4:310–315
8. Luan L, Evans RD, Jokerst NM, Fair RB (2008) Integrated optical sensor in a digital microfluidic platform. IEEE Sens J 8:628–635
9. Global In Vitro Diagnosis Market Analysis, PRLog Free Press Release, http://www.prlog.org/10080477-global-in.vitro-diagnostic-market-analysis.html
10. World Malaria Day (2009) Key Figures, http://www.rollbackmalaria.org/worldmalariaday/keyfigures
11. Semiconductor Industry Association, International Technology Roadmap for Semiconductors (ITRS), (2007) [Online]. Available: http://www.itrs.net/Links/2007ITRS/Home2007.htm
12. Pollack MG, Fair RB, Shenderov AD (2000) Electrowetting-based actuation of liquid droplets for microfluidic applications. Appl Phys Lett 77:1725–1726
13. Fair RB, Srinivasan V, Ren H, Paik P, Pamula VK, Pollack MG (2003) Electrowetting-based on-chip sample processing for integrated microfluidics. In: Proceedings of IEEE international electron devices meeting (IEDM), pp. 32.5.1–32.5.4
14. Cho SK, Moon HJ, Kim CJ (2003) Creating, transporting, cutting, and merging liquid droplets by electrowetting-based actuation for digital microfluidic circuits. J Microelectromech Syst 12:70–80
15. Wheeler AR, Moon H, Bird CA, Loo RRO, Kim C-J, Loo JA, Garrell RL (2005) Digital microfluidics with in-line sample purification for proteomics analyses with MALDI-MS. Anal Chem 77:534–540
16. Gong J, Kim CJ (2005) Two-dimensional digital microfluidic system by multi-layer printed circuit board. In: Proceedings of IEEE MEMS, 726–729
17. Chatterjee D, Hetayothin B, Wheeler AR, King DJ, Garrell RL (2006) Droplet-based microfluidics with nonaqueous solvents and solutions. Lab Chip 6:199–206
18. Berthier J, (2007) Microdrops and Digital Microfluidics: processing, development, and applications (micro & nano technologies), William Andrew Publishing
19. Ren H, Fair RB, Pollack MG (2004) Automated on-chip droplet dispensing with volume control by electro-wetting actuation and capacitance metering. Sens Actuators B 98:319–327
20. Srinivasan V, Pamula VK, Fair RB (2004) Droplet-based microfluidic lab-on-a-chip for glucose detection. Anal Chim Acta 507:145–150
21. Advanced Liquid Logic, http://www.liquid-logic.com
22. http://www.ultimatepcb.com/index.php
23. Xu T, Thwar P, Srinivasan V, Pamula VK, Chakrabarty K (2007) Digital microfluidic biochip for protein crystallization. In: IEEE-NIH Life science systems and applications workshop, Bethesda, MD

24. Xu T, Chakrabarty K, Pamula VK (2008) Design and optimization of a digital microfluidic biochip for protein crystallization. In: Proceedings of IEEE/ACM international conference on computer-aided design, pp 297–301
25. Zhao Y, Chakrabarty K (2009) Cross-contamination avoidance for droplet routing in digital microfluidic biochips. In: Proceedings of IEEE/ACM design, automation and test in europe conference, pp 1290–1295
26. Fan SK, Hashi, C, Kim CJ (2003) Manipulation of multiple droplets on N × M grid by cross-reference EWOD driving scheme and pressure-contact packaging. In: Proceedings of MEMS, pp 694–697
27. Cho M, Pan DZ (2008) A high-performance droplet router for digital microfluidic biochips. Proceedings of international symposium on physical design (ISPD)
28. Yuh PH, et al (2008) A progressive-ILP based routing algorithm for cross-referencing biochips. In: Proceedings of DAC, pp 284–289
29. Yuh PH, et al (2007) BioRoute: A network flow based routing algorithm for digital microfluidic biochips. In: Proceedings of ICCAD, pp 752–757
30. Yuh PH et al (2007) Placement of defect-tolerant digital microfluidic biochips using the T-tree formulation. ACM J Emerg Tech Comput Sys 3:13.1–13.32
31. Xu T, Chakrabarty K (2008) Broadcast electrode-addressing for pin-constrained multi-functional digital microfluidic biochips. In: Proceedings of IEEE/ACM design automation conference, pp 173–178
32. Xu T, Chakrabarty K (2007) A cross-referencing-based droplet manipulation method for high-throughput and pin-constrained digital microfluidic arrays. In: Proceedings of design, automation and test in europe (DATE) conference, pp 552–557
33. Xu T, Chakrabarty K (2007) Integrated droplet routing in the synthesis of microfluidic biochips. In: Proceedings of IEEE/ACM design automation conference, pp 948–953
34. Maftei E, Pop P, Madsen J, Stidsen T (2008) Placement-aware architectural synthesis of digital microfluidic biochips using ILP. In: Proceedings of the international conference on very large scale integration of system on chip, pp 425–430
35. Su F, Chakrabarty K (2004) Architectural-level synthesis of digital microfluidics-based biochips. In: Proceedings of IEEE international conference on cad, pp 223–228
36. Su F, Chakrabarty K (2005) Unified high-level synthesis and module placement for defect-tolerant microfluidic biochips. In: Proceedings of IEEE/ACM design automation conference, pp 825–830
37. Su F, Chakrabarty K (2005) Design of fault-tolerant and dynamically-reconfigurable microfluidic biochips. In: Proceedings of the date conference, pp 1202–1207
38. Su F, Chakrabarty K (2005) Reconfiguration techniques for digital microfluidic biochips. In: Proceedings of IEEE design, test, integration and packaging of mems/moems symposium, pp 143–148
39. Su F, Chakrabarty K (2005) Defect tolerance for gracefully-degradable microfluidics-based biochips. In: Proceedings of IEEE VLSI test symposium, pp 321–326
40. Su F, Hwang W, Chakrabarty K (2006) Droplet routing in the synthesis of digital microfluidic biochips. In: Proceedings of design, automation and test in europe (date) conference, pp 323–328
41. Su F, Chakrabarty K (2006) Module placement for fault-tolerant microfluidics-based biochips. ACM Trans Des Autom Electron Syst 11:682–710
42. Su F, Chakrabarty K (2008) High-level synthesis of digital microfluidic biochips. ACM J Emerg Tech Comput Syst, 3, Article 16
43. Su F (2006) Synthesis, Testing, and Reconfiguration Techniques for Digital Microfluidic Biochips. Ph.D. thesis, Duke University, Durham, NC, USA
44. Xu T, Chakrabarty K (2006) Droplet-trace-based array partitioning and a pin assignment algorithm for the automated design of digital microfluidic biochips. In: Proceedings of IEEE/ACM international conference on hardware/software codesign and system synthesis, pp 112–117
45. De Micheli G (1994) Synthesis and optimization of digital circuits, McGraw-Hill, New york

46. Kramer ME, van Leeuwen J (1984) The complexity of wire routing and finding the minimum area layouts for arbitrary VLSI circuits. In: Advances in computing research 2: VLSI theory, JAI Press, London
47. Diestel R (2005) Graph Theory. Springer, Berlin
48. Kerkhoff HG (2007) Testing of microelectronic-biofluidic systems. IEEE Des Test Comput 24:72–82
49. Su F, Ozev S, Chakrabarty K (2003) Testing of droplet-based microelectrofluidic systems. In: Proceedings of IEEE international test conference, pp 1192–1200
50. Su F, Ozev S, Chakrabarty K (2005) Ensuring the operational health of droplet-based microelectrofluidic biosensor systems. IEEE Sens 5:763–773
51. Xu T, Chakrabarty K (2007) Parallel scan-like test and multiple-defect diagnosis for digital microfluidic biochips. IEEE Trans Biomed Circuits Syst 1:148–158
52. Su F, Ozev S, Chakrabarty K (2006) Test planning and test resource optimization for droplet-based microfluidic systems. J Electron Test: Theory Appl 22:199–210
53. Su F, Hwang W, Mukherjee A, Chakrabarty K (2007) Testing and diagnosis of realistic defects in digital microfluidic biochips. J Electron Test: Theory Appl 23:219–233
54. Xu T, Chakrabarty K (2007) Functional testing of digital microfluidic biochips. In: Proceedings of IEEE international test conference
55. Zhao Y, Xu T, Chakrabarty K (2008) Built-in self-test and fault diagnosis for lab-on-chip using digital microfluidic logic gates. In: Proceedings of IEEE international test conference
56. Xu T, Chakrabarty K (2009) Design-for-testability for digital microfluidic biochips. In: Proceedings of IEEE VLSI test symposium, pp 309–314

Part IV
Reconfigurable Systems

Chapter 17
FPGA Startup Through Sequential Partial and Dynamic Reconfiguration

Joachim Meyer, Michael Hübner, Lars Braun, Oliver Sander, Juanjo Noguera, Rodney Stewart and Jürgen Becker

Abstract Dynamic and partial reconfiguration of Xilinx FPGAs is a well known technique for runtime adaptive system design. The technique enables to substitute parts of a configuration while other regions stay operative without any disturbance. The advantage is the fact, that the spatial and temporal partitioning can be exploited with the goal to increase the performance and to reduce the power consumption due to the re-use of chip area. However, the feature of dynamic and partial reconfiguration can be further exploited. This novel kind of exploitation is described in this chapter. FPGAs still have to compete with other solutions like e.g., processor based designs if e.g., a restricted power budget is available for a specific application domain. In order to reduce the power consumption to a minimum, many devices use different kinds of power saving modes, called sleep-or hibernating modes. In these modes, the power supply of the device is reduced or even fully down. Taking this idea to the extreme, many devices in systems are only powered at run-time when it is necessary. If not, they are released from their power supply and do not drain current at all. Due to the fact that such a power down mode leads to a complete loss of data in SRAM based FPGAs, special techniques for such designs needs to be developed. The configuration has to be reloaded into the device every time when reattaching the power to the FPGA. This circumstance leads to restrictions for the device deployment in some electronic systems since in many cases the time a device may use to wake up is strictly limited. In several use cases, the configuration time of a SRAM based FPGA exceeds this limitation and forces designers to use processors instead of FPGAs. This chapter describes a way to decrease the configuration time of a design by exploiting the method of dynamic

J. Meyer · M. Hübner (✉) · L. Braun · O. Sander · J. Becker
Karlsruhe Institute of Technology (KIT), karlsruhe, Germany
e-mail: Michael.Huebner@kit.edu

J. Noguera · R. Stewart
Xilinx Inc, Dublin, Ireland

N. Voros et al. (eds.), *VLSI 2010 Annual Symposium*,
Lecture Notes in Electrical Engineering 105, DOI: 10.1007/978-94-007-1488-5_17,

and partial reconfiguration in order to enable the usage of a sleep mode. With the presented method, the configuration time of any Xilinx SRAM based FPGA from the identical series (e.g., Spartan 3) is independent from the size of the used device.

17.1 Introduction

Field programmable Gate Arrays found their niche in a variety of academic and industrial applications. The usage reaches from embedded electronics in e.g., automotive domain to accelerators in computer systems e.g., in graphic cards. However, the reconfigurable technology still suffers from the fact, that ASIC solutions of the functional elements for an application including microprocessors, have significant better performance and power specifications. The current improvements for FPGA based technology target more and more these benchmarks due to the usage of latest technology in transistor design and e.g., in dynamic threshold voltage scaling for a stable point of operation. However, power consumption can also be reduced tremendously through the deployment of a sleep mode realized on the computing elements. This method is well established in processor and microcontroller designs. Such a technique is also applicable in FPGA based system. A critical drawback for this is the fact that SRAM based FPGAs needs to be re-configured after they have been in a sleep mode. The SRAM loses its information if the device is not powered. Therefore the wake up time for such a device is directly proportional with its configuration memory size, which has to be reinitialized. The reconfiguration time after the power-on can be up to some hundreds of milliseconds depending on the size of the FPGA. This time frame affects the decision of hardware developers for the choice which device is used in a hardware system. Especially if the device has to process data immediately after the wake up, the choice must be to use a microprocessor which has currently much faster wake up times than FPGAs have if they are in a certain size. This scenario can be found e.g., in the automotive domain where electronic control units (ECUs) are connected in a network with a number of other ECUs in a car. Especially at standby time it is important, that only the required ECUs are active, which are used to monitor some critical functionality like e.g., theft protection. Certainly it is important, that these devices have a fast wakeup time in order to receive and process the information in the messages which are currently on the bus. It can easily be shown, that modern SRAM based FPGAs cannot overcome this hurdle and fail in such a use case. Furthermore the current trend of FPGA vendors is to include more and more logic on the devices which comes with an increased amount of configuration data and therefore an increased startup time. Many other use cases can be found e.g., for other mobile devices like PDAs or in measurement systems with a restricted battery capacity. A novel method for a fast sequential start up for SRAM based FPGA was developed and will be presented in this chapter. The idea here is to start the FPGA with only the required IP blocks through the exploitation of dynamic and partial reconfiguration. The Xilinx Spartan and Virtex SRAM based FPGAs support this feature and are therefore highly relevant for this approach. In the next sections the general idea and the detailed realization will be presented. The chapter is organized as follows: Sect. 17.2 introduces the basics of dynamic and partial reconfiguration. In Sect. 17.3 the general method for the sequential startup of FPGAs is described. Section 17.4 describes an appropriate tool flow for Spartan 3 devices and presents exemplarily a test scenario with real implementation results. The chapter is closed in Sect. 17.5 with the conclusions and the outlook.

17.2 Dynamic and Partial Reconfiguration

The SRAM-based Xilinx FPGAs consist of configurable components, such as configurable logic blocks (CLBs), Block RAM, hardware multiplier, input–output (I/O) elements, and switch matrixes for the connection of routing resources (see Fig. 17.1).

In comparison to the coarse grained reconfigurable architecture described in [2, 3], this fine grained architecture allows path width down to one bit. Changing the content of the SRAM implies changes to the architecture, which represents the actual configuration (function) of the hardware. A detailed overview of reconfigurable hardware and the related methodologies is described in [4]. Design tools, such as Xilinx Integrated System Environment, enable the generation of the actual data needed to program such components, for example, from a Very High Speed Integrated Circuit Hardware Description Language (VHDL) description after synthesis, translation, mapping, and place and route processes. This programming data can be downloaded as bitstream to the FPGA which is then configured. After configuration, the designed architecture starts working immediately.

The idea of dynamic and partial reconfiguration describes the possibility to change parts of hardware on the configurable device while all other parts stay unaffected and operative. This is done by programming the respective cells of the SRAM while other memory areas stay unaffected. The memory of Xilinx FPGAs is organized into columns. The smallest changeable unit is a column with a width of one bit, a so-called frame, which therefore contains configuration information of a complete column. Accordingly, for partial reconfiguration, writing configuration data is only possible in full configuration frames.

There exist two concepts for dynamic partial reconfiguration. Difference based partial reconfiguration is used for changing configuration properties of the running system. Therefore usually only few configuration data needs to be changed. The technique to create bitstreams for this is based on a comparison of the actual and the target design. Afterwards the differences between them are extracted into a partial bitstream.

The second concept is the module based partial reconfiguration. It is used to replace whole blocks of functionality of an actual design. Therefore the design is

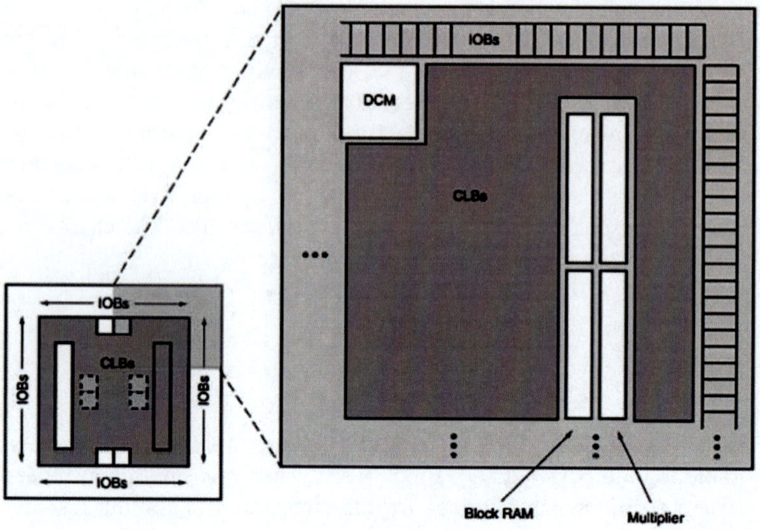

Fig. 17.1 Schematic view on Xilinx FPGA resources [1]

divided into static areas and areas for partial reconfigurable modules. The standard flow to create the needed bitstreams creates each of the partial modules in separate runs, taking into account resources which are used by the static module.

Like for a standard configuration, for partial reconfiguration the corresponding configuration data has to be transferred to FPGA as well. This transmission of reconfiguration data can be achieved, with different interfaces. The Serial Peripheral Interface (SPI) allows a low cost configuration over an external flash memory. Since this single bit connection suffers from low data throughput, the next generation of the Xilinx devices support Quad SPI mode where 4 parallel data lines are used to transfer the configuration to the FPGA. The Byte Peripheral Interface (BPI) is a parallel configuration interface in which the FPGA also acts as the "Master" device. With this interface, the highest data throughput for configuration data can be achieved due to the fact, that 8 bits in parallel are transferred to the FPGA device. Novel FPGAs like Spartan 6 provide a BPI interface with 16 bits. Furthermore, the quantity of date to be transferred plays a major role for the startup time.

17.3 General Method of Sequential Startup

The general idea for the sequential start up is derived from the state of the art method of dynamic and partial reconfiguration (see [5]). As described in Sect. 17.2, parts of the active design on an FPGA are substituted by another functionality while the other parts of the design stay operative. The first step for the realization of a sequential start up design is the analysis of the complete design

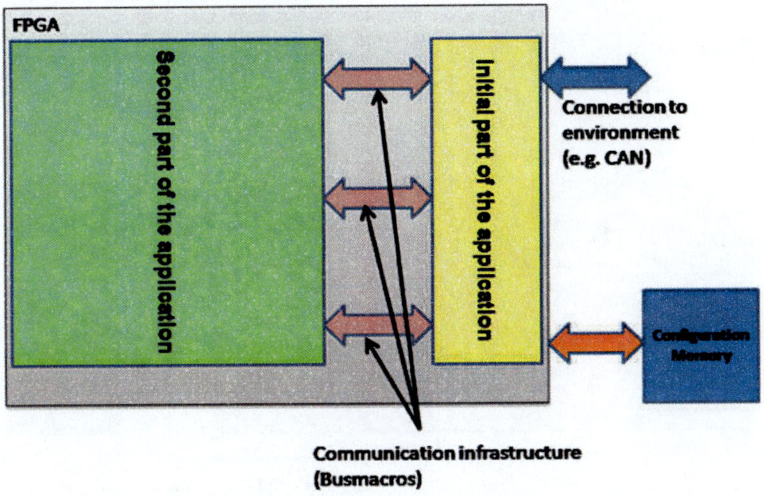

Fig. 17.2 Schematic view to a partition application on a FPGA

in terms of two parts which can be loaded in 2 separate phases. The first one in the first phase, in order to receive and process data very early after the power up, and the second part in the second phase, in order to complete the design to the full functionality.

Figure 17.2 shows a schematic view to a partitioned design. The initial design of the complete application is positioned on the right side of the FPGA and connected over a communication infrastructure to the second part. The two parts together are the complete application with its full functionality. The initial design needs to be connected to the environment via the I/O pins of the FPGA in order to process incoming data very early after powering up the device. Certainly the configuration data needs to be provided from an external memory. For this reason boot memory is connected over the standard interfaces (e.g., SPI) as described in Sect. 17.2.

It is obvious, that the size of the initial design affects the start-up time of the device due to the required data transfer of the configuration information. It is therefore beneficial, to include only the required components for first data processing. Such components could be e.g., a small microprocessor or simple logic for preprocessing of the data. In the next section, a more concrete example for an initial design as well as the tool flow to implement it on Spartan 3 (E) devices are described.

After power on, the FPGA reads the configuration data from the external memory after a specific and fixed delay. Due to the fact that FPGA internal processes have to be initiated, the data transfer is delayed by a fixed time which is called power-on-reset time. This time is device dependent and cannot be optimized by the user.

Figure 17.3 shows the first phase of the sequential startup. The initial design is loaded and ready to process data received from an external device. The initial

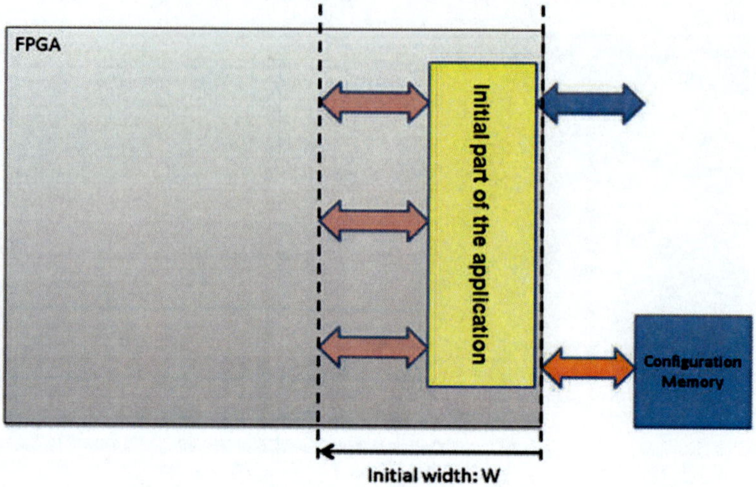

Fig. 17.3 Phase 1 of sequential startup

width W shown in the figure includes the area utilized by the initial design and the communication infrastructure. It can be seen that this width has to be minimized in order to reduce the initial configuration time. The communication infrastructure has to be kept as close as possible to the initial design due to the fact that each utilized logic block or any resource on an additional column of the FPGA configuration memory has a high impact to the size of the bitstream which has to be transferred after power on.

In Fig. 17.4, the schematic view to the FPGA based system after the second phase is presented. The FPGA now contains the complete design with its full functionality. The initial design pre-processes the incoming data and forwards the results to the second design for further operations. The communication infrastructure connects both parts of the design through standard interfaces. As shown in the figure, the complete configuration for width C of the FPGA would be necessary to load the design after power on. Due to the sequential start up, the delay time after power on is now reduced proportional to the quotient of C and W. After describing the procedure to build such a design, the next section evaluates exactly this result and shows that this assumption is valid also in real realizations on a Spartan 3 FPGA.

17.4 Implementation

The following chapter describes the implementation of the sequential startup using a Spartan 3E FPGA. It also identifies the constraints and techniques of this configuration methodology for this specific FPGA architecture. A device of the

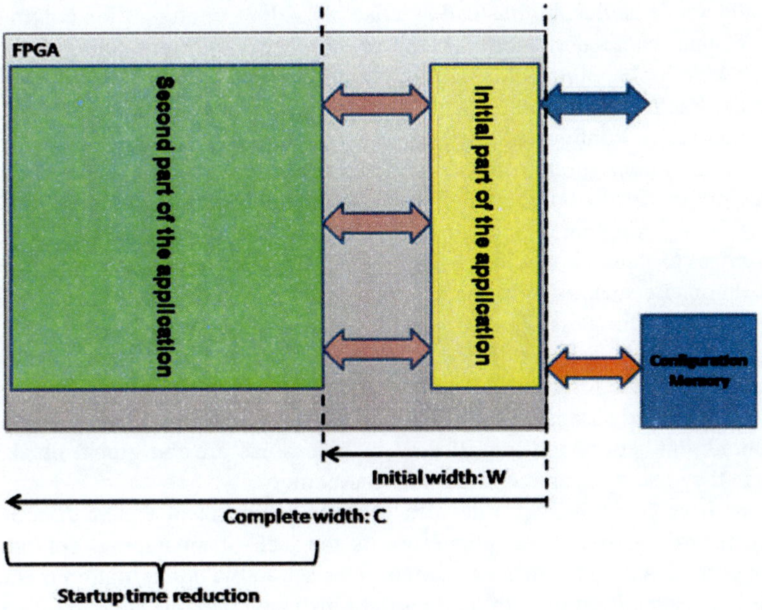

Fig. 17.4 Phase 2 with complete design on the FPGA

Spartan 3 family was chosen as target architecture since the low cost and low power characteristics of this family suits the requirements for the chosen embedded systems scenario much better than the high end Virtex series. Although Xilinx does not support partial reconfiguration for Spartan 3 devices, research groups successfully implemented this technique also for this device family [6].

17.4.1 Spartan 3 Configuration Memory Architecture

The use of partial reconfiguration for Spartan 3 FPGAs is possible but brings along some drawbacks compared to the high end Virtex devices. Those drawbacks are caused by specific characteristics of their configuration memory architecture.

Like for all Xilinx FPGAs, the configuration memory of Spartan-3 FPGAs can be visualized as a two-dimensional matrix of bits. Those bits can be grouped into vertical frames which are one-bit wide and span the whole device. Such a configuration frame is the smallest amount of data which can be read or can be written. One frame does not directly map to any single hardware resource of the FPGA, it rather contains the configuration information for one part of several logic resources as well as routing.

The next bigger configuration structure is a configuration column which is composed of several configuration frames. Therefore a configuration column also spans the whole device, but instead of a one bit width a configuration column has a width of one resource element. Therefore there exist different configuration columns, for example columns of CLBs, columns of BRAM blocks, or columns of IO elements. Each of those different configuration columns are composed by a different number of configuration frames.

Although a configuration frame is the smallest accessible unit of the configuration memory, for partial reconfiguration, the smallest accessible unit of Spartan 3 FPGAs is a configuration column. This is a result of the Spartan 3 specific requirement to delete a whole configuration column before writing any frames of that column. Because of this characteristic, partial reconfiguration for Spartan 3 devices is not glitchless. Even if the design stays exactly the same the corresponding columns get first deleted and then rewritten. Therefore, for partial reconfiguration, the logic as well as the routing of different configuration modules hast to be strictly separated into different configuration columns. The only exceptions for nets which are allowed in both areas are the global clock lines, because they use a separated configuration memory.

Apart from Spartan 3A(N) devices, another drawback of the Spartan 3 architecture regarding partial reconfiguration is the lack of an internal configuration access port (ICAP). The missing internal interface to the configuration memory of the device makes it difficult to load partial bitstreams without the help of external intelligence. A possible solution to this problem is an external loopbacks to one of the external configuration interfaces, like presented in [7].

17.4.2 Design Flow for Fast Sequential FPGA Startup with Spartan 3

The first step for implementing the sequential startup approach on an FPGA is to separate the design into boot-time critical components and boot-time tolerant components. Since the size of the initial partial bitstream depends on the amount of utilized columns to be configured, the initial design should be strictly limited to the boot-time critical components and also optimized for minimal area consumption by placing all those components into a contiguous area as small as possible. This area needs to provide enough resources for all necessary components (e.g., Input Output resources for communication interfaces) of the initial design but should include not more resources than needed since those cannot be used by the second design anymore and would be wasted.

Because of the characteristics of the Spartan 3 configuration memory architecture the partitioning options of the physical resources of the device into two areas, one for the initial and one for the second part of the application, are very limited. The obvious partitioning would be to place one part on the left hand side and the other part on the right hand side. To realize the initial part in a vertical stripe located in the middle of the device and realize 2 application areas, one on the left hand side, the other on the right hand side of this stripe is another option.

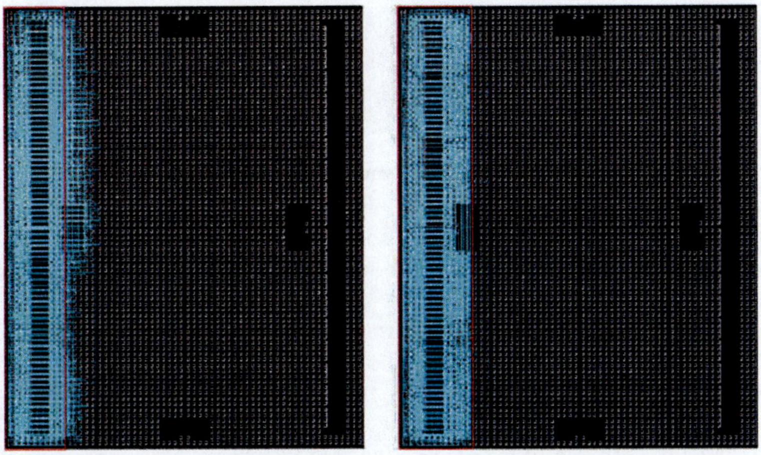

Fig. 17.5 Compared to the standard implementation tools (*left hand side*), the EAPR tool flow (*right hand side*) provide improved capabilities to constraint the routing. This is useful to minimize the initial area and thus the initial bitstream size

But since the majority of Spartan 3 devices do not provide special resources like BRAM block in the middle of the device, this partitioning will rarely provide the right set of resources for the initial design part.

The key technique for using the Fast Sequential Startup is to create appropriate partial bitstreams. Although not officially supported by Xilinx, there are three ways to create valid partial bitstreams for Spartan 3 FPGAs using Xilinx tools. The first possibility is given by the BitGen option for difference based partial reconfiguration [8]. But since this option is supposed to be used rather for small design changes than for the exchange of functionality for whole modules, it is not suitable for the Fast Sequential FPGA Startup. The second way is another option of BitGen, called Partial Mask [9]. It allows determining exactly which configuration column should be included in the resulting partial bitstream. Therefore it provides a high amount of freedom by the cost of the requirement of a detailed knowledge on the memory architecture and the design to implement. To generate appropriate partial bitstreams for Fast Sequential Startup we used this option. The last option is the Early Access Partial Reconfiguration flow [10] which can be used by patching the standard tools with a special EAPR-patch. While this flow is adequate to create the partial bitstream for the second design, it is not able to create the partial bitstream for the initial configuration. This is because of the flow assumes a full configuration bitstream to configure the global clock resources and thus cannot be used to include such resources in partial bitstreams.

Nevertheless the EAPR flow is used in the proposed tool flow. The reasons for this are the modified implementation tools which allow the user to set advanced constraints regarding the routing. This is necessary to strictly separate the two design parts but still be able to utilize as much resources as possible.

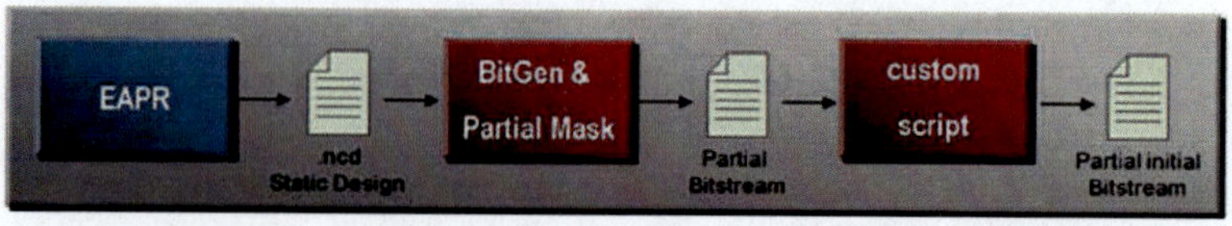

Fig. 17.6 Proposed flow for fast sequential startup using spartan 3 devices

Figure 17.5 compares the implementation of the initial design part using the standard tools from ISE 10.1 on the left and an implementation using the ISE 9.2 sp4 tools with the EAPR-patch. For the standard tools many nets leave the area which the logic for the initial design part is constraint to. When using the EAPR tools only global clock routes leave the constrained area. This is no problem because of a separated configuration memory for those lines.

The overall proposed flow to implement a design using Fast Sequential Startup for Spartan 3 is shown in Fig. 17.6. When using the Partial Mask option of BitGen in order to include only the appropriate columns for the initial design, make sure to also use the option for active reconfiguration. This makes sure the bitstream includes a valid startup sequence. The resulting bitstream still will not work immediately for an initial configuration, first the proper value for the Frame Length Register (FLR) has to be included. The value determines the length of a frame and is different for different Spartan3 device types. After inserting this value either the CRC values in the bitstream has to be recalculated and replaced, or the CRC-check feature has to be disabled during BitGen. A custom script was used to automate these actions, compare Fig. 17.6. For the second part of the bitstream the partial bitstream resulting from the EAPR flow can be used.

17.5 Experiments and Results

To verify the idea of Fast Startup and the proposed flow for Spartan 3, the test design presented in Fig. 17.7 was implemented and the configuration time was measured. The goal was to bring up a realistic initial subsystem as fast as possible and configure the remaining part of the FPGA with additional logic afterwards. Therefore a MicroBlaze microprocessor and a Controller Area Network (CAN) interface from the automotive domain were combined with several minor modules in order to form a flexible microprocessor based, initial system.

Since sequential startup is based on dynamic partial reconfiguration, all special requirements of this technique are also given for the sequential startup. Therefore, well-defined interfaces had to be used for nets which connect the second part of the application with the initial parts of the design. This was done by so called busmacros [11]. Busmacros must be placed at the border of the initial and secondary design. It is crucial that the signals of the interfaces between both design fit exactly to another. To keep the sequential startup as flexible as possible, a data bus like the Processor Local Bus (PLB) should be routed through such busmacros in

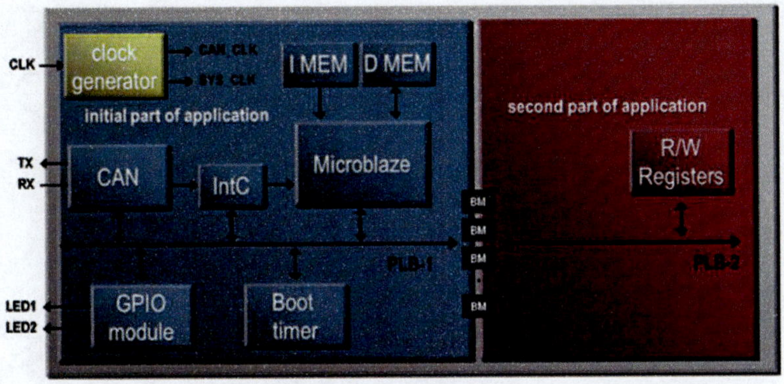

Fig. 17.7 Hardware design used for the experiments

order to make the logic of the second design accessible to the microprocessor of the first design. Those busmacros provide the possibility to disable the connection between the two design parts. On the one hand this avoids an undefined bus state when there is no second design configured yet. On the other hand glitches resulting from the dynamic partial reconfiguration of the second design part cannot reach the initial part if the busmacros are disabled.

The second design part which was used for the experiments were several 32-bit registers implemented as PLB-slave to provide read and write access to the Microblaze. The FPGA Editor view for the whole design is shown in Fig. 17.8. The initial design part (P1) is located at the left hand side, the second design part at the right hand side.

The picture in Fig. 17.8 gives a hint, that the initial design has to be designed area efficient in order to keep the area to be configured after startup of the FPGA as small as possible. Furthermore it is obvious, that larger FPGAs will benefit enormously from the reduced startup times due to the fact, that the quotient between the time for the initial design and the time for a complete configuration is higher than for smaller FPGAs.

For the partial initial configuration the SPI interface was used, the second dynamic partial configuration was done by using iMPACT with the external JTAG interface. Measured configuration times of the initial partial bitstream and of a full bitstream can be seen in Table 17.1.

The size of the partial initial bitstream for the sequential startup was about 72% of the full bitstream size which confirms the expected proportionality between the configuration time and configuration data. The reason for the small difference in percentage of the measured times compared to percentage of the bitstream sizes is the Power-on-Reset time which is the same for all test scenarios and is therefore a static bias for the calculation.

To get further values for calculations and to show the benefit of the sequential startup, the same design was ported to an XC3S1600E device, which is the device in the Spartan3E series with the highest number of logic gates. For this design the

Fig. 17.8 FPGA Editor view of the sample design. P1: Initial design, P2: second design

Table 17.1 Measured results using SPI interface

XC3S500E	SPI					
	Configrate 6		Configrate 12		Configrate 25	
	ms	%	ms	%	ms	%
Traditional Startup	431	100	218	100	112	100
Sequential Startup	320	74	163	75	84	75

initial partial bitstream had only a size of 31% of a full bitstream. Tables 17.2, 17.3 show worst case calculations assuming SPI as configuration interface for both de-signs. The FPGA Editor view of the initial design part for this device can be seen in Fig. 17.4.

The calculation basis for the values in Tables 17.2, 17.3 is described in Eq. 17.1. The worst case numbers for the power on reset time and the configuration clock frequency generated by the FPGA can be found in the Spartan 3 datasheet [1].

$$T_{conf} = T_{Power-On} + \frac{\text{Bitstream size}}{f_{conf}} \tag{17.1}$$

T_{conf} : Configuration time

Table 17.2 Calculated worst case numbers for a XC3S500E device

XC3S500E	SPI			
	CCLK: 12.8 MHz		CCLK: 25.6 MHz	
	ms	%	ms	%
Traditional startup	182.36	100	93.68	100
Sequential startup	131.72	72	68.36	73

Table 17.3 Calculated worst case numbers for a XC3S1600E device

XC3S1600E	SPI			
	CCLK: 12.8 MHz		CCLK: 25.6 MHz	
	ms	%	ms	%
Traditional startup	473.38	100	240.19	100
Sequential startup	152.28	32	79.64	33

$$T_{Power-On} : \text{Power on reset Time}$$

$$f_{conf} : \text{Configuration Clock Speed}$$

As the calculations show, the benefit of sequential startup is higher for large devices. This is due to the fact that the size in terms of required logic resources on the FPGA of the initial design stays constant. Therefore the size of data to be transferred to the FPGA for the initial design is also constant while the complete amount of configuration data increases if the device is larger. A startup of a large device therefore can be accelerated dramatically as the calculations show.

17.6 Conclusions and Outlook

The work presented in this chapter introduces the general idea of sequential startup for configuring an FPGA with an initial and followed by a second bitstream. By exploiting the small sizes of partial bitstreams the sequential startup enables FPGAs to provide boot time critical components independent to the remaining components of the design or the size of the used device. Whenever it is possible to separate the full FPGA design into at least two temporal partitions and whenever there is only a limited time to boot, the sequential startup brings tremendous advantages. The advantages with this approach are obvious, FPGAs with a high startup time can now be used in a wider field of application scenarios. Especially when pointing to the next generation FPGAs with bitstream sizes bigger than 22 Megabyte, it is obvious that the importance of this technique will even increase in the future (as also described in [12]). However, with all its advantages, sequential startup also brings drawbacks in terms of additional requirements and limitations to the system design. Most of these requirements depend on the used method to generate the partial bitstreams and thus they depend highly on the device architecture. Therefore the next steps are an integration of the sequential startup method in the standard tools for Xilinx FPGAs.

Furthermore it has to be investigated which functionality the initial design includes for different application scenario. For example it could be a small state machine which gets initially loaded on the FPGA for checking the status of the environment and to decide if a full wakeup is needed, maybe even decide dynamically between different designs. Intelligent wakeup mechanisms play a major role if additional power needs to be saved in several applications e.g., in the automotive domain.

References

1. Spartan-3E FPGA Family: Data Sheet DS312 (v3.8), August 26, 2009, Xilinx Inc
2. Baumgarten V, May F, Nuckel A, Vorbach M, Weinhardt M (2003) PACT XPP A self-reconfigurable data processing architecture. In: International conference engineering of reconfigurable systems and algorithms (ERSA 2001), Monte Carlo Resort, Las Vegas
3. Thomas A, Becker J (2004) Dynamic adaptive routing techniques in multigrain dynamic reconfigurable hardware architectures. In: 14th International conference field programmable logic and applications (FPL 2004), Antwerp, Belgium
4. Compton K, Hauck S (2002) Reconfigurable computing: A survey of systems and software. ACM Comput Surv 34(2):171–210
5. Blodget B, McMillan S, lysaght P (2003) A lightweight approach for embedded reconfiguration of FPGAs. Design, automation and test in Europe, Date 2003
6. Gonzalez I, Aguayo E, Lopez-Buedo S (2007) Self-reconfigurable embedded systems on low-cost FPGAs. Micro, IEEE 27(4):49–57. doi:10.1109/MM.2007.72
7. Paulsson K, Hübner M, Auer G, Dreschmann M, Becker J (2007) Implementation of a virtual internal configuration access port (jcap) for enabling partial self-reconfiguration on xilinx spartan III FPGAs. FPL 2007, 351–356
8. Difference-based partial reconfiguration: Application note XAPP290 (v2.0), December 3, 2007, Xilinx Inc
9. Virtex-II Pro, Virtex-II Pro X FPGA user guide: User guide ug012 (v4.2), 364–367, 5 November, 2007, Xilinx Inc
10. Early access partial reconfiguration user guide: User guide UG208 (v1.2), September 9, 2008, Xilinx Inc
11. Huebner M, Becker T, Becker J (2004) Real-time LUTbased network topologies for dynamic and partial FPGA selfreconfiguration. In: Proceedings of the 17th symposium on Integratedcircuits and system design (SBCCI 04), pages 28–32, 25–29 2004
12. Nelson BE, Wirthlin MJ, Hutchings BL, Athanas PM, Bohner S (2008) Design productivity for configurable computing. ERSA 57–66

Chapter 18
Two Dimensional Dynamic Multigrained Reconfigurable Hardware

Lars Braun and Jürgen Becker

Abstract Partial dynamic reconfigurable (PDR) systems designed with state-of-the-art tool chains, like the Early Access Partial Reconfiguration (EAPR) Flow from Xilinx, does not exploit the full flexibility and all features which a state of the art FPGA chip offers. For example the utilized chip area and the position of a region which can be reconfigured dynamically is traditionally specified during design-time. Thereby the shape and the size of the reconfigurable area are set by the size of the largest module to be reconfigured. The consequence is that if a smaller module is placed on that region, chip area stays unused as so called black silicon. This drawback is only one example for the limitation of development tools of reconfigurable hardware architectures. In this book section, a new approach for exploiting the capability of reconfigurable hardware architectures is presented. It allows exploiting the reconfigurable architectures more efficient than other solutions introduced before. This is achieved through a novel concept of using micro blocks for the communication infrastructure as well as for the functional elements on the FPGA. The granularity and the online versus offline tradeoff for the usage of the micro blocks for building up more complex structures on the FPGA will be presented in this chapter.

L. Braun (✉) · J. Becker
Karlsruher Institut für Technologie (KIT) Institut für Technik der Informationsverarbeitung (ITIV), 76131 Karlsruhe, Germany
e-mail: lars.braun@kit.edu

J. Becker
e-mail: becker@kit.edu

N. Voros et al. (eds.), *VLSI 2010 Annual Symposium*,
Lecture Notes in Electrical Engineering 105, DOI: 10.1007/978-94-007-1488-5_18,

Fig. 18.1 Overview of the complete system

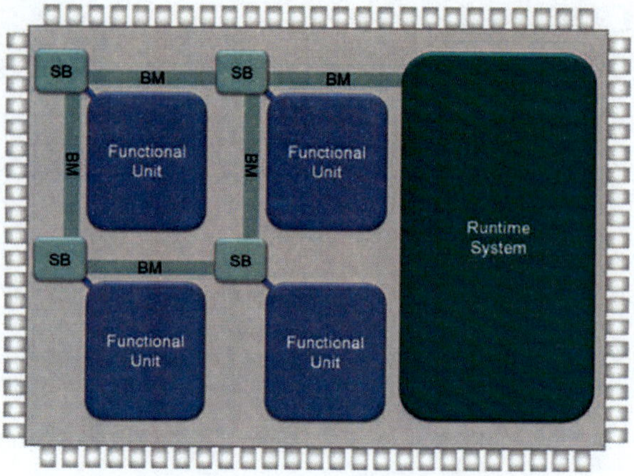

18.1 Introduction

The goal of the presented work is to bridge the hurdle of known restrictions of PDR systems which are well known for example from the EAPR flow from Xilinx [1]. These restrictions are e.g., the fixed size of areas for the reconfigurable modules during run-time, the fixed amount of regions for reconfigurable modules, a fixed communication structure (e.g., buses) and, last but not least, the algorithms realized within the modules. An overview of the proposed design is presented in Fig. 18.1. Due to the rectangular placement of the FPGA resources on the die, obviously a mesh based network is one the most suitable communication structure. To achieve this goal, the achievements are divided in two goals. Each of them examines a part of the addressed limitations. The first one is a scalable, flexible and extendable Network-on-Chip (NoC). The network is the backbone of the system. The limitations of the NoC influences the overall system in terms of performance, due to a loss of data throughput [2]. The limitation e.g., through the bandwidth or the used topology, has a direct effect on the performance of the complete system and certainly the attached modules. Also the routing scheme and strategy of the communication structure has obviously a direct effect on where and how the modules can be placed. In Fig. 18.1, the communication modules are named as *BM* and *SB*. The positions named BM show the communication lines which have to be established during runtime and SB stands for the communication nodes which represents the links to the functional units. The second part of this book section examines the reconfigurable modules which are fixed in their size if traditional design methodologies are used to develop the system. As already mentioned, the consequence of this is a fixed functionality run-time. In case the algorithm of a module has to be changed due to the requirement of the application, a module has to be generated completely new during design-time and made available to be configured on the chip. This leads to a large amount of reconfigurable modules which are normally stored on an external configuration memory.

The required memory capacity can be enormous due to the high number of required modules. The presented work handles these limitations with the approach described in the following manner: Sect. 18.2 presents the novel development of a NoC and the online routing of its communication structures. Section 18.3 describes the online generation of functional modules. The tool for generating the necessary bitstreams is introduced in Sect. 18.4 and the final conclusions are presented in Sect. 18.5.

18.2 Online Adaptable Mesh Based NoC

The on-chip NoC has to fulfill different requirements. It has to be area efficient because of the limited resources an FPGA offers. Furthermore, the NoC could beneficially offer the possibility to be expandable during run-time in term of bitwidth and therefore data throughput. In the following chapter, the NoC switch and its elements are introduced. Furthermore, the novel concept and realization of the on demand online routing for the communication structures for building up a run-time extensible mesh structured network-on-chip [3] are presented.

18.2.1 Switch for 2D Mesh Based NoC Approach

In this approach, the term of a switch describes a module which changes the data flow path in the system. Thereby it is possible to choose different sinks for a source and hence to control the data path. This mechanism manages the switch (*SB* in Fig. 18.1) and thereby the data stream which has to be routed via the switch. One challenge in the development of such switches is the controlling of the switch itself. One way to control the switch is to use a global master controller, which handles the whole communication on the chip and controls all involved switches. This approach has the benefit that the resource utilization of the switches is decreased since no controlling logic has to be implemented on the switch node. But this solution would require connecting all switches with the master. This connection requires additional FPGA resources which are only available in a limited amount on the device. The second approach is to deploy a packed based routing. In this case every packet which has to be routed via the network has a destination address included in the packet itself. The routing will be done directly within the switches. Therefore the switch needs additional logic, which decodes the destination address and chooses the next neighbor for the data transmission. Hence there is no master needed and also no additional connections, but the resource utilization of the switch are highly increased since the routing algorithm can be very complex. In our approach the first method will be modified to meet the requirements of an FPGA design. The goal is to minimize the resource utilization and to increase the speed of a switching mechanism. Another fact which argues for

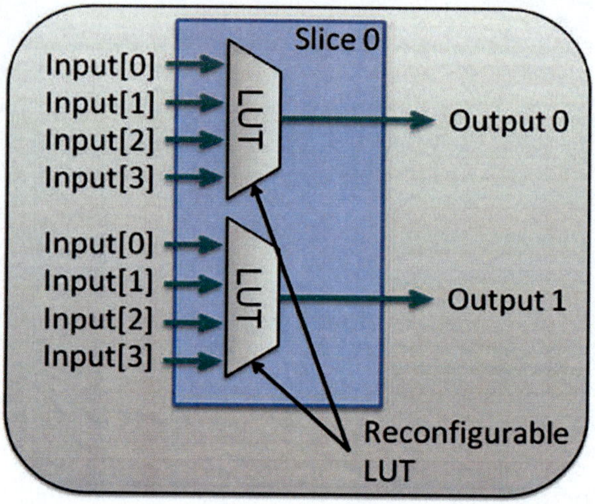

Fig. 18.2 Slice implementation of the presented switch

the use of simple switches is, that the design is easy to implement on FPGAs. The network switch will be controlled by the configuration mechanism of the FPGA. To control a switch it is necessary to know the exact position of the content of the look up tables (LUTs) within the bitstream. Only with this information it is possible to modify the bitstream and to change the content of a specific LUT. Equally important is the ability to locate the correct frame which has to be changed and written back. These tasks have been evaluated in prior work by [4].

The switch was designed using the LUTs as multiplexers as shown in Fig. 18.2. These multiplexers are controlled by the content of the LUTs which is stored in the configuration memory of the FPGA. The innovation of the presented approach lies in the fact, that the content of these LUTs can be changed by a master without being directly connected by using the configuration mechanisms of the FPGA itself. To connect each node with its neighbors (as shown in Fig. 18.1), four channels must be provided as well as the connection between the switch and the corresponding functional unit (FU). Hence, a switch with five interfaces has to be generated as shown in Fig. 18.3. The content of a configuration frame represents the data which must be edited within the bitstream to control the switch. In this system a Virtex-II Pro FPGA from Xilinx is used, but the bitstream format of the Virtex-II and II Pro families are quite similar. Also the switch is tested and evaluated on Virtex4 and Virtex5 FPGAs. The configuration stream is divided into several different kind of bitstreams. Each of them configures a different hardware block of the FPGA. For this approach only the configuration stream of the Configurable Logic Blocks (CLBs) is relevant. To explain the configuration mechanism more briefly, the Virtex-II FPGA is chosen as example. For Virtex 4 and Virtex 5 FPGAs the mechanism can be derived. The configuration of a Virtex-II FPGA is performed CLB wise, in a way that every CLB column is divided into 22 single frames with a bit width of one bit per frame. The reconfiguration time is

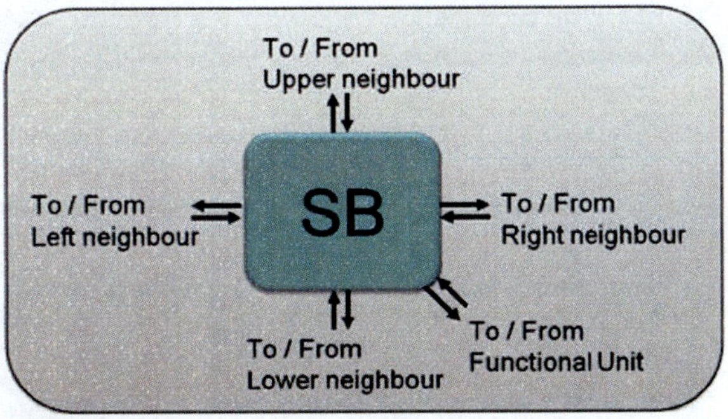

Fig. 18.3 Possible routing directions of the presented switch

proportional to the number and the length of the frames which are reconfigured, so for a fast reconfiguration, the number of modified frames should be minimized, since the length of a frame is fixed for a given FPGA architecture and cannot be modified. Within a CLB bitstream the information for the FPGA global routing switch matrices is also included. This can be exploited by direct modification of the routing resources of the FPGA. Because of the enormous combinatorial possibilities and the difficult and computation intensive way to handle them, this technique will not be used in this approach. The required information for the LUTs is coded into two frames in the bitstream of a CLB. Each frame stores the content of the LUTs of two slices within one CLB and the other LUTs which are included in the same CLB are stored in a second bitstream.

18.2.1.1 Switch Layout

The switches are implemented through a combination of LUTs which are available in the CLB blocks. The on chip implementation of such a switch box is presented in Fig. 18.4. This example shows an implementation of an 8 Bit Switch. The calculation of the resource usage should show how less the on chip resource utilization of this implementation is. As described before a switch with 4 inputs has to be realized to connect each possible input to one output. So one LUT with its four inputs can be used to do this. Hence one LUT is used to route 1 Bit from four directions. For 8 bit 8 LUTs are needed. This means each input needs eight LUTs to route eight bit from four directions to this eight bit input. Hence eight LUTs are within a CLB one CLB per output is needed. In order to generate such a five output switch, five CLBs are needed. This example shows that only five CLBs are needed to fabricate a eight bit switch. In Fig. 18.4, these CLBs are in the middle of the design. The utilized CLBs surrounding them are responsible for the connection to the routing resources and the functional unit. The multiplexers of the switch are controlled with the method described above. If additional multiplexers are connected in series, each stage needs a reconfigurable LUT in order to control the multiplexers of this stage.

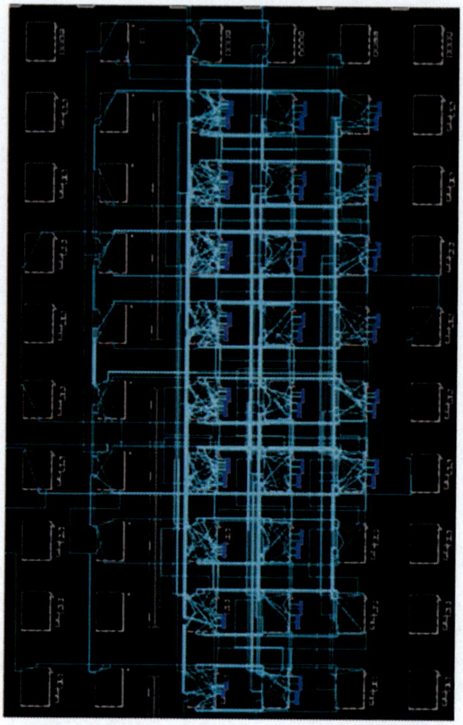

Fig. 18.4 Switch design in FPGA editor

18.2.1.2 Controlling the Switch

There are several possibilities to change the content of the reconfigurable LUTs and thereby the switch itself. If the system uses the ICAP in combination with a microprocessor, such as MicroBlaze or the PowerPC, the LUTs can be reconfigured directly over this interface. The glue logic for this is included in the package of Xilinx's HWICAP [5] and can be used to modify specific LUTs. With the provided software subroutine XHwIcapSetClbBits it is possible to write a new content into a specified LUT. This makes it simple to change the content and thereby control the data path of the switch. A method which is independent from special functions provided from Xilinx to control the switch will use the read-modify and write-back method. This technique can be performed internally by a microprocessor using the ICAP also externally using the configuration port of the FPGA. This requires to read back the bitstream of the specific frames, which holds the data from the switch, into the memory of the controlling system. In the system the frames will be modified at the required positions and then written back to the FPGA. The next approaches to modify the content of the LUTs are only exploitable if the involved LUTs are placed in a static system, that means that the area above and below the switch does not change during runtime. Therefore the configuration bitstream of the frames which should be modified to control

Table 18.1 Comparisons of the switch configuration time of different ICAP controller

ICAP Controller	Virtex 4 FX 20	Virtex II Pro 30	Virtex II Pro 100
XPS_HWICAP [5]	19 μs	168.4 μs	244.8 μs
PLB_ICAP [6]	556 ns	9.2 μs	13.6 μs
FSL_ICAP [7]	5.8 μs	n.a.	n.a.

the switch are stored in the system. When the switch shall toggle, the stored bitstream is modified at the position where the content of the corresponding register is stored and is written into the configuration memory on the FPGA. If the switch is placed in the static part of the system it is also possible to generate a set of partial bitstreams which contain the different contents of the involved LUTs. This method has the benefit that the bitstream does not have to be modified before the reconfiguration. But for every possible LUT content a bitstream must be provided on an internal or external memory. Thereby a large amount on pre-stored bitstreams must be generated and stored. The following calculation will give the number of bitstreams that have to be stored. If the system has N switches with M possible interconnections, the designer has to generate and store NxM bitstreams if the Reconfigurable LUTs are not in the same column, if they are in the same column N^M bitstreams have to be generated and stored. This calculation shows that this approach for modifying the LUTs is only reasonable for a small design. Hence other approaches should be exploited as already mentioned in this chapter. The approximate time the switch needs to change the dataflow depends on the speed of the internal ICAP or the JTAG interface if an external reconfiguration is performed. The following calculation is based on measurements of the XPS_ICAP controller of a Virtex-II FPGA. The measured configuration speed of this interface is about 5 Mbyte per s. One frame of a Virtex-II Pro V2P100 will be written into the configuration memory within approximately 244.8 μs. This will also be the time the switch needs to change the dataflow. This enables to change the switches aligned in one frame with a frequency of 4 kHz. Table 18.1 shows the switching speed for different FPGAs and ICAP controllers.

18.2.2 Physical Online Routing of Communication Structures

18.2.2.1 Motivation

Actual state-of-the-art systems require communication structures which are determined completely at design time. For this purpose, bus macros are used which are implemented fix in the system and cannot be moved around from their physical location. That is why modules can be placed at run time only on predetermined positions, thus reducing the degree of freedom of the placement area. To avoid this problem, a methodology has been developed that takes care of the physical realization of the communication lines between the individual modules and the static

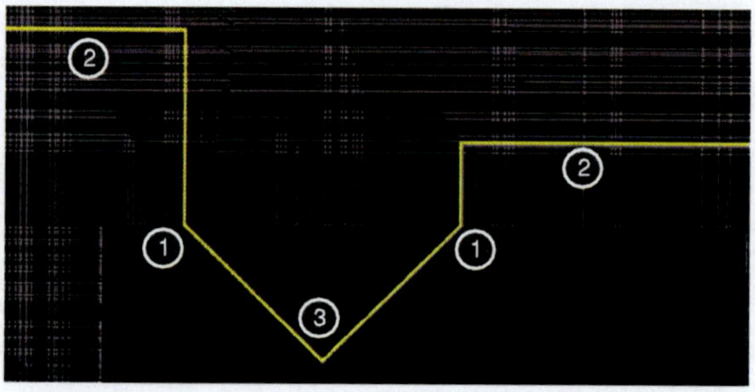

Fig. 18.5 Switch design in FPGA editor

area during runtime. The extended system approach aims for a reduction of the restrictions for communication structure in a way that the modules can be placed independently and flexible on the chip area. Therefore a technique has been developed to route communications structures on the FPGA during run time in order to react on the varying positions of the modules. In addition to this, a method has been developed to create connections on the FPGA. This extends the degree of freedom for the 2-dimensional placement compared to the slot based systems.

18.2.2.2 Method

In order to make use of the degrees of freedom described above the read-modify-writeback (RMW) technique was used [8]. This means that the configuration memory of the part of the FPGA that has to be changed is read back and saved in the local memory of the run time system. Based on former analysis of the configuration bit streams, the construction and content of the data packets is well known. Hence it is possible to apply well-directed manipulations on the configuration bitstream. After the insertion of the modifications, the bitstream is written back on the FPGA (write back). We assume that the position of the modules respectively the positions of the I/O ports are known by the run time system. Thus the run time system has the information of the position of the start and end points and of the communication lines that has to be established on the FPGA. Using these parameters the routing algorithm of the run time system tries to find a suitable path for this point-to-point connection. Here the A*-algorithm showed up to be well suited for this problem. The advantage of this algorithm is its low complexity compared to other algorithm like Dijkstra where it still delivers very good results with low computation time. In addition, this algorithm can also take care of blocked areas and bypasses them. This is extremely important for on-line routing as other previously placed blocks can be bypassed.

Now that the path to be routed is known, the run time system calculates the position of the routing templates to be placed in order to establish a communication link. The communication resources of the FPGA are constructed in a way, that every end of a communication line is connected to one port of a switch box. This is shown in Fig. 18.5. Thereby a connection point (1) describes a connection

Fig. 18.6 Online routing system overview

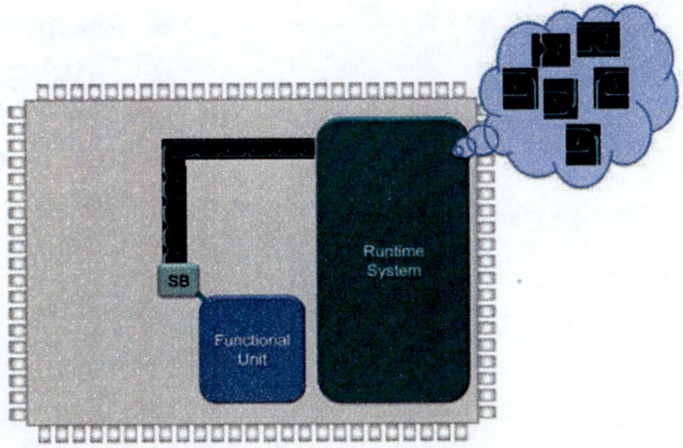

to a communication line (2) of the FPGA. If a connection between two lines has to be established, the connection points of the two lines have to be connected in the switch box (3). This is shown as highlighted line in the figure. To establish a routing or to connect two lines with each other, only the information of the connection in the switch matrix (1) also called Programmable Interconnect Point (PIP) has to be stored permanently. Using this methodology it is possible to connect individual lines in order to establish a communication path based on individual components at run time. This is done by using routing templates that are configured to the adequate position on the FPGA. These routing templates include the information which PIPs in a switch box have to be enabled. The information differs depending on the direction and the length of the routed line. Here, different routing templates for every orientation are used as well as 90 degree connections to change the orientation.

18.2.2.3 Implementation

The method described above has been used to develop a system that enables routing from the static area to any point on the FPGA. Therefore routing templates for every orientation as well as 90 degree connections have been generated. The bit width of every single macro is 8 bit which is caused by the hardware structure of the FPGA. This limitation can be handled by cascading the templates. An overview of the systems is shown in Fig. 18.6.

At Fig. 18.7 the textual output of the online routing system is visualized. As input for the system the starting point and the target point are provided. Afterwards the system calculates the route using the presented A*-Algorithm. The output of this is shown in Fig. 18.8. In this picture the 1's represent blocked areas, the 8 represent the starting point and the 9 the target point. After the A*-Algorithm finds a way the corners and the start- / endpoint is given to the placing algorithm. With this information this algorithm decides which templates have to be placed where in order to establish the calculated route.

- Startpoint : (1 , 2)
- Targetpoint : (17 , 15)

- the number of Corners : 3
- Corner : 0 , (1 , 9)
- Corner : 1 , (11 , 9)
- Corner : 2 , (11 , 15)

- the number of Busmacros :8
- Busmacro :0 , 77 , (1 , 2)
- Busmacro :1 , 20 , (1 , 8)
- Busmacro :2 , 41 , (1 , 9)
- Busmacro :3 , 5 , (7 , 9)
- Busmacro :4 , 48 , (13 , 9)
- Busmacro :5 , 65 , (13 , 15)
- Busmacro :6 , 4 , (19 , 15)
- Busmacro :7 , 78 , (20 , 15)

Fig. 18.7 Textual output of the online-routing system

18.2.2.4 Innovation

By usage of this technique it is now possible, to place modules with higher degree of freedom since the limitation to routing channels placed at design time does not longer apply. Therefore the chip area of the FPGA can be used more efficient then before. In addition this technique avoids the designer from the need to placed modules in fixed slots. This results from the fact that the online routing modules can be placed based on the actual situation and efficiently on the chip area. Because of the usage of templates, this technique is very flexible in respect to different bit widths based on the applications needs. Also the porting of the systems to other FPGAs is feasible. This has already been done for the Virtex-2 architecture also practically for the Virtex-4 architectures. For this purpose, the templates have to be adapted to the physical conditions of the FPGA and the algorithm has to be adapted to the new hardware structure.

18.3 Physical Online Routing of Modules

18.3.1 Motivation

For every functional unit, which has to be provided in state of the art FPGA systems, the corresponding bitstream must be available. In systems in where a

Fig. 18.8 Output of the online A*-algorithm

larger amount of modules are used, a tremendously higher number of bitstreams must be available in the memory. This is obviously accompanied by a larger requirement of memory capacity. Furthermore, such a system is limited to the modules stored in the memory and it is impossible to generate further modules during runtime. Now a method has been developed which allows to generate simple and complex functions out of a set of basic blocks. These blocks are available to the run-time system in bitstream fragments called snippet and can be newly composed according to the connecting rule provided by the system during runtime. Thus, it is possible to create a new module out of these small snippets. Those way modules can be assimilated very flexible to the requirements. With this method it is possible to create a multitude of modules out of a subset of sub-functions. Moreover, the memory requirements, which are caused by the bit-streams that has to be stored, can be minimized. A further advantage of this method is the possibility to adapt the size of the modules created to the available chip surface. As a result, the geometrical form of the modules is no longer pre-defined. This makes it possible for the system to use the available chip surface much more efficiently as the modules can adapt to the surfaces given.

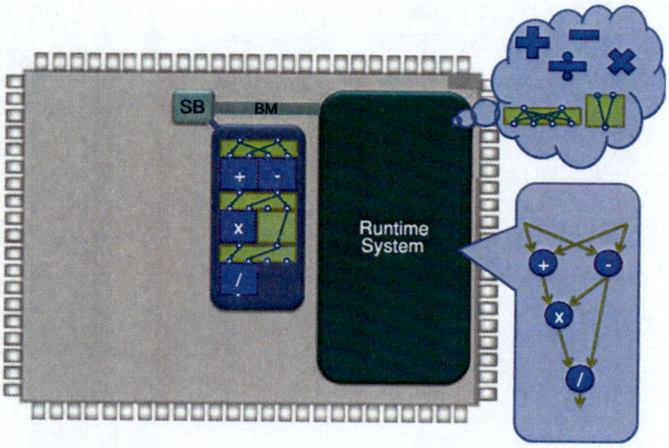

Fig. 18.9 Overview of the system for online module routing

18.3.2 Method

As already mentioned modules consist of single fine-granular sub-modules and are composed according to the connecting rule by the runtime system. Therefore these sub-modules must have defined connection points. These connecting points must be at the same relative position for all the different sub-modules and must be identical whatever their function may be. This guarantees that the sub-modules are compatible and exchangeable and can be put together like a plugging system. To connect the single modules to each other and to realize any different types of connections necessary, additional routing blocks are placed between the modules. Via these modules it is possible to connect the exit of the sub-module to the succeeding modules. The system described above is shown schematically in Fig. 18.9. As shown in the figure, sub-blocks like adders, subtractors, dividers and multipliers are available to the run-time system in form of bitstreams. Furthermore, a set of connecting macros is also available for the run-time system. In Fig. 18.9 the data flow graph shown at the bottom on the right is now to be configured onto the FPGA. Therefore the bitstream packer analyses a textual form of this dataflow graph. After this the graph will be composed of the corresponding sub- and connecting-blocks to create a bitstream. This bitstream generated during run-time according to the connecting rule is now written onto the FPGA via Read-Modify-Write back (RMW). Through the picture in the figure it becomes quite obvious that the standardized interfaces at the sub-modules as well as at the routing blocks are necessary to connect them to each other without any problems. They must therefore always be in the same place to guarantee communication.

18.3.3 Implementation

The system described above has already been physically implemented onto a Virtex-4 FX Board. In doing so, sub-modules like the ones described above were

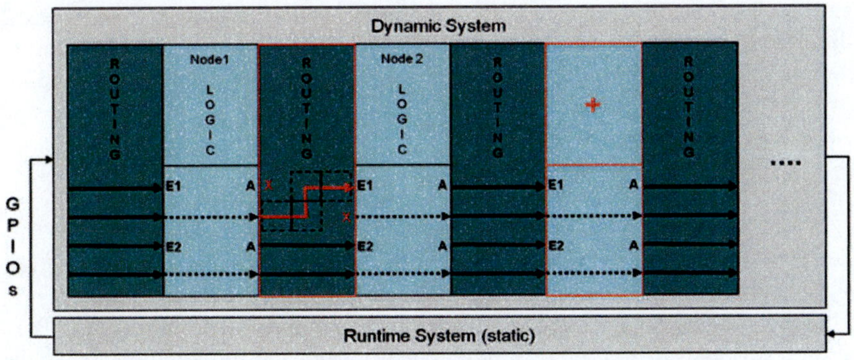

Fig. 18.10 Configuration of the module online routing shown on a virtex 4 FPGA

generated and made available for the run-time system as bitstreams. These blocks hold defined I/O ports, which allow to connect the modules to routing templates. Moreover, a method has been created to generate these routing templates during runtime. Hence the bitstreams which has to be stored could be reduced once more. These routing templates act as connecting unit between the computation-modules. The system is designed as co-processor for the Power-PC, which has eight registers with the width of 16 bits. Four of those registers serve as entry for the accelerator, four as register for the results. The run-time system writes the data into the appropriate entry registers. From here the accelerator takes the data, processes it and writes the result into the register from where the run-time system can read in the result afterwards.

In Fig. 18.10 the test system described above is shown schematically for a Virtes-4 FPGA. On the bottom of the picture, the area which contains the fixed modules can be seen which will not alter during run-time. Because of the rough vertical partition of the FPGA into reconfigurable blocks, a two-dimensional fragmentation of the configuration area is already given. Therefore the height of the sub-modules and the routing shown in Fig. 18.10 will fit in this limitation given by the Virtex 4 architecture and hold per default the height of the configuration area. The alteration of the blocks within such a configuration area can be carried out during run-time without influencing modules located above or below. The bus-macros, which connect the accelerating system to the static area, are named as GPIO on the right and left side of the figure. Those connect the eight registers to the static area to the inputs and the output of the accelerator. The registers in these systems are fixed and cannot be altered as they serve as fix docking points for the accelerator. The blocks named as routing are the routing templates used to connect the sub-modules. They are arranged in a way that the interfaces for the sub-modules are always in the same place. In-between the routing templates are the computational sub-modules. In this approach an adder, subtractor, multiplier, divider, square root and an xsum. Parallel to the activities described above, the feasibility of the technology on Virtex-2 FPGAs has been examined. The handling of the configuration mechanism existing on the Virtex-2 is different to the ones for the Virtex 4. To achieve a similar degree of freedom in reconfiguration, the above described readback-modify write method was used.

18.3.4 Innovation

Because of the possibility to place the functional modules flexible within the given limits, the method described in this paper offers a higher flexibility. Furthermore, the functional units can not only be adapted to the geometrical requirements in size and form but also to the altering requirements from the environment (e.g., the user). A further advantage is the description of the function blocks without dependencies to the architecture. This means, that the abstraction of the module representation makes it possible to alter the hardware enabled by a simply exchanging the templates. Merely the sub-modules given in bitstream format have to be provided for the particular platform. In contrast an alteration of the type of block within the model range is possible without further ado, as the bitstream fragments are identical for all blocks. Thus, a change from e.g., a Virtex-4 FX 20 FPGA to a Virtex-4 FX 100 block is possible without adapting the sub-modules. Another benefit of this technology is the possibility to alter the granularity of the sub-modules. In the system described above the sub-modules were simple functions to test the technology. Nevertheless, it is possible to design these blocks in a more complex way. Moreover, the bit width of the I/Os and the routing templates can be widened to achieve a higher performance. However, with rising complexity the flexibility of the module design and the probability between the types of blocks gets lost. Certainly the biggest advantage of this technology is the simple way of adapting the functional units to changing requirements during run-time.

18.4 Universal Tool for Working With Bitstreams

The methods described in Sects. 18.2 and 18.3 underline and strength the need of a specialized tool to manipulate the bitstream. Also it is advantageous to be able to visualize the FPGA content without having access to the hardware [4]. Because of this a software package was designed to perform offline visualization and manipulation of Xilinx Virtex II and 4 bitstreams. The software takes a bitstream as its content and produces a graphical representation of the utilized logic cells including routing resources of the FPGA. The tool offers also the possibility to cut out areas of the FPGA configuration to be used as bit stream snippets for the methods described before. Another advantage is to check and debug the bit streams generated by the tools before they will be programmed on the FPGA. This makes it easy to detect faults during the development. Figure 18.11 shows the different windows the tool offers. The *Resource View* shows the utilization of

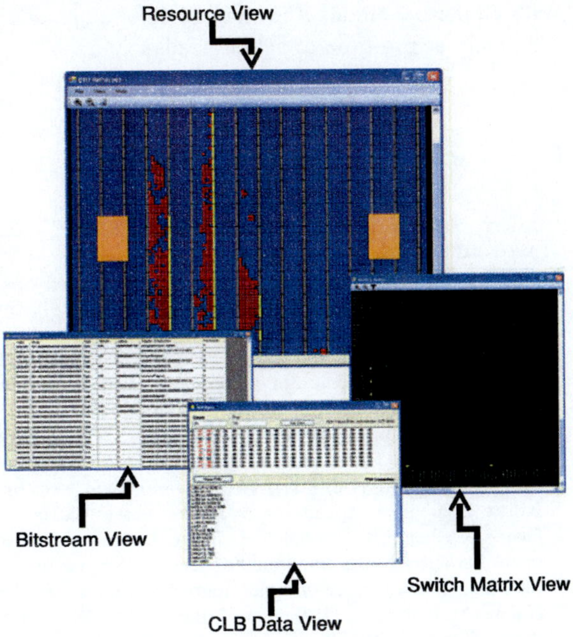

Fig. 18.11 Windows of the universal tool for working with bitstreams

the available FPGA resources. The *Bitstream View* shows the bitstream in a more human readable form, including description of header and packets. The *CLB Data View* shows the bitstream data relating to a specific CLB including the interconnections set in the according switch matrix. The *Switch Matrix View* shows the interconnections set inside the switch matrix of a specific CLB.

18.5 Conclusion

The research work described in this paper offers the possibility of a two-dimensional, run-time adaptable, reconfigurable system. Therefore three components and tools have been designed. The introduced mesh based network has the possibility to be build and extend during run-time. For that reason a switch was designed which is cascadable during run-time. To meet the requirements, the resource utilization of the switch is as low as possible and no additional control signals are needed. The signal routing to connect the switches and thus the functional units can also be established during run-time. It holds also the possibility to be cascadable to extend the bit width of the communication channels. Also the length and direction of the channels are adaptable to the placement of the functional units. The functional units described here offer the possibility to be placed and, under certain constraints, to be generated during run-time. Moreover, the functional units are adjustable in terms of the bit width and the functionality during run-time.

With all these methods it is now possible to create a system which can be adjusted adaptively at run-time to actual needs.

References

1. Xilinx, Early access partial reconfiguration user guide, ug208. http://www.xilinx.com, Accessed Mar 2005
2. Benini L, De Micheli G (2002) Networks on chip: a new paradigm for systems on chip design. In: Proceedings of Design, automation and test in Europe conference and exhibition, pp 418–419
3. Braun L, Goehringer D, Perschke T, Schatz V, Huebner M, Becker J (2008) Adaptive real-time image processing exploiting two dimensional reconfigurable architecture. J Real-Time Image Process. http://dx.doi.org/10.1007/s11554-008-0095-8
4. Huebner M, Braun L, Becker J, Claus C, Stechele W (2007) Physical configuration on-line visualization of xilinx virtex-ii fpgas. In: Proceedings of IEEE computer society annual symposium on VLSI ISVLSI, 07, pp 41–46, 9–11 Mar 2007
5. Xilinx, Logicore ip xps hwicap (v5.00a), Access, 2010
6. Claus C, Zhang B, Stechele W, Braun L, Hubner M, Becker J (2008) A multi-platform controller allowing for maximum dynamic partial reconfiguration throughput. In: Proceedings international conference on field programmable logic and applications FPL 2008, pp 535–538
7. Hubner M, Gohringer D, Noguera J, Becker J (2010) Fast dynamic and partial reconfiguration data path with low hardware overhead on xilinx fpgas, 2010 IEEE international symposium on parallel and distributed processing, workshops and PhD forum (IPDPSW), pp 1–8 http://ieeexplore.ieee.org/lpdocs/epic03/wrapper.htm?arnumber=5470736
8. Huebner M, Schuck C, Kuehnle M, Becker J (2006) New 2-dimensional partial dynamic reconfiguration techniques for real-time adaptive microelectronic circuits. In: Schuck C (ed) Proceedings of IEEE Computer society annual symposium on emerging VLSI technologies and architectures, p 6

Chapter 19
Design for Embedded Reconfigurable Systems Using MORPHEUS Platform

Paul Brelet, Philippe Millet, Arnaud Grasset, Philippe Bonnot, Frank Ieromnimon, Dimitrios Kritharidis and Nikolaos S. Voros

Abstract This chapter is related to the paper "System Level Design for Embedded Reconfigurable Systems using MORPHEUS platform" (Brelet et al. (2010) System level design for embedded reconfigurable systems using MORPHEUS platform). It presents a novel approach for designing embedded reconfigurable systems. Reconfigurable systems bring a significant importance for their highly attractive mix of performance density, power efficiency and flexibility. In this chapter, we present a toolset that abstracts the heterogeneity and benefits of a dynamically reconfigurable heterogeneous platform called MORPHEUS (Voros et al. (2009) Dynamic system reconfiguration in heterogeneous platforms, the MORPHEUS approach. This platform consists of a System-on-Chip made of a regular system infrastructure

---

Partners of the project: Thales Research & Technology (France), Deutsche THOMSON OHG (Germany), INTRACOM Telecom Solutions S.A. (Greece), ALCATEL-LUCENT Deutschland AG (Germany), Thales Optronics SA (France), STMicroelectronics SRL (Italy), PACT XPP Technologies AG (Germany), M2000 (France), Associated Compiler Experts bv (The Netherlands), CriticalBlue (United Kingdom), Universitaet Karlsruhe (Germany), Technische Universiteit Delft (The Netherlands), Commissariat à l'Energie Atomique (France), Université de Bretagne Occidentale (France), Universita di Bologna (Italy), ARTTIC SAS (France), Technische Universitaet Braunschweig (Germany), Technische Universitaet Chemnitz (Germany).

Start–End Date: January 2006-September 2009.

Global Budget/Funding by EU: 16.57 M€/: 8.24 M€.

P. Brelet (✉) · P. Millet · A. Grasset · P. Bonnot
Thales Research and Technology, 45 Rue de Villiers, 92200 Neuilly-sur-Seine, France
e-mail: paul.brelet@thalesgroup.com

F. Ieromnimon · D. Kritharidis
INTRACOM Telecom Solutions S.A, Athens, Greece

N. S. Voros
Department of Telecommunication Systems and Networks (consultant to Intracom Telecom Solutions S.A), Technological Educational Institute of Mesolonghi, Athens, Greece

N. Voros et al. (eds.), *VLSI 2010 Annual Symposium*,
Lecture Notes in Electrical Engineering 105, DOI: 10.1007/978-94-007-1488-5_19,

hosting different kinds of heterogeneous reconfigurable engines accelerating some operations. Integrated mechanisms simplify the utilization of these reconfigurable accelerators at design time and minimize the time to fetch and reconfigure a function dynamically at run time. Implementing an application on the platform is made easier and faster by a comprehensive design environment. Industrial use cases from various application domains are also presented and used to evaluate the performance of the platform and assess the MORPHEUS concept.

Keywords Reconfigurable computing · Systems-on-Chip · Heterogeneous architectures · Dynamic reconfiguration · Toolset · Embedded systems

19.1 Introduction

Taking into account the nature of modern embedded systems and their application domains, one can deduce that two of the most important requirements are cost effectiveness and flexibility. This means the processing components used in such systems must provide the high density performance requires by the embedded market, enable inexpensive development with fast modification/validation loop process and be flexible enough to support several application domains.

In that context, the approach presented in this chapter is described with these requirements by defining a reference platform for dynamic reconfigurable computing that is efficiently used in different application domains. Real-time processing is in particular in the focus of this platform. It is obvious that flexibility, modularity, and scalability of such a platform are key requirements in order to allow an efficient adaptation of the platform architecture to the specific requirements of a given application.

Most competing state-of-the-art System-on-Chip (SoC) are based on a single CPU mono or multi-core or on a combined CPU and enhanced with a DSP or some dedicated hardware accelerators (Nomadik by STMicroelectronics [3], OMAP by Texas Instruments [4], and PXA by Intel [5]). Such architectures are very efficient but also highly linked to an application domain: movie codec, mobile phone's signal processing and so on. With MORPHEUS, application's demands can be distributed on different units at runtime. Its heterogeneous reconfigurable engines (HREs) and I/Os remain runtime configurable. It is therefore not linked to an application domain but still offers high performance density. New MultiProcessor System-on-Chips (MPSoCs) is also highly programmable with very high performances but still homogeneous. Existing heterogeneity in multimedia and other data intensive applications cannot be mapped as efficiently on MPSoC with uniform means of computation of a single granularity, similar I/O bandwidths and access patterns as on the MORPHEUS HREs.

On one hand, reconfigurable platforms come with specific languages and tools. Application specific instruction processors and accompanying tools can be developed with the help of Mescal [6], ArchC [7], LisaTek [8], Chess/Checkers [9]. C-based languages like CatapultC [10], Mitrion-C [11] and ImpulsC [12] are available to target specific architectures. A more stream-oriented approach is available with tools like Matlab/Simulink [13] or Scilab [14].

On another hand, a trend to standardise the APIs for parallel programming has converged to a reduced set of languages like VSIPL++ [15], MPI [16] or OpenMP [17]. Several modelling languages like UML [18] and sysML [19] are used to hide lower system levels and improve code consistency.

In MORPHEUS, a fundamental and novel idea is the integration of Heterogeneous Reconfigurable computation Engines (HREs). They bring different but complementary types of reconfigurable computing in one platform. Three states-of-the-art dynamically reconfigurable computation engines (a fine-grain, a mid-grain and a coarse-grain kind of reconfigurable) have been selected and integrated into the MORPHEUS platform making it a very flexible architecture.

The associated toolset combines the benefits of the different styles of reconfigurable computing using the high flexibility and scalability to address various application domains. This toolset is swayed by state-of-the-art software like OpenMP, Matlab/Simulink, Scilab and CatapultC merged in one design flow to bring an efficient way to get the best performances from the chip.

Section 19.2 and Sect. 19.3 present respectively an overview of the MORPHEUS architecture and toolset. The experimental results using the MORPHEUS platform with three different case studies are shown in Sects. 19.4, 19.5 and 19.6 while Sect. 19.7 concludes this study.

19.2 Architecture

The MORPHEUS architecture (Fig. 19.1) is based on an ARM9 embedded RISC processor which is responsible for data, control and configuration transfers between all resources in the system, memory, IO peripherals, and a set of heterogeneous reconfigurable engines (HREs) each residing in its own clock domain with a programmable clock frequency.

As dynamic reconfiguration might imposes a significant performance demand for the ARM processor, a dedicated reconfiguration control unit is foreseen to serve as a respective offload-engine. All system modules are interconnected via multilayer AMBA busses. Each HRE is composed of a reconfigurable IP seen as a memory-mapped co-processor or peripheral.

Additionally, a NoC infrastructure has been added to extend inter-HRE communication capabilities. It speeds up data delivery among HREs and enables stream data processing; thus removing a bottleneck for parallel data transmission.

MORPHEUS is built around three HREs: XPP-III [20] is a coarse-grained reconfigurable stream processor featuring an array of 16-bit computational elements communicating through a matrix of configurable data channels. It mainly

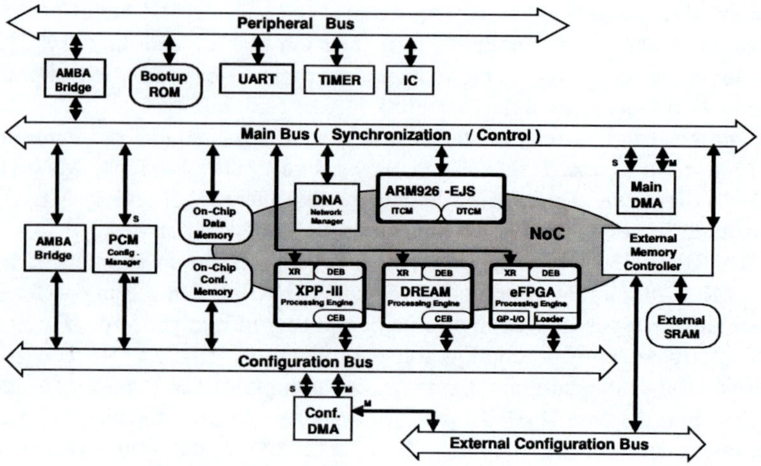

Fig. 19.1 MORPHEUS hardware architecture

targets streaming applications with huge computational densities and regular dataflow structures; DREAM [21] is a reconfigurable processor composed of a RISC processor coupled with a mid-grain reconfigurable. DREAM is aimed at exploiting instruction level parallelism for a wide range of applications (e.g. multimedia, telecom, cryptography and so on); FlexEOS [22] is an embedded Field Programmable Gate Array (eFPGA). It is suitable for fine-grain algorithm, arbitrary logic implementation or peripherals interface due to its 30 GPIO pins exported to the pad frame.

HREs are chosen and they are sized to provide a good trade-off between hardware resources and computation performance but this is not a limitation since the basic frame is designed to support any additional HREs (e.g. ASIC accelerators or DSP cores) or different sizing of the proposed ones. The main design challenge resides in the definition of the top-level architecture especially the way data is moved to feed all the computation resources in the system. The interconnect system must be efficient and flexible to support the requested bandwidth, to avoid resource starvation and to support application requirements.

Each HRE comes with its specific language and programming styles that require deep knowledge of inner hardware details. The toolset abstracts the complexity of the whole hardware architecture and unifies the HREs' languages.

19.3 The Toolset

The toolset (Fig. 19.2) is composed of three parts: (1) HREs accelerated functions design, (2) main ARM application programming and (3) runtime libraries and operating system.

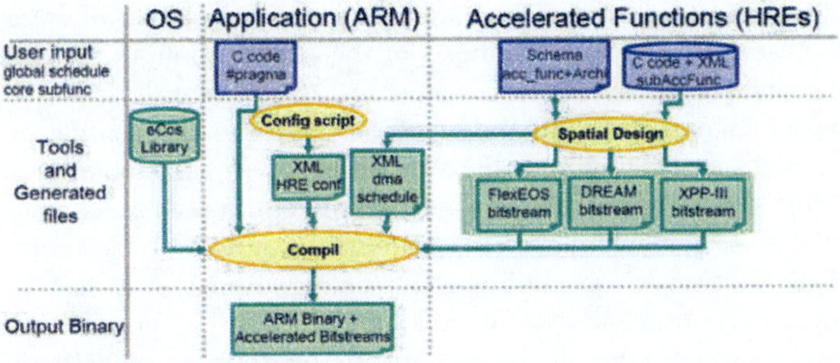

Fig. 19.2 MORPHEUS tool chain

Fig. 19.3 Annotation of application code

```
int pin[10], pout[10];

#pragma MOLEN 1
void func(int*in,int*out){
    *out=*in++;
}

int main(int argc, char*argv[]){
    func(pin,pout);
    return 0;
}
```

Pragma that gives this function the ID number 1 in MOLEN paradigm

Accelerated function running on the targeted HRE

Main control function running on the ARM

It reduces the application design time and improves the code quality. Shortening the application design time is targeted via the utilisation of high level programming solutions for the platform. Code quality is increased thanks to code generation. Code quality is considered here as code readability, absence of bug with respect to the system level implementations (address mapping, etc.), and the possibility to easily implement upgrades and fixes as maintenance.

The MOLEN paradigm [23] targets this objective. It abstracts the tedious management of configurations, execution calls, data communications and process synchronisations. During application implementation, the designer splits its application into control parts (executed on the ARM processor) and computation intensive parts (mapped on the HREs) by following these steps: (1) the application is written in standard C language. During this step, one validate the application against use cases and test benches on host workstation; (2) the programmer selects the functions that must be accelerated by allocating an ID to each of them through MOLEN *pragma* (Fig. 19.3); (3) these accelerated functions are captured inside SPEAR [5], a graphical environment, by connecting and assembling some building blocks called elementary sub-functions and written in C language; (4) the programmer selects which HRE executes which accelerated function. The targeted HRE code for each accelerated function is then generated.

SPEAR provides a common programming interface for the different reconfigurable units using a high level synthesis of the HREs' configurations. This hides the heterogeneity of the architecture and eases its utilisation.

The toolset manages the flexibility of the platform by facilitating run-time dynamic allocation of function to the various HREs. It contributes to the performance of the platform by quickly generating optimised executable code by combining the implementation on reconfigurable units and the communication aspects. Thus, optimised scheduling shall be performed at compilation. Here, configuration, execution and communication are sources of possible performance optimisation at system level in such a complex platform as MORPHEUS.

A Control Data Flow Graph (CDFG) format is used as an intermediate and technology independent format inside the framework. MADEO [24] uses it with high-level synthesis techniques to target HRE's proprietary tools and generates the configuration binaries. Provided the compliancy of a tool with the CDFG format, any design capture tool, code generator or compiler can be integrated in the toolset. It allows us to integrate the specific compilers of the three HREs and ARM.

With SPEAR, the user manipulates the C code of the sub-functions, abstracted Models to combine these sub-functions, and Makefiles to generate the binary for each HRE. Functions are specified at a high-level of abstraction, improving design time and flexibility, without sacrificing performances. They are modelled with a directed acyclic graph, in a formalism called Array-OL [25] which is well suited to represent deterministic, data intensive and data-flow applications such as the accelerated functions on HREs. SPEAR automatically generates a CDFG skeleton for each accelerated function and the communication parameters to move data from one accelerated function to another, between HREs and to/from the shared memories. The design flow contributes in this way to seamless hardware/software integration. The Cascade tool [26] fills the CDFG skeleton with the operational code of each function.

Three weeks was required to implement the video surveillance application on the MORPHEUS platform while the double was needed on a dedicated SIMD platform and 6 month is estimated in the case of a classical hardware implementation on FPGA [27].

19.4 Wireless Telecommunications Application

The first implemented case study comes from the wireless communication domain. In a standard environment one finds several wireless possibilities, 802.11 (WiFi), 802.16 (WiMax)..., to access core Networks like Internet (Fig. 19.4).

With MORPHEUS we design a unique device that can adapt automatically to the best available Wireless Network. The reconfigurable units deliver the required services with high computation performances at low power consumption. The DREAM is selected to implement a word-level processing block, a 128-point Fast Fourier Transform (FFT128) of the wireless application, followed by a

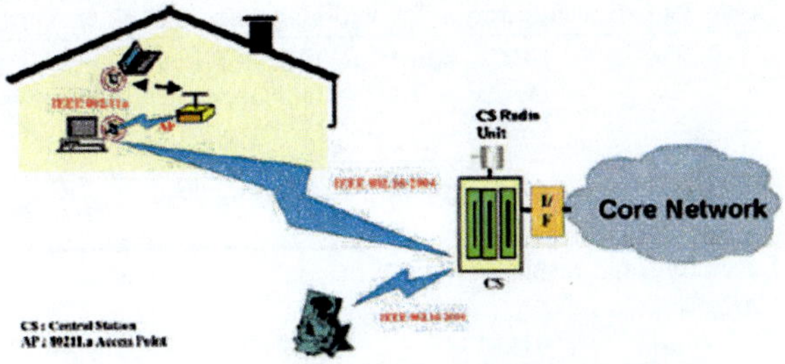

Fig. 19.4 Broadband wireless access system architecture

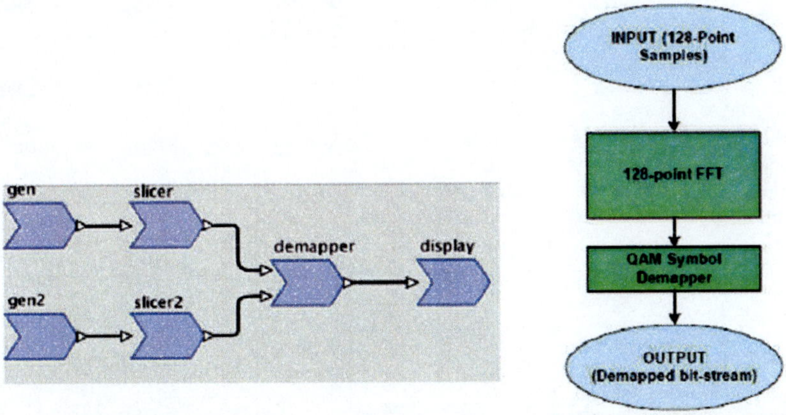

DEMAPPER accelerated function Graph from SPEAR

Application scenario for case study

Fig. 19.5 Wireless telecommunication case study

Quadratic Amplitude Modulation (QAM) symbol demapper, capable of supporting modulation schemes ranging from a 4-point Quadratic Phase Shift Keying (QPSK) to 64-point QAM. The DREAM application dataflow is depicted in Fig. 19.5.

It is a cascade of the above two blocks (FFT and QAM demapper). Due to the size constraints, the mapping of the accelerated functions is done in two steps: (1) a CDFG for the FFT (2) and a separate one for the QAM demapper. The resulting CDFGs are used by the MADEO tool downstream for the generation of the configuration streams targeting the DREAM array, as well as the control code running on the ARM processor, responsible for downloading the configuration streams and for managing the data transfers to and from the DEBs used for communication.

A number of discrete simulation scenarios were run on the PC hosting the tool chain. Simulated execution time can be broken-down to DMA configuration and transfer time, bitstream load/deload time and actual execution time. As can be seen in Table 19.1, the ratio of actual execution time for the ARM + DREAM complex over the corresponding time for the application running on ARM alone is generally favourable to MORPHEUS, indicating speedups ranging from 2.07 to 6.55 (Fig. 19.6).

Table 19.1 Speedup ratios for various test programs vs Morpheus configuration

	FFT + Slicer + Demappe r QPSK	FFT + Slicer + Demappe r QAM16	FFT + Slicer + Demappe r QAM64	Slicer + Demappe r QAM64	FFT128
	ARM cycles	ARM cycles	ARM cycles	ARM cycles	ARM cycles
Execution time result					
ARM only results					
Execution (ARM only)	918617	927578	940197	26982	913215
Total	925487	931847	947109	28730	918379
Accelerated results					
Total conf_DMA	3745	3745	3745	1835	1910
Total load/deload	5279	5278	5278	3126	2152
Execution (with HRE)	6584	8408	10301	6108	4193
Total	19751	21574	23467	12754	10713

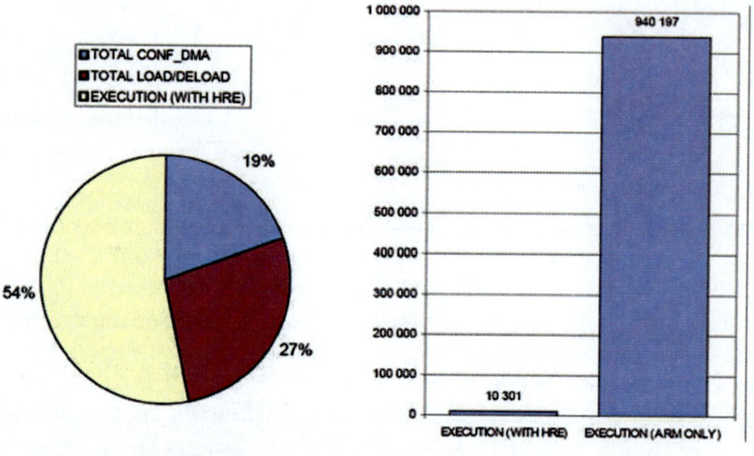

Time spent in the component during the acceleration **Application Execution Time**

Fig. 19.6 Wireless telecommunication Execution Time

The speedup for the overall application is particularly impressive, the Slicer + QAM64 Demapper and the 128-point FFT are very well adapted to such an implementation. The pipelined operations taking place in the chip and the huge amount of data to be processed, due to a high locality of the computation, mask the overhead for configuration and data-transfer. For smaller components, with less computation locality, the results discussed above are different. To have a positive impact on the computation performance, the HW-accelerated functions that are taken off the ARM execution environment and mapped onto the HREs must meet the two following conditions: (a) The configuration/DMA/bitstream load/deload time, added to the actual execution time must not exceed execution time on the ARM alone; (b) Reconfiguration intervals must be spaced sufficiently far apart for the configuration/DMA overhead to be absorbed by the gains in execution time.

Among the two conditions, the first is almost trivial to meet, provided that the dataset processed by the accelerated functions is large enough. To get more performance by saving HRE configuration, one can statically load the HW accelerated functions so that there is no reconfiguration at each run. In this study, we need to adapt the chip to the network availability, which does not need to be at each run, since it does not exceed a few symbols' worth of data.

MORPHEUS can support applications that include dynamically configurable components, provided certain timing constraints are met. These constraints are variable, i.e. dependable on each particular application. For wireless application, the approach gives very good results with ability to adapt the processing elements to the available environment.

19.5 Systems for Intelligent Cameras

An intelligent camera can be viewed as a large collection of real-time algorithms which are activated (or not) depending on non predictable events such as the content of the image, an external information or a request from the user.

This case presents a motion detection application embedded in a smart camera. It computes first an absolute pixel-to-pixel difference between the current frame and the background. The background is periodically refreshed but this will not be considered here. This step isolates the object to be analysed, if any, but with a greyscale it remains too complex to be analysed. A binarisation step will lower the complexity of the frame thanks to a threshold conventionally fixed to $0.3\times$ the maximum value of the pixels.

The "cleaning" step removes isolated pixels and enhances connections between pixels by an opening phase. It is implemented with two operators: the erosion and the dilatation which consist in replacing the central pixel of a 3×3 matrix by the minimum and the maximum values of this matrix.

The next step is the edge detection which finds the boundaries of the shape in the monitored area. This operation is implemented with a simple convolution applied to 3×3 pixel matrices using the Sobel algorithm. Here, the Sobel edge detector uses two convolution kernels to detect vertical and horizontal edges.

This gives the result as a binarised shape, since the aim of the application is not to detect the magnitude of the gradient but the presence of a gradient. Finally the detected edge is merged with the original image. For that goal, inverse binarisation is applied: the background is filled by 1 s and moving image edges by 0 s, thus allowing to implement the merge operation with a multiplication.

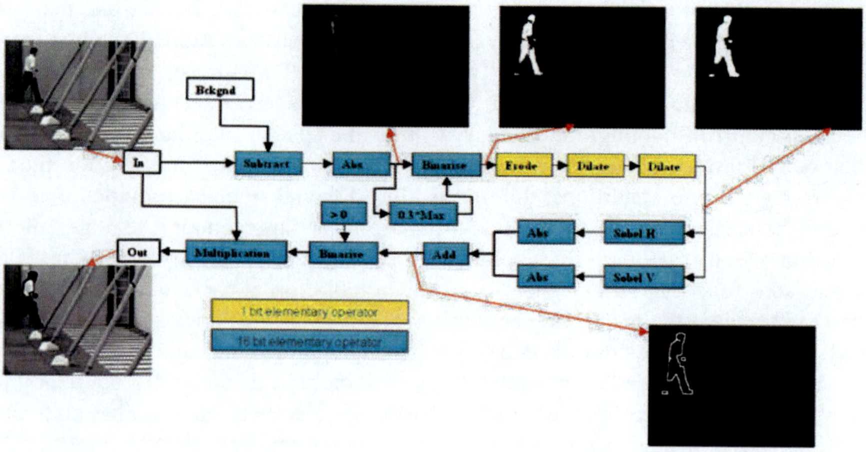

Fig. 19.7 Simple motion detection application

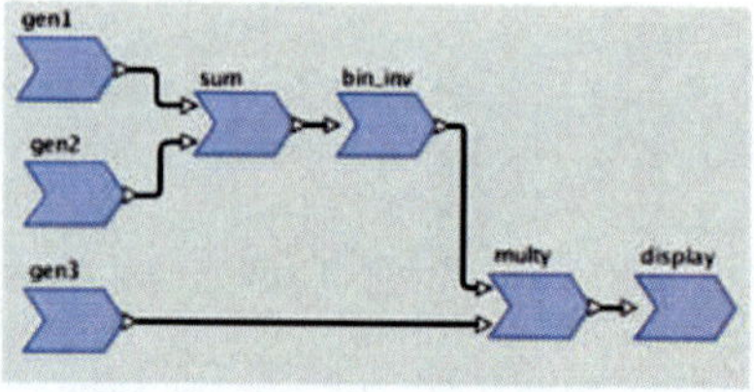

Last accelerated function Graph from SPEAR

ARM only results	ARM cycles
EXECUTION (ARM ONLY)	613 131
TOTAL	666 134

Accelerated results	ARM cycles
CONF DMA	8 038
DMA	67 721
EXECUTION (WITH HRE)	103 978
TOTAL	179 737

Accelerated results	DREAM cycles
TOTAL DREAM EXECUTION	459 816

EXECUTION TIME RESULT

Fig. 19.8 Spear graph with results

Figure 19.7 shows the different steps of the image processing application for an intelligent camera.

All the application is generated by the toolchain in 4 accelerated functions: (1) the subtraction, the absolute value and the binarisation between two images (the background and the current image); (2) the opening (erode + dilate) and a second dilate; (3) the convolution H/V from a SOBEL matrix; (4) the sum, the absolute value and the multiplication between the current image and the result of the third accelerated function.

Figure 19.8 was obtained with a manually implemented application on the DREAM unit only. The MORPHEUS's DREAM frequency is 250 MHz. The measured performance shows a speedup of $\times 3.7$ with the utilisation of the DREAM.

By using different HREs available on the MORPHEUS platform, the performance could even reach 1.27 cycles per pixel (20% performance increase). For this implementation, the critical kernel is the Erosion/Dilatation/Edge Detection. It has been implemented on the DREAM engine whose frequency is mentioned above.

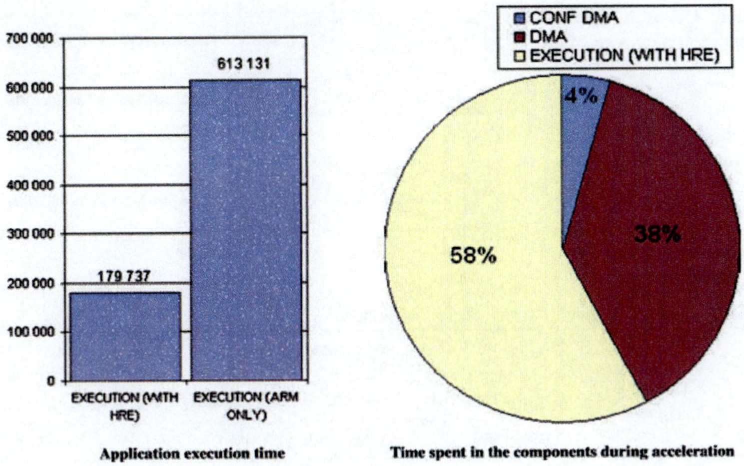

Fig. 19.9 Execution time for intelligent camera application

Figure 19.9 shows the repartition of the time spent by the components during execution. The DMA communications take about the same time as the HRE execution. The DMA transfers can be masked using pipelined communications and computations. In a case where the DMA and the HRE work would take place in parallel (e.g. with the pipeline), we would gain about half the total time. The DMA time would then no longer be added to the HRE's computation time.

Moreover, the higher level programming of the toolset gives a programming accessibility to a larger range of programmers. Since it is not needed to get specific skill corresponding to low level architectural considerations, even non-specialists can program the chip.

19.6 Discrete Wavelet Transformation Application

The wavelet transform was borne out of a need for further developments from Fourier transforms.

Discrete wavelet transforms (DWT) are applied to discrete data sets and produce discrete outputs. Transforming signals and data vectors by DWT is a process that looks like the fast Fourier transform (FFT), the Fourier method applied to a set of discrete measurements.

The DWT is being increasingly used for image compression today since it supports features like progressive image transmission (by quality, by resolution), ease of compressed image manipulation, region of interest coding, etc.

In our case, the DWT is applied for an image processing application composed by FIR filter bank structures.

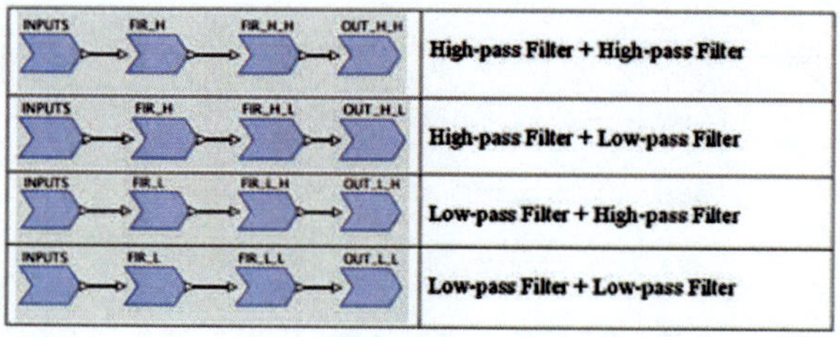

Fig. 19.10 Discrete wavelet transformation decomposition

Table 19.2 Discrete wavelet transformation results

	High-pass Filter + High-pass Filter	High-pass Filter + Low-pass Filter	Low-pass Filter + High-pass Filter	Low-pass Filter + Low-pass Filter
	ARM cycles	ARM cycles	ARM cycles	ARM cycles
Execution time result				
ARM only results				
Execution (ARM only)	57025	50725	49805	49503
Total	66687	60387	59467	50304
Accelerated results				
Total conf_DMA	6025	5948	5947	5947
Total load/deload	5689	5699	5884	5880
Execution (with HRE)	6404	8749	11713	13051
Total	27738	30016	33164	35325

The 2D-DWT transform is repeated 3 times (for the 3 levels of compression). Each level is made of vertical and horizontal filtering. In each direction a low pass filter and a high pass filter are computed.

The code for the high pass filter and the low pass filter has been written in C. These two sub-functions are used for the capture of the whole 2D-DWT transform as shown in the SPEAR capture figure. This application is composed by four main accelerated functions (Fig. 19.10): (1) a filter combination (High-pass Filter + High-pass Filter); (2) a combination of Filter (High-pass Filter + Low-pass Filter); (3) a Low-pass Filter and a High-pass Filter; (4) Two Low-pass Filters.

To simulate the DWT application, we have resized the initial 5200 × 5200 High Definition Image (HDI) to an 80 × 80 Low Resolution image (LR).

Each Filter function is an elementary function and has been tested separately. Their implementations on the HREs take less than a day effort for each of them and do not require any HREs knowledge. Unlike standard developments the only requirement is to write the elementary C code of the functions. The design on the SPEAR graphical interface can easily be built and requires a few minutes.

Globally, the implementation of the application including C code for ARM and SPEAR capture takes a few days. The application can be implemented within 4 days on the MORPHEUS platform where the non-integrated approach requires several weeks.

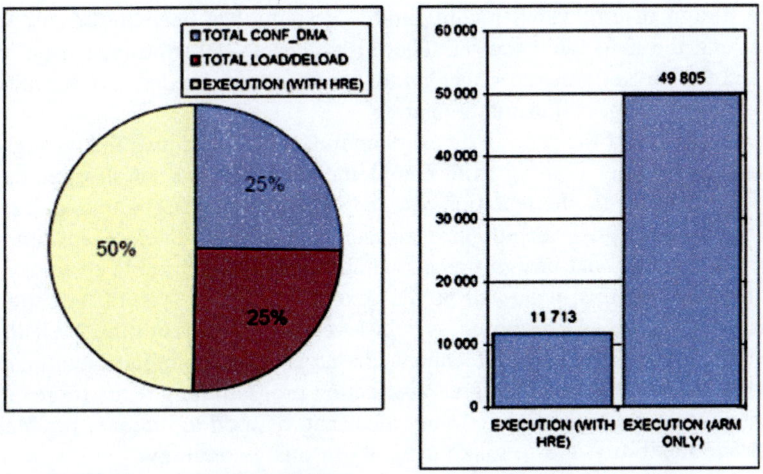

LOW-Pass Filter + High-pass Fillter **Application execution time**

Fig. 19.11 Discrete wavelet transformation execution time

The high pass and low pass filters composing the 2D Discrete Wavelet Transformations have been implemented on the DREAM and the eFPGA. Table 19.2 shows the implementation results. In our case studies, these filters are applied on an array of 80 × 80 integers.

For this application, the complete tool flow has been involved (compilation, RTOS, Spatial Design). The discrete wavelet decomposition execution time results indicate a better performance with HRE than with ARM processor only.

Figure 19.11 shows the performance given with MORPHEUS toolset for the DWT image processing application. They are in the same order than in the previous case (×4 gain).

Regarding the development effort, the implementation time of the film grain removal algorithm on the MORPHEUS platform was approximately the same as with a state of the art FPGA-based platform. But modification, code reusability is improved with the MORPHEUS toolset. Moreover, computation performance gains have been observed (×4). In this case, the DMA configuration and the LOAD/DELOAD times are equal to the execution times on a HRE. Implementing a pipeline stage would also increase the performances. The MORPHEUS toolset and platform overrides a previous manually programmed SIMD implemented on a FPGA.

19.7 Conclusions

In the previous sections, we presented the concept of a novel heterogeneous platform, called MORPHEUS, and the associated toolset for the design of reconfigurable embedded systems.

The design of embedded reconfigurable systems has been introduced in this chapter and the associated toolset. The initial system description is usually in C code and the toolset generates accelerator target specific codes and optimises the schedule of the application implemented.

The feasibility of the proposed design approach has been justified through three different case studies coming from complementary domains. In all three cases, it has been proven by implementation results that the MORPHEUS approach enables the design of embedded reconfigurable systems in shorter development times than compared to traditional design flows of embedded reconfigurable systems.

A question remains: what can be the next steps for this type of reconfigurable SoC in the future? The proposed system-level principles from the MORPHEUS approach can be extended and enhanced: larger number of accelerators, more powerful allocation schemes, higher abstraction programming level, more efficient memory interfaces, and so on... Many ideas can be used to increase productivity, to improve flexibility and to make progress in this technology.

It is very easy to change the architecture components and to modify the toolset implementation to adapt the whole hardware/software approach from MORPHEUS to a novel reconfigurable engine. For example, just by changing the ARM9 to ARM11, the gain of performance is increased without taking into account additional improvements of the heterogeneous reconfigurable engines.

Moreover, the toolset can be used easily by both beginners and experienced developers in the field of reconfigurable computing.

Acknowledgments The authors would like to thank all the partners of the project consortium who were involved in studying and providing the required technologies, specifying the requirements and assessing the results. This research was partially funded by the European Community's 6th Framework Program.

References

1. Brelet P, Grasset A, Bonnot P, Ieromnimon F, Kritharidis D, Voros NS (2010) System level design for embedded reconfigurable systems using MORPHEUS platform. In: Proceedings of the 2010 IEEE annual symposium on VLSI 5 July 2010. ISVLSI. IEEE computer society, Washington, DC, pp 500–505. http://dx.doi.org/10.1109/ISVLSI.2010.13
2. Voros N, Rosti A, Hübner M (2009) Dynamic system reconfiguration in heterogeneous platforms, the MORPHEUS approach. Springer, Berlin
3. Mair H, Wang A, Gammie G, Scott D, Royannez P, Gururajarao S, Chau M, Lagerquist R, Ho L, Basude M, Culp N, Sadate A, Wilson D, Dahan F, Song J, Carlson B, Ko U. A 65-nm mobile multimedia applications processor with an adaptive power management scheme to compensate for variations. Digital Signal Processing, pp 8–9
4. Clark L, Hoffman E, Miller J, Biyani M, Strazdus S, Morrow M, Velarde K, Yarch M (2001) An embedded 32-b microprocessor core for low-power and high-performance applications. IEEE J Solid State Circ 36:1599–1608
5. Lenormand E, Edelin G (2003) An industrial perspective: pragmatic high-end signal processing environment at Thales. In: Proceedings of the 3rd international workshop on synthesis, architectures, modeling and simulation (SAMOS)

6. Gast N, Gaujal B (2010) A mean field approach for optimization in discrete time. J Discrete Event Dyn Syst
7. Klein F, Leao R, Araujo G, Santos L, Azevedo R (2007) A multi-model power estimation engine for accuracy optimization. ISLPED '07 Proceedings of the 2007 international symposium on Low power electronics and design
8. Muhammad R, Apvrille L, Pacalet R (2008) Evaluation of ASIPs design with LISATek: Springer Volume 5114/2008
9. Clarke P (2002) Chess/checkers tool flow brings verification into play. EETimes article
10. SystemC Modeling, Synthesis, and Verification in Catapult C, Mentor Graphics Corporation, 8005 SW Boeckman Road, Wilsonville, Oregon, USA. [online] Available: http://www.mentor.com/products/esl/techpubs/
11. Mitrionics (2008) Low power hybrid computing for efficient software acceleration. White paper
12. Pellerin D, Thibault EA (2005) Practical FPGA programming in C, Prentice Hall
13. University of Newcastle upon Tyne (2003) Matlab/Simulink tutorial
14. Campbell SL, Nikoukhah R (2004) Auxiliary signal design for failure detection. Princeton University Press, Princeton
15. Bergmann J, McCoy D (2004) Sourcery VSIPL++ HPEC benchmark performance. HPCMP-UGC '06 Proceedings of the HPCMP Users Group Conference
16. Snir M, Otto S, Huss-Lederman S, Walker D, Dongarra J, MPI the complete reference. [online] Available: http://www.netlib.org/utk/papers/mpi-book/mpi-book.html
17. Chapman B, Jost G, van der Pas R, Kuck DJ (2008) Using openMP: portable shared memory parallel programming. MIT Press, Cambridge
18. Fowler M (2008) UML distilled: a brief guide to the standard object modeling language. Published September 25th 2003 by Addison-Wesley Professional
19. Weilkiens T (2006) Systems engineering with SysML/UML: modeling analysis, design. Hüthing, Heidelberg
20. PACT XPP Technologies (2005) PACT software design system XPP-IIb (PSDS XPP-IIb)—programming tutorial. Version 3.2, November 2005
21. Stitt G, Grattan B, Villarreal J, Vahid F (2002) Using on-chip configurable logic to reduce embedded system software energy. IEEE symposium on field-programmable custom computing machines, Napa Valley, USA
22. Baron M (2004) M2000's spherical FPGA cores. MicroProcessor report, Dec 2004
23. Coppola M, Locatelli R, Maruccia G, Pieralisi L, Scandurra A (2004) Spidergon: a novel on-chip communication network. Proceedings of the international symposium on system-on-chip, pp 16–18
24. Whitty S, Ernst R (2008) A bandwidth optimized SDRAM controller for the MORPHEUS reconfigurable architecture. In: Proceedings of the IEEE parallel and distributed processing symposium (IPDPS)
25. Amar A, Boulet P, Dumont P, Projection of the array-OL specification language onto the Kahn process network computation model. [online] Available: http://hal.archives-ouvertes.fr/docs/00/07/04/91/PDF/RR-5515.pdf
26. CRITICALBLUE (2005) Boosting software processing performance with co-processor synthesis. White paper
27. Whitty S, Sahlbach H, Hurlburt B, Putzke-Röming W, Ernst R (2010) Application-specific memory performance of a heterogeneous reconfigurable architecture. In: Proceedings of design, automation and test in Europe (DATE)

Chapter 20
New Dimensions in Design Space and Runtime Adaptivity for Multiprocessor Systems Through Dynamic and Partial Reconfiguration: The RAMPSoC Approach

Diana Göhringer and Jürgen Becker

Abstract Embedded high performance computing applications have two requirements which hardly can be achieved simultaneously: high performance and low energy consumption. One solution is the exploitation of the low-level parallelism of a field programmable gate array (FPGA). Due to the manifold parameters, such as the adaptation of the clock frequency in relation to the application requirements, a better energy efficiency compared to traditional processor-based platforms can be achieved. However, the FPGA programming is time consuming until today and requires a very good understanding of the underlying hardware. There exist C-to-FPGA tools, which leverage the traditional FPGA programming using HDL-languages. However, C-to-FPGA tools can only be used for submodules and accelerators, because they do not handle the communication with the environment, e.g., camera interfaces, PCI-interfaces, etc. Furthermore, the results of an automatic code transformation are until today suboptimal in comparison to a hand coded design. Due to this fact, the interfaces have to be either programmed by hand, which is very time consuming, or they have to be bought from IP suppliers. Furthermore, these C-to-FPGA tools often have some restrictions on the input C, C++ language. In this book chapter a novel holistic approach called RAMPSoC (Runtime Adaptive Multiprocessor System-on-Chip) is presented. RAMPSoC provides a meet-in-the middle solution by combining the hardware flexibility and low power consumption of FPGAs with the software flexibility and the high-level programming paradigms of multiprocessor systems-on-chip. The RAMPSoC approach consists of a flexible and energy efficient hardware architecture, consisting

D. Göhringer (✉)
Fraunhofer IOSB, Gutleuthausstr. 1, 76275 Ettlingen, Germany
e-mail: diana.goehringer@iosb.fraunhofer.de

J. Becker
Karlsruhe Institute of Technology (KIT), 76128 Karlsruhe, Germany
e-mail: becker@kit.edu

N. Voros et al. (eds.), *VLSI 2010 Annual Symposium*,
Lecture Notes in Electrical Engineering 105, DOI: 10.1007/978-94-007-1488-5_20,

of heterogeneous processing elements connected over a heterogeneous Star-Wheels Network-on-Chip, a user-guided design methodology and a new operating system for runtime resource management. RAMPSoC provides new dimensions for design space and runtime adaptivity by exploiting the features of dynamic and partial reconfiguration in FPGA-based designs. Using an object recognition algorithm, it was shown that the RAMPSoC is more energy efficient than a standard CPU and an NVIDIA Tesla GPU.

Keywords Field Programmable Gate Array (FPGA) · Dynamic and Partial Reconfiguration · Multiprocessor System-on-Chip (MPSoC) · Network-on-Chip (NoC) · Operating System · Design Methodology · High Performance Computing

20.1 Introduction and Motivation

Embedded high performance computing systems, such as used e.g., in surveillance applications, have two contrary requirements: high computing performance and low power consumption. Furthermore, they need to be able to adapt to the requirements of the environment. These requirements are, e.g., different image processing algorithms, depending on the objects and their distance to the camera; varying computing requirements, depending on the number of objects to be tracked.

Field programmable gate arrays (FPGAs) are one possible hardware solution for embedded high performance computing systems, due to their low-level parallel architecture. By exploiting the inherent parallelism of the embedded high performance applications a high energy efficiency can be achieved. Furthermore, the hardware architecture can be adapted at design-time and for some FPGAs, such as Xilinx FPGAs, even at runtime. This runtime adaptivity is called dynamic and partial reconfiguration. With this feature, a part of the modules configured on the FPGA device can be exchanged at runtime, while the other modules are still executing and are not disturbed. With this feature the FPGA cost and the power consumption can be reduced, because only the currently required functionality needs to be configured on the device. All unused functionality is removed. Therefore, a smaller FPGA device can be used, which also results in lower power consumption. An additional benefit is that components, which always need to be present on the device, are not disturbed, while part of the functionality is exchanged. This is beneficially, e.g., for image processing applications, where the camera interface needs to remain fully functional, while the filter algorithm needs to be adapted depending on the image data. This way the function is brought to the data instead of the traditional von Neumann approach, where a program counter points to the current instruction and directs the data to the corresponding function (see [1]). This means, the function is configured at runtime onto the device at the position, where it is needed. After processing the function can be removed from the device configuration memory or it can be exchanged by a different function on-demand without disturbing the other executing functions on the device. Xilinx offers a special toolflow for designing partial bitstreams. For Xilinx tools version 9.2 and older it was called Early Access Partial Reconfiguration Flow (EAPR flow) [2] and since version 12.1 it is part of the standard tools and it is called Partial Reconfiguration Flow [3].

However, FPGA programming is very complex and time consuming, as a hardware description language (HDL) such as VHDL or Verilog has to be used. Therefore, the applications, which mostly are designed in languages such as C or C++, need to be translated into VHDL or Verilog. Furthermore, the debugging and the modification of an algorithm written in an HDL are more time-consuming than the imperative programming languages such as C, C++. There exist commercial C-to-FPGA tools, such as CatapultC [4], ImpulseC [5] or Handel C [6], which can transform an algorithm written in C, C++ into a HDL description for the FPGA. However, the C, C++ Code has to be adapted to fulfill the requirements of the corresponding tool. Also, most of these tools, except Handel C, can only generate HDL-code for a single module, not for a full FPGA configuration. Even more difficult in terms of development and debugging is the design of a dynamically and partially reconfigurable FPGA system. The selection of the partially reconfigurable modules has to be done with care and a runtime system has to be developed for managing the reconfigurations depending on the changing application requirements.

Another solution for embedded high performance applications are Multiprocessor Systems-on-Chip (MPSoCs), which allow to exploit the task-level parallelism of an application. Programming MPSoCs is easier than programming FPGAs, because parallel programming models can be used, which are an extension of imperative languages such as C, C++ and are therefore more familiar to application developers than an HDL language. The most well known parallel programming models are based on threads, such as PosixThreads [7], OpenMP [8], or on message passing, such as the Message Passing Interface (MPI) [9]. Special types of multiprocessors are general purpose graphic processing units (GPGPU), such as the NVIDIA Tesla GPU [10]. They are programmed with C-for-CUDA [11], which is C with NVIDIA extensions and was specifically designed to program NVIDIA GPUs.

However, the hardware architecture of MPSoCs, such as the number of processors, the bandwidth of the communication infrastructure and the memory, is fixed and cannot be modified at design- or runtime. This means, that the hardware of MPSoCs cannot be adapted at runtime in relation to the performance and power requirements of the application, like it is possible in FPGA platforms.

The work presented in this chapter combines the flexibility and low power consumption strategies of FPGA-based design with the simpler programming model of multiprocessor systems-on-chip (MPSoCs) within a holistic approach called RAMPSoC (runtime adaptive MPSoC) [12]. Also, RAMPSoC supports the possibility to combine microprocessors with application specific architectures obtained by hardware compilation tools. Using an image processing application, it

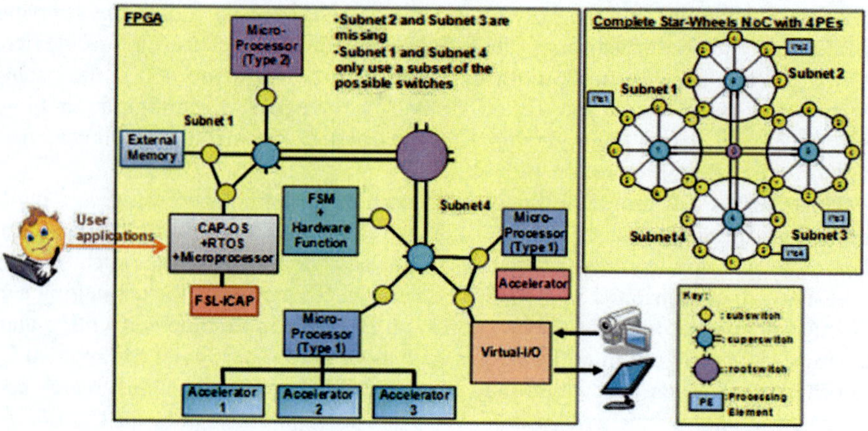

Fig. 20.1 A RAMPSoC system connected over an incomplete Star-Wheels Network-on-Chip at one point in time. On the top right an example for a maximal established Star-Wheels NoC is shown

is proven, that the RAMPSoC system has a higher energy efficiency than an NVIDIA GPU and a standard CPU.

This chapter is organized as follows: In Sect. 20.2, the holistic RAMPSoC approach is presented. This means, the hardware architecture, the Star-Wheels Network-on-Chip, the RAMPSoC design methodology and the configuration access port-operating system (CAP-OS) are described. In Sect. 20.3, the implementation of an image processing application onto the RAMPSoC is described and the results of the RAMPSoC system in terms of performance and energy consumption are compared against a standard CPU and an NVIDIA Tesla C1060 GPU. Finally, the chapter is closed by presenting the conclusions and an outlook Sect. 20.4.

20.2 The Holistic RAMPSoC Approach

The holistic RAMPSoC approach consists of a heterogeneous and runtime adaptive multiprocessor architecture, a design methodology for architecture development and application partitioning and a special purpose operating system called CAP-OS, which is responsible for runtime scheduling, task mapping and resource management.

Figure 20.1 shows a RAMPSoC system at one point in time. It supports different types of processing elements (PEs) (e.g., processors, finite state machines (FSMs)). Each PE can be connected to one or more accelerators.

Different types of communication infrastructure (such as bus, Network-on-Chip (NoC), Point-to-Point) or a heterogeneous combination of those are supported. In the example of Fig. 20.1, the PEs are connected over the novel heterogeneous and runtime adaptive Star-Wheels NoC [13]. As this NoC is runtime adaptive, it

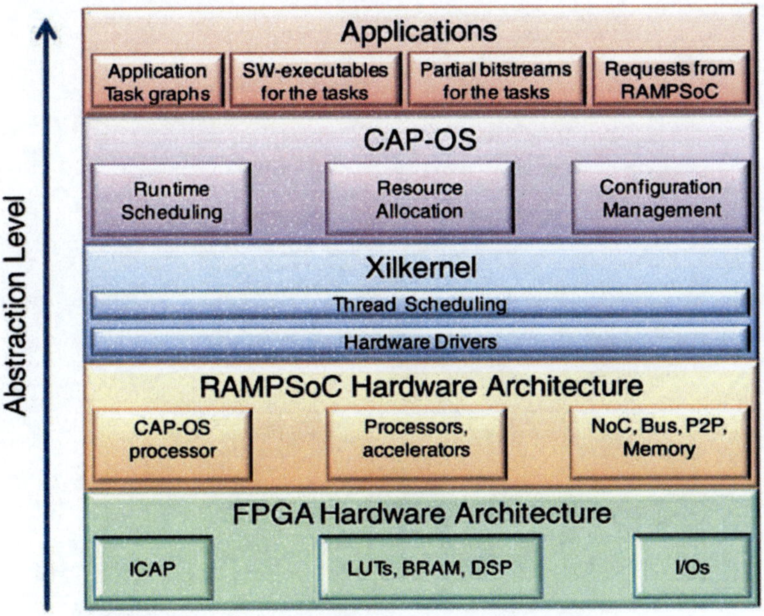

Fig. 20.2 Abstraction levels of the CAP-OS

can detect if new PEs were added, or if existing ones have been removed or exchanged. The top right picture in Fig. 20.1 shows an example for a maximal implemented Star-Wheels NoC. Not only the structure, but also the communication paradigm is heterogeneous, since the routing is based on a combination of packet-switching and circuit-switching. An additional benefit of this NoC is that it supports different clock frequencies for different PEs.

A special communication gateway is the Virtual-I/O component. It is used to connect peripherals such as cameras, monitors, PCI-interfaces, etc. with the communication infrastructure of the PEs. Moreover, the Virtual-I/O is used for exploiting data parallelism, as it can split the incoming frames of a camera into tiles for multiple PEs and it also can be used to collect the computation results and display them on a monitor.

The most important PE is the microprocessor with the Xilkernel Real-Time Operating System (RTOS) from Xilinx on top of which the Configuration Access Port-OS (CAP-OS) [14] is executed. Figure 20.2 shows the different abstraction layers of CAP-OS.

While the CAP-OS is responsible for runtime scheduling and resource management of the overall RAMPSoC, the Xilkernel is responsible for scheduling the different CAP-OS threads on the microprocessor and for providing the hardware drivers to the CAP-OS. The processor is connected to a special component called Fast Simplex Link-Internal Configuration Access Port (FSL-ICAP) [15], through which the CAP-OS can access the configuration memory on the FPGA with high

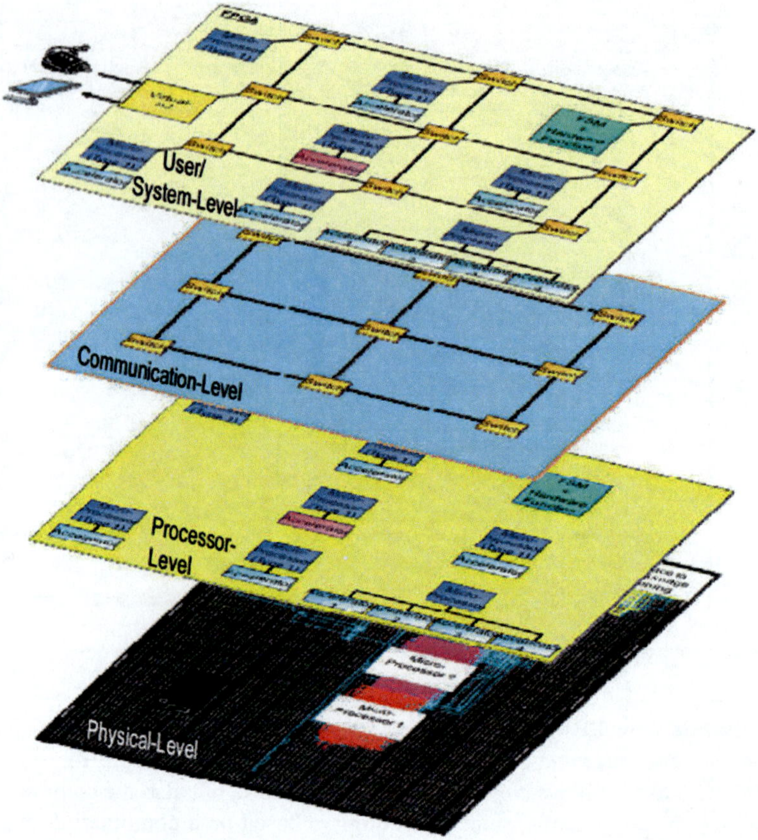

Fig. 20.3 Four abstraction layers used within the design methodology and the CAP-OS to hide the complexity of the RAMPSoC architecture

throughput and therefore can modify the hardware structure of the overall MPSoC to achieve a high energy efficiency. This processor receives the algorithmic dependencies of the user application in form of a task graph at runtime. In addition, it receives partial configuration bitstreams for processors and accelerators and software executables for the different threads to be executed on the processors. CAP-OS then schedules the different software tasks to the available PEs. If sufficient processing elements are not available, it modifies the configuration memory of the FPGA via the FSL-ICAP to add new PEs. Currently unused PEs are substituted or shut down. To ensure that real-time constraints are met, CAP-OS can also modify the clock frequency of some or all PEs by dynamically reconfiguring the corresponding digital clock manager (DCM) of the Xilinx FPGA.

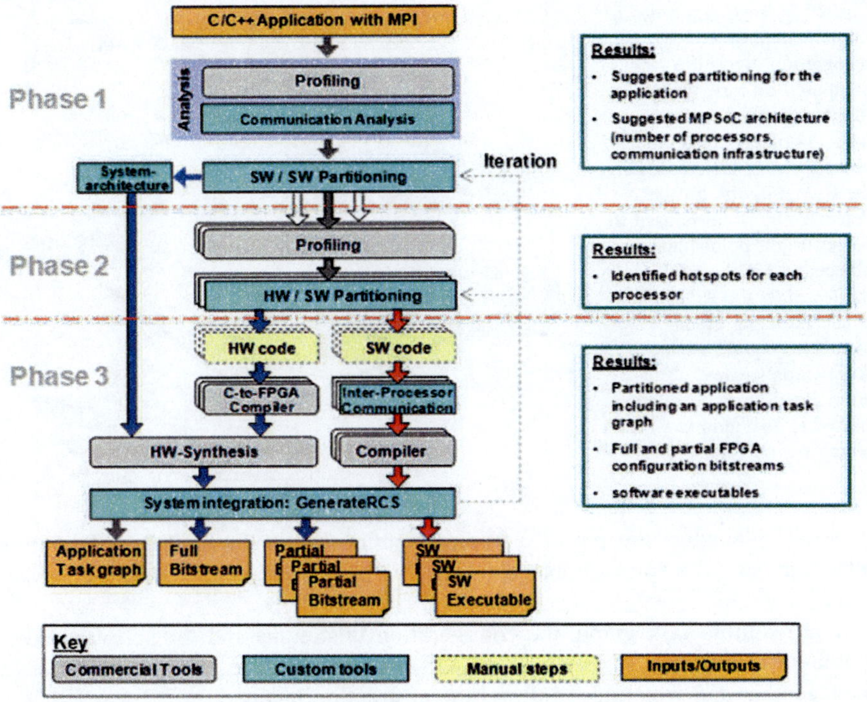

Fig. 20.4 Design Methodology of RAMPSoC

To close the gap between the software applications written in C, C++ and the physical RAMPSoC hardware architecture, four abstraction layers have been introduced as shown in Fig. 20.3.

The top level is the User/System-level, which is seen by the user. This means the user knows, that an MPSoC system is being programmed, but the structure of the system e.g., types of processors, communication infrastructure is hidden from the user. The lowest level of abstraction is the physical level of the FPGA. The levels in between are the Communication- and the Processor-level.

All abstraction layers are used within the novel design methodology [16]. Figure 20.4 shows the different phases of the design methodology, each of which correspond to a specific abstraction layer.

In Phase 1 the design flow analyses the user applications written in C, C++ and generates a task graph of the application. Hierarchical Clustering is used to map specific tasks onto the PEs. This phase corresponds to the User/System- and the Communication-level.

In Phase 2 each task is profiled and a Hardware-Software Codesign is done to find possible computation intensive blocks, which can be outsourced into accelerators. This phase corresponds to the Processor-level.

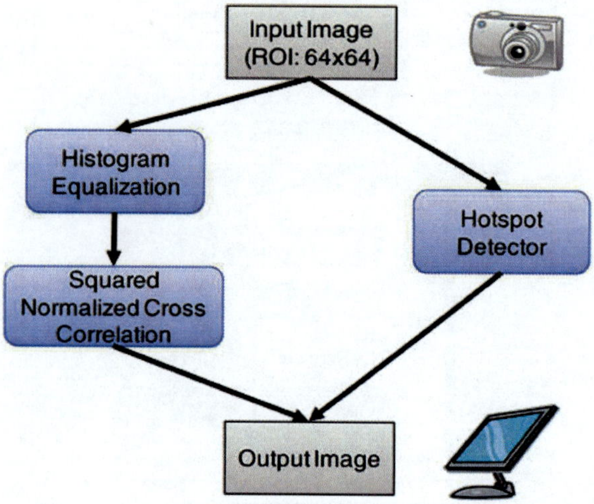

Fig. 20.5 Task graph of the implemented object recognition algorithm computed on a region of interest (ROI) of size 64 × 64 pixels. Two kinds of objects can be detected. On the *right side* the hotspot detector algorithm is used to detect bright point-like objects within the ROI. On the *left side*, the ROI is enhanced using global histogram equalization. Afterwards squared normalized cross correlation is used to do a template-based search for objects

Phase 3 handles the physical implementation. It generates the configuration bitstreams and the software executables. This phase corresponds to the Physical-level.

The resulting task graph, the configuration bitstreams and the software executables are then passed over to CAP-OS, which is responsible for the runtime management and which also follows the abstraction layers.

20.3 Implementation and Results

All elements of the RAMPSoC approach have been exemplarily implemented on different FPGA platforms (such as Virtex-4 and Virtex-5) and have been evaluated using different application scenarios from image processing [13] and bioinformatics [17].

An object recognition algorithm was selected for a comparison of the performance, the power and energy consumption as well as the energy efficiency of the RAMPSoC with a standard CPU and a modern NVIDIA GPGPU (general purpose graphic processing unit). Figure 20.5 shows an abstract representation of the object recognition algorithm. The algorithm is computed on a region of interest (ROI) of 64 × 64 pixels within the frames acquired from a video source. Bright point-like objects are detected using the hotspot detector algorithm. Bigger objects are detected using template matching. To improve the detection rate, the ROI is enhanced using global histogram equalization. Afterwards, squared normalized cross correlation is used to calculate the similarity between the template and a subregion of the ROI in the size of the template. Here a template size of 10 × 10 pixels was used.

Fig. 20.6 Implemented RAMPSoC system consisting of three processors and two hardware accelerators. For the hotspot detector no hardware accelerator was implemented. Instead, the image was partitioned to speed up the execution by using two processors, one for each partition of the image

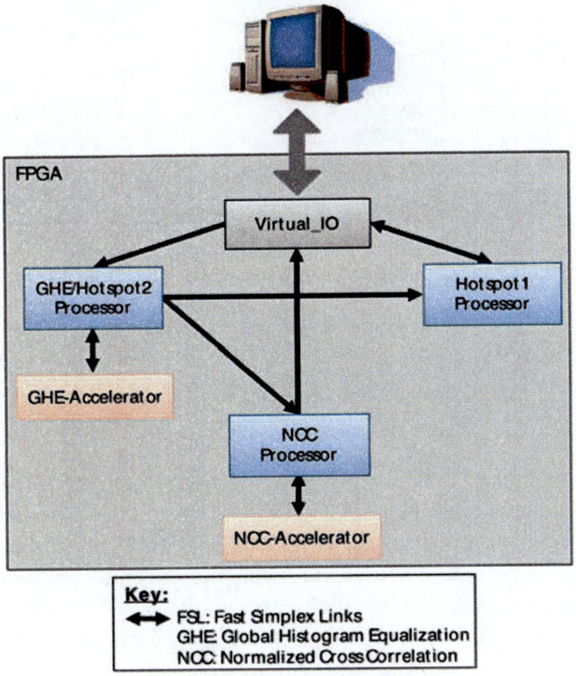

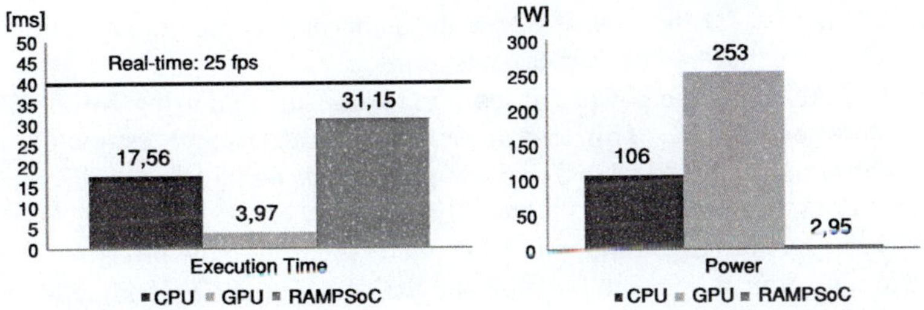

Fig. 20.7 Execution time and power consumption for an object recognition algorithm on a RAMPSoC system running at 125 MHz, on a CPU running at 2 GHz and on an NVIDIA Tesla C1060 GPGPU running at 1.3 GHz

This object recognition algorithm was implemented in C for the CPU and with CUDA for the NVIDIA Tesla C1060 GPU. For the RAMPSoC, a heterogeneous system consisting of a Virtual-IO, three processors and two hardware accelerators (one for the Histogram Equalization and one for part of the squared normalized cross correlation) were used as shown in Fig. 20.6. Instead of increasing the performance of the hotspot detector using an additional hardware accelerator, the ROI was partitioned into two overlapping tiles and the histogram equalization processor was reused for calculating the hotspot detector for the second partition of the input image. This way resources and therefore power consumption can be kept low.

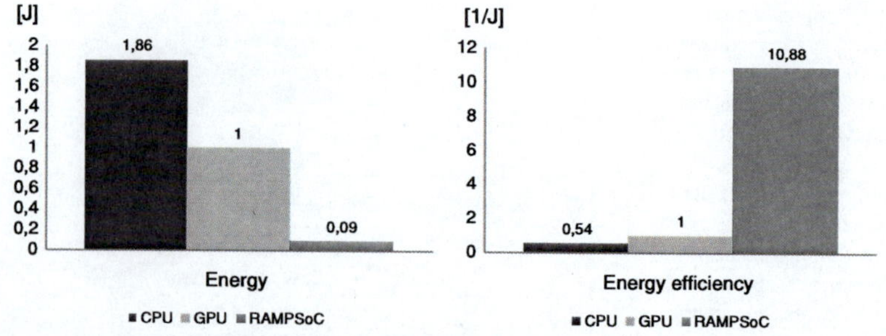

Fig. 20.8 Energy consumption and energy efficiency for an object recognition algorithm on a RAMPSoC system running at 125 MHz, on a CPU running at 2 GHz and on an NVIDIA Tesla C1060 GPGPU running at 1.3 GHz

Of course a hardware only solution would be the fastest one, but for all three implementations (CPU, GPU, RAMPSoC) the goal was to optimize each implementation, only until it was fast enough to fulfill the real-time constrains.

Figure 20.7 shows the measured execution time and the measured / estimated power consumption for the CPU, the GPU and the RAMPSoC. The GPU is the fastest platform, followed by the CPU and the last one is the RAMPSoC, but all platforms fulfill the real-time constraints. All performance values were measured on the implemented platforms. For the power consumption it is vice versa. The RAMPSoC has by far the lowest power consumption. The CPU is in the middle and the GPU has the highest power consumption.

For the RAMPSoC the power consumption was estimated using Xilinx XPower tool, as described in [18]. This tool allows an accurate enough estimation of the power consumption. RAMPSoC was implemented on a large Xilinx Virtex-4FX100 FPGA, but only uses less than 30% of the available resources. The power consumption of the CPU and the GPU were measured by inserting a power measuring instrument between the plug and the power outlet. Of course in the case of the GPU, the GPU power consumption was measured together with the Host-CPU power consumption. But as the Host-CPU also computed parts of the algorithm together with the GPU, the GPU could not have calculated the algorithm without the Host-CPU. Therefore it was decided to be fair to measure both together.

The energy consumption and the energy efficiency were calculated using the performance and power consumption results. Figure 20.8 shows the results for the three systems.

The RAMPSoC execution time has been 8 times slower, than the one of the GPU. On the other hand, the power consumption of RAMPSoC was 86 times lower than the one of the GPU. This results into the fact, that the RAMPSoC has the lowest energy consumption, followed by the GPU and then by the CPU. The GPU has a lower energy consumption compared to the CPU due to its much faster execution time.

As a consequence for this object recognition algorithm the RAMPSoC is much more energy efficient than the GPU and the CPU. This behavior will be evaluated with more applications in the future.

20.4 Conclusions and Outlook

The holistic RAMPSoC approach consists of the novel runtime adaptive hardware architecture, a newly developed design methodology and the special purpose operating system called CAP-OS. Several abstraction layers are used within the complete approach to hide the complexity of the system from the user. It provides the flexibility, high performance and low power consumption required by embedded high performance computing applications. The energy efficiency of the RAMPSoC was evaluated using an object recognition application. The results were compared against the energy efficiency of a standard CPU and a high performance NVIDIA Tesla C1060 GPU. All platforms fulfilled the real-time requirements of 25 frames per second (fps), but the RAMPSoC had by far the lowest power consumption and has therefore the best energy efficiency for the given application.

Future steps are to evaluate the advantages of the RAMPSoC approach with further high performance embedded applications and with other multiprocessor platforms, such as the Intel Single Chip Cloud Computer (SCC) [19]. Based on the results of the evaluation the hardware architecture, the design methodology and CAP-OS will be improved to further ease the programming of the RAMPSoC architecture and to support a greater library of processing elements, communication infrastructures and hardware accelerators.

References

1. Hartenstein R (2006) Why we need reconfigurable computing education. RC-Education workshop, Karlsruhe, Germany
2. Lysaght P, Blodget B, Mason J, Young J, Bridgford B (2006) Invited paper: Enhanced architectures, design methodologies and CAD tools for dynamic reconfiguration of Xilinx FPGAs. In: Proceedings of FPL 2006, August 2006
3. PlanAhead Software Tutorial (2010) Overview of the partial reconfiguration flow, UG 743 (v 12.1). Available at: http://www.xilinx.com
4. Fingeroff M (2010) High level synthesis blue book. Xlibris Corporation
5. Pellerin D, Thibault S (2005) Practical FPGA programming in C. Prentice Hall Professional Technical Reference
6. Kamat RK, Shinde SA, Shelake VG (2009) Unleash the system on chip using FPGAs and Handel C. Springer, Berlin
7. Butenhof DR (1997) Programming with POSIX threads. Addison-Wesley, Reading
8. OpenMP. Available at: http://openmp.org
9. MPI (2009) A message-passing interface standard, version 2.2. Message passing interface forum, September 4. Available at: www.mpiforum.org

10. NVIDIA 739 ® Tesla™, GPU Computing technical brief, Version 1.0.0, May 2007
11. NVIDIA CUDA™ (2009) Programming guide, Version 2.3.1, August Available at: http://www.nvidia.com/object/cuda_home.html
12. Göhringer D, Becker J (2010) FPGA-based runtime adaptive multiprocessor approach for embedded high performance computing applications. In: Proceedings of the IEEE Computer Society annual symposium on VLSI (ISVLSI 2010), Lixouri, Kefalonia, Greece
13. Göhringer D, Liu B, Hübner M, Becker J (2009) Star-wheels network-on-chip featuring a self-adaptive mixed topology and a synergy of a circuit- and a packet-switching communication protocol. In: Proceedings of FPL 2009, September 2009, pp 320–325
14. Göhringer D, Hübner M, Nguepi Zeutebouo E, Becker J (2010) CAP-OS: Operating system for runtime scheduling, task mapping and resource management on reconfigurable multiprocessor architectures. In: Proceedings of RAW at the IPDPS 2010, April 2010
15. Hübner M, Göhringer D, Noguera J, Becker J (2010) Fast dynamic and partial reconfiguration data path with low hardware overhead on Xilinx FPGAs. In: Proceedings of RAW at the IPDPS 2010, April 2010
16. Göhringer D, Hübner M, Benz M, Becker J (2010) A design methodology for application partitioning and architecture development of reconfigurable multiprocessor systems-on-chip. In: Proceedings of FCCM 2010, May 2010
17. Göhringer D, Hübner M, Hugot-Derville L, Becker J (2010) Message passing interface support for the runtime adaptive multi-processor system-on-chip RAMPSoC. In: Proceedings of the 10th international conference on embedded computer systems: Architectures, modeling and simulation (SAMOS X), July 2010, Samos, Greece
18. Development System Reference Guide, v9.2i, Chap. 10 XPower. Available at: http://www.xilinx.com
19. Howard J, Dighe S, Hoskote Y et al (2010) A 48-Core IA-32 message-passing processor with DVFS in 45 nm CMOS. In: Proceedings of IEEE international solid-state circuits conference (ISSCC 2010), February 2010, San Francisco